Studies in Computational Intelligence

Volume 512

Series Editor

Janusz Kacprzyk, Warsaw, Poland

Subseries

Learning, Optimization and Interdisciplinary Applications

Subseries Editors

Natalio Krasnogor, Nottingham, United Kingdom
David Pelta, Granada, Spain

For further volumes:
http://www.springer.com/series/7092

Germán Terrazas · Fernando E.B. Otero
Antonio D. Masegosa

Editors

Nature Inspired Cooperative Strategies for Optimization (NICSO 2013)

 Springer

Editors
Germán Terrazas
School of Computer Science
University of Nottingham
Jubilee Campus
Nottingham, NG8 1BB
United Kingdom
gzt@cs.nott.ac.uk

Antonio D. Masegosa
Center for Research on ICT (CITIC)
C/ Periodista Rafael Gómez Montero, 2
University of Granada
18071 Granada, Spain
admase@decsai.ugr.es

Fernando E.B. Otero
School of Computing
University of Kent
Medway Building
Chatham Maritime, ME4 4AG
United Kingdom
F.E.B.Otero@kent.ac.uk

ISSN 1860-949X   ISSN 1860-9503   (electronic)
ISBN 978-3-319-03347-1   ISBN 978-3-319-01692-4   (eBook)
DOI 10.1007/978-3-319-01692-4
Springer Cham Heidelberg New York Dordrecht London

Preface

The ability of computer scientist, engineers and practitioners to find sources of inspiration in nature to abstract metaphors suitable for problem solving has proven limitless. For instance, the rationale behind computational platforms such as cloud computing, whereby users access distributed computing power to exploit data sharing and information processing, has been observed in the lysogenic cycle which is one of the ways virus operate within cell DNA machinery to reproduce itself. Besides the already well-known applications on classification, learning and robotics to name but a few, nature inspired strategies became increasingly important to address challenges ranging at all scales from molecular reactions to social networks. As examples, we might mention that evolutionary algorithms already represent indispensable tools for the design optimisation of biochemical signalling pathways in synthetic biology and that cooperative spatio-temporal architectures are suitable approximations for modelling complex emergent behaviour seen in micro-scaled biological systems.

This book is a collection of research works presented in the VI International Workshop on Nature Inspired Cooperative Strategies for Optimization (NICSO) held in Canterbury, UK. Previous editions of NICSO were held in Granada, Spain (2006 & 2010), Acireale, Italy (2007), Tenerife, Spain (2008), and Cluj-Napoca, Romania (2011). The aim of NICSO 2013 is to provide a place where state-of-the-art research, latest ideas and emerging areas of nature inspired cooperative strategies for problem solving are vigorously discussed and exchanged among the scientific community. The contributions of this volume have undergone a strictly peer reviewed process by members of the international Programme Committee. The breadth and variety of articles report on nature inspired methods and applications such as Swarm Intelligence, Hyper-heuristics, Evolutionary Algorithms, Cellular Automata, Artificial Bee Colony, Dynamic Optimisation, Support Vector Machines, Multi-Agent Systems, Ant Clustering, Evolutionary Design Optimisation, Game Theory and other several Cooperation Models. In this edition, NICSO had three plenary lectures delivered by Dr. Alex A. Freitas, *Automating the Design of Data Mining Algorithms with Genetic Programming*, Dr. Angel Goñi Moreno, *Bacterial micromachines – Living logic circuits for computing*, and Dr. Leonardo Vanneschi, *Applications of Genetic Programming to Drug Discovery and Pharmacokinetics*.

Of course, neither NICSO 2013 or this book would exist without the help of many people and institutions. We wish to thank the authors for contributing with valuable articles to publication and each of the Programme Committee members for their dedicated time, suggestions and advice. In addition, we thank the financial support received from the Spanish Ministry of Economy and Competitiveness (project TIN2011-27696-C02-01), the Andalusian Government (project P11-TIC-8001), the European Regional Development Fund (ERDF), the School of Computing, University of Kent, and The European Commission FP7 Future and Emerging Technologies initiative (FP7/2012–2015) under grant agreement number 318235 Training and Research in Unconventional Computation in Europe (TRUCE). Finally, we thank the School of Computing, University of Kent, for providing both administrative and logistical assistance.

UK   Germán Terrazas
UK   Fernando E.B. Otero
Spain   Antonio D. Masegosa
September 2013

Organization

Steering Committee

David A. Pelta — University of Granada, Spain
Natalio Krasnogor — University of Nottingham, UK

Organizing Committee

Germán Terrazas — University of Nottingham, UK
Fernando E.B. Otero — University of Kent, UK
Antonio D. Masegosa — University of Granada, Spain

Programme Committee

Alberto Moraglio — University of Birmingham, UK
Alex A. Freitas — University of Kent, UK
Alvaro García Piquer — Institute of Space Sciences (ICE-CSIC), Spain
Andrés R. Masegosa — University of Granada, Spain
Belén Melian — University of La Laguna, Spain
Blas J. Galván — University of Las Palmas de Gran Canaria, Spain
Carlos Cruz Corona — University of Granada, Spain
Carlos García Martínez — Univeristy of Cordoba, Spain
Cecilio Angulo — Technical University of Catalunya, Spain
Christof Teuscher — Portland State University, USA
Colin Johnson — University of Kent, UK
Dario Landa-Silva — University of Nottingham, UK
Davide Anguita — University of Genova, Italy
Enrique Onieva — University of Deusto, Spain
Evelyne Lutton — INRIA, France
Francisco Herrera — University of Granada, Spain
Gabriela Ochoa — University of Stirling, UK
Gianluigi Folino — Istituto di Calcolo e Reti ad Alte Prestazioni, Italy
Gisele Pappa — Universidade Federal de Minas Gerais, Brazil
José Marcos Moreno — University of La Laguna, Spain
Jaume Bacardit — University of Nottingham, UK
Jean-Louis Giavitto — Université d'Evry, France
Jiawei Li — University of Nottingham, UK

Jim Smith	University of the West of England, UK
Jon Timmis	University of York, UK
Jorge Casillas	University of Granada, Spain
José A. Moreno Perez	University of La Laguna, Spain
José A. Castillo	Instituto Nacional de Investigaciones Nucleares, Mexico
José Manuel Cadenas	University of Murcia, Spain
Juan José Merelo	University of Granada, Spain
Julia Handl	University of Manchester, UK
Marco Dorigo	Université Libre de Bruxelles, Belgium
Mario Pavone	University of Catania, Italy
Oliver Korb	Cambridge Crystallographic Data Centre, UK
Pablo José Villacorta	University of Granada, Spain
René Doursat	Complex Systems Institute, France
Shengxiang Yang	De Montfort University, UK
Stefano Pizzuti	Energy, New Technologies & Envir. Agency, Italy
Thomas Stibor	GSI Helmholtz Centre for Heavy Ion Research, Germany
Vittorio Maniezzo	University of Bologna, Italy

Plenary Lectures

Dr. Alex A. Freitas

University of Kent, United Kingdom

Automating the Design of Data Mining Algorithms with Genetic Programming

Rule induction and decision-tree induction algorithms are among the most popular types of classification algorithms in the field of data mining. Research on these two types of algorithms produced many new algorithms in the last 30 years. However, all the rule induction and decision-tree induction algorithms created over that period have in common the fact that they have been manually designed, typically by incrementally modifying a few basic rule induction or decision-tree induction algorithms. Having these basic algorithms and their components in mind, we describe the use of Genetic Programming (GP), a type of evolutionary algorithm that automatically creates computer programs, to automate the process of designing rule induction and decision-tree induction algorithms. The basic motivation is to automatically create complete rule induction and decision-tree induction algorithms in a data-driven way, trying to avoid the human biases and preconceptions incorporated in manually-designed algorithms. Two proposed GP methods (one for evolving rule induction algorithms, the other for evolving decision-tree induction algorithms) are evaluated on a number of datasets, and the results show that the machine-designed rule induction and decision-tree induction algorithms are competitive with well-known human-designed algorithms of the same type.

Dr. Angel Goñi Moreno

National Center for Biotechnology, Spain

Bacterial micromachines – Living logic circuits for computing

Engineering Boolean logic circuits in bacteria is a major research theme of synthetic biology. By using living technology as DNA blocks or molecular wires we can mimic the behaviour of electronic devices. Examples of this engineering, such as logic gates, clock signals, switches, multiplexers or half adders have been successfully built inside bacteria. Just as the pioneers of computer technology quickly incorporated the early transistor into larger circuits, researchers in synthetic biology merge this genetic devices to achieve distributed computations within a microbial consortia. The rapid development of bacterial-based devices is accompanied by a need for computational simulations and mathematical modelling to facilitate the characterisation and design of such systems. Therefore, computer sciences are mixed with synthetic biology (i.e. computational biology and bioinformatics) for two closely related purposes: 1) the design of devices, where the knowledge acquired in electronics since the beginning of computation is crucial; and 2) the desire to better understand the underlying biological substrate, giving answers to questions impossible to solve in a wet-lab. Up to now, bacterial micromachines are only in the form of basic computing devices. However, this machines are alive. Where is this engineering going?

Dr. Leonardo Vanneschi

ISEGI, Universidade Nova de Lisboa, Portugal

Applications of Genetic Programming in Drug Discovery and Pharmacokinetics

The success of a drug treatment is strongly correlated with the ability of a molecule to reach its target in the patients organism without inducing toxic effects. Moreover the reduction of cost and time associated with drug discovery and development is becoming a crucial requirement for pharmaceutical industry. Therefore computational methods allowing reliable predictions of newly synthesized compounds properties are of outmost relevance. In this talk, I discuss the role of Genetic Programming (GP) in predictive pharmacokinetics, considering the estimation of adsorption, distribution, metabolism, excretion and toxicity processes (ADMET) that a drug undergoes into the patients organism. In particular, I discuss the ability of GP to predict oral bioavailability (%F), median oral lethal dose (LD50) and plasma-protein binding levels (%PPB). Since these parameters respectively characterize the percentage of initial drug dose that effectively reaches the systemic blood circulation, the harmful effects and the distribution into the organism of a drug, they are essential for the selection of potentially effective molecules. In the last part of the talk, I show and discuss how recently defined geometric semantic genetic operators can dramatically affect the performances of GP for this kind of application, in particular on out-of-sample test data.

Table of Contents

Extending the ABC-Miner Bayesian Classification Algorithm

Khalid M. Salama and Alex A. Freitas

School of Computing, University of Kent,
Canterbury, CT2 7NF, UK
{kms39,A.A.Freitas}@kent.ac.uk

Abstract. ABC-Miner is a Bayesian classification algorithm based on the Ant Colony Optimization (ACO) meta-heuristic. The algorithm learns Bayesian network Augmented Naïve-Bayes (BAN) classifiers, where the class node is the parent of all the nodes representing the input variables. However, this assumes the existence of a dependency relationship between the class variable and *all* the input variables, and this relationship is a type of "causal" (rather than "effect") relationship, which restricts the flexibility of the algorithm to learn. In this paper, we propose ABC-Miner+, an extension to the ABC-Miner algorithm which is able to learn more flexible Bayesian network classifier structures, where it is not necessary to have a (direct) dependency relationship between the class variable and each of the input variables, and the dependency between the class and the input variables varies from "causal" to "effect" relationships. The produced model is the *Markov blanket* of the class variable. Empirical evaluations on UCI benchmark datasets show that our extended ABC-Miner+ outperforms its previous version in terms of predictive accuracy, model size and computational time.

Keywords: Ant Colony Optimization (ACO), Data Mining, Classification, Bayesian Network Classifiers.

1 Introduction

Ant Colony Optimization (ACO) is a meta-heuristic for solving combinatorial optimization problems, inspired by the observation of the behavior of biological ant colonies [6]. One of the fields in which ACO has been successfully applied is data mining, which involves finding hidden patterns and constructing analytical models from real-world datasets [20]. Classification is one of the widely studied data mining tasks, where the aim is to discover, from labeled cases (instances), a model that can be used to predict the class of unlabeled cases. While there are several types of classification methods, such as decision tree and rule induction, artificial neural networks and support vector machines [20], our focus is on Bayesian network (BN) classifiers.

BN classifiers model the (in)dependency-relationships between the input domain variables given the class variable by means of a probabilistic network [7],

G. Terrazas et al. (eds.), *Nature Inspired Cooperative Strategies for Optimization*
(NICSO 2013), Studies in Computational Intelligence 512,
DOI: 10.1007/978-3-319-01692-4_1, © Springer International Publishing Switzerland 2014

which is used to predict the class of a case by computing the class with the highest posterior probability given the case's predictor attribute values. Since learning the optimal BN structure from a dataset is $\mathcal{NP}$-hard [4], stochastic heuristic search algorithms - such as ACO – can be a good alternative to build high-quality models, in terms of predictive accuracy and network size, within an acceptable computational time. Developing ACO-based algorithms to learn BN classifiers is the research topic addressed in this work.

We have recently introduced ABC-Miner [19], as an Ant-based Bayesian Classification algorithm that learns the structure of a Bayesian network Augmented Naïve-Bayes (BAN), where the class node is the parent of all the input variables, and at most k parents are allowed for each variable in the network. The ABC-Miner algorithm showed predictive effectiveness compared to other Bayesian classification algorithms, namely: Naïve-Bayes, TAN and GBN [19].

In this paper, we propose ABC-Miner+, which extends our ABC-Miner algorithm to learn more flexible BN classifier structures, where it is not necessary to have a (direct) dependency relationship between the class variable and each of the input variables. In addition, ABC-Miner+ allows the dependency between the class and the input variables to vary from "causal" to "effect" relationships, where the class variable can be a parent or a child of an input node. The produced model is called the *Markov blanket* of the class variable. Empirical results on 18 UCI datasets show that ABC-Miner+ improves the performance of ABC-Miner by producing simpler (smaller) BN classifiers that have higher predictive accuracy in less computational time.

Note that we use the word "causal" in a loose sense in this work, simply to refer to a direction of the dependency relationship between two variables. The issue of whether or not Bayesian networks learned from observational data represent truly causal knowledge is controversial (depending on how we define causality) [15], and is out of the scope of this paper.

The rest of the paper is structured as follows. The next section gives some background on BN classifiers. We briefly review the previously introduced ABC-Miner algorithm in Section 3, to make this paper more self-contained. Our proposed extension, ABC-Miner+ is described in detail in Section 4. We describe our experimental methodology and show the results in Section 5. Finally, we conclude with some general remarks and future research in Section 6.

2   Bayesian Network Classifiers

Bayesian networks (BNs) are knowledge representation and reasoning tools that model probabilistic dependence and independence relationships amongst variables in a specific domain [5]. Learning a BN from a dataset (in which the attributes are referred to as variables) consists of two phases: learning the network structure, and then learning the parameters of the network. Parameter learning is relatively straightforward for any given BN structure with specified dependencies between variables. The task is to estimate a conditional probability table (CPT), one for each variable, by computing the relative frequencies of the variable with respect to its parents directly from the dataset. The CPT of variable X_i encodes the likelihood of each value of this variable given each combination of values of the parents of variable X_i in the network.

There are two paradigms for learning the structure of a BN. The first one is referred to as *CI-based* (Conditional Independence-based, or constraint-based) algorithms [5, 8], which suggests learning the BN structure by identifying the conditional independence relationships among the nodes, according to the concept of *d-separation*. The second paradigm views the BN as a structure that encodes the joint distribution of the attributes. Hence, the aim is to find the graph that best fits a given dataset in terms of maximizing the value of a scoring function, which led to the scoring-based algorithms [5, 8]. In the context of data mining, the scoring-based approach has been (overall) more popular and it is somewhat easier to be used than the CI-based approach, partly because the former views the problem as a well-defined optimization task, where various search and meta-heuristic techniques can be employed [3]. K2, MDL, KL, BDEu and several other scoring functions can be used for this task [8].

A recent, very comprehensive review on BN-learning approaches and issues is presented by Daly et al. in [5]. For further information about BNs, the reader is referred to [8].

While BNs should perform inference to answer probabilistic queries about any node(s) in the network, BN classifiers are a special kind of the probabilistic networks, which focus on answering queries about the probability of a specific node: the class attribute. Thus, the class node is treated as a special variable in the network. The purpose is to compute the probability of each value c in the class variable C given a case $\mathbf{x}$ (an instance of the input attributes $\mathbf{X}$) using classifier BNC, then label the case with the class having the highest probability, as in the following formulas:

$$C(\mathbf{x}) = \underset{\forall\ c \in C}{\arg\max}\, P(C = c | \mathbf{x} = x_1, x_2, ..., x_n, BNC), \qquad (1)$$

$$\overbrace{P(C = c | \mathbf{x} = x_1, x_2, ..., x_n)}^{\text{posterior probability}} \propto \overbrace{P(C = c)}^{\text{proir probability}} \prod_{v=1}^{n} \overbrace{P(x_v | \mathbf{Parents}(X_v), BNC)}^{\text{likelihood}}, \qquad (2)$$

where $C \in \mathbf{Parents}(X_v) \forall\ X_v \in \mathbf{X}$.

Naïve-Bayes is the simplest kind of BN classifiers; it has a network structure where the class node is the only parent node of all other nodes (input variables). This structure assumes that all attributes are independent of each other given the class. However, in many real-world application domains this assumption is not satisfied, and more sophisticated types of BN classifiers, which consider dependencies between the predictor attributes, can lead to higher predictive accuracy. This led to the development of a general type of BN classifiers called BN Augmented Naïve-Bayes (BAN).

In a BAN classifier, each node representing an input attribute not only has the class node as a parent, but it is also allowed to have other parent nodes which are also input attributes. Hence, the edges representing dependencies among input attributes can be regarded as a kind of BN, which justifies the name "BN-augmented" Naïve-Bayes. Usually, however, each node representing a input attribute is allowed to have a maximum number (k) of parents, in order to reduce computational time and reduce the chances of over-fitting the BN to the data, and in this case the algorithm is often referred to as a k-dependency BAN. Note that when the maximum number of parents k is set to 1, the BAN is usually referred to as a TAN (Tree-Augmented Naïve-Bayes), because in that type of classifier each node representing an input attribute can have at most one parent node (in addition to the class node), so that the dependencies among input attributes are represented as a tree.

Figure 1 illustrates the various kinds of the BN classifiers. Friedman et al. provided an excellent study of these algorithms in [7]. A comprehensive comparison of these various Bayesian classifiers by Cheng and Greiner is found in [3]. Surveys on improving Naïve-Bayes for classification are found in [10, 11].

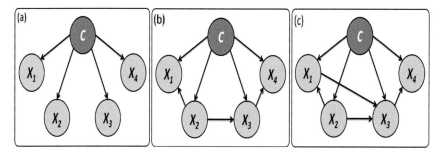

Fig. 1. Different types of BN classifiers: (a) Naïve-Bayes, (b) TAN, and (c) BAN

3   An Overview of the ABC-Miner Algorithm

ACO algorithms have been successful in solving several combinatorial optimization problems, including classification rule discovery [12–14, 17, 18] and general purpose BN construction [2, 16, 21]. However, ABC-Miner, introduced by the authors in [19], is the first ACO algorithm that learns the structure of BAN classifiers [19].

In ABC-Miner, the decision components in the construction graph (which define the search space that an ant uses to construct a candidate solution) are all the edges of the form $X \rightarrow Y$ where $X \neq Y$ and X, Y belong to the set of input attributes. These edges represent the attribute dependencies in a constructed BN classifier – i.e., an edge $X \rightarrow Y$ means that the value of Y depends (probabilistically) on the value of X.

In order to build the structure of a BN classifier, the maximum number of parents for a node is typically specified by the user. However, the selection of the optimal number of parents (dependencies) that a variable in the network can have (in addition to having the class as a parent node) is automatically carried out in ABC-Miner [19]. To create a candidate solution, an ant starts with the network structure of Naïve-Bayes, where every variable has only the class variable as its parent. Then the ant expands that structure into a BAN structure by adding edges to the network. The selection of the edges is performed according to a probabilistic state transition formula that involves the pheromone amount and the heuristic function value – measured by the conditional mutual information [19] – of the edges. An edge is valid to be added to the BN classifier being constructed if its inclusion does not create a directed cycle and does not exceed the limit of k parents (chosen by the current ant).

After the ant adds a valid edge to the current candidate solution (BN classifier), all the invalid edges are eliminated from the construction graph. The ant keeps adding edges to the current solution until no valid edges are available. When the structure is finished, the CPT of each variable is computed, producing a complete BN classifier. Then the quality of the solution is evaluated and all the edges become available for constructing further candidate solutions. The ABC-Miner algorithm evaluates the quality of the candidate constructed BN classifier using a measure of predictive accuracy [19], since the goal is to build a BN only for predicting the value of a specific class attribute, unlike conventional BN learning algorithms whose scoring function does not distinguish between the input (predictor) and the class attributes.

4   The Proposed ABC-Miner+ Extension

The motivation behind our proposed extension is the following. As mentioned in the previous section, the structure of the BAN models constructed by ABC-Miner has two limitations. First, it assumes that the class variable has dependency relationships with *all* the input variables (the case's attributes), which means that the state of each input variable affects the posterior probability of the class values, and consequently the class prediction. This assumption is not necessarily valid in all applications domains. In some domains, some attributes are irrelevant, or at least not directly related, to the prediction of the target class. Including these irrelevant attributes in the computation of the posterior probability of the class values, according to Equation 2, can be disadvantageous, and may lead to incorrect predictions.

Second, in the BAN classifier constructed by ABC-Miner, the relationship between the class and all the input variables is always a type of "causal" relationship, that is, the class variable can only be a parent of an input variable. Such a property limits the flexibility of the algorithm to learn. Nonetheless, in real-world domains, some input variables are "causes" (parents) of the class variable, whereas others are "effects" (children) of the same class variable. For example, in a cancer diagnosis domain, the state of the *smoker* variable can be considered a cause of the state of the *Cancer* class variable, while the state of the *X-Ray* variable can be considered an effect of the class variable.

Accordingly, we propose ABC-Miner+, which extends the ABC-Miner algorithm to learn more flexible BN classifier structures, where it is not necessary to have a (direct) dependency relationship between the class variable and each of the input variables. This means that an input variable may not have a direct connection (edge) to the class node in the network, or an input variable may not even be presented in the network. In this case, our proposed ABC-Miner+ performs an embedded feature selection during the construction of the BN classifier. In addition, ABC-Miner+ allows the type of dependency (edge) between the class and the input variables to vary from "causal" to "effect" relationships, where the class variable can be a parent or a child of an input node.

The advantage of allowing this kind of edges in the BN model is the possibility of capturing new conditional (in)dependency-relationships. For example, if X and Y are input variables that are unconditionally independent of the class variable C, then X and Y should be parents to C. This kind of (in)dependency-relationship cannot be modeled by a BAN structure. Such a flexible BN classifier structure should better represent the dependency-relationships between the input variables, with respect to the class variable, and lead to higher classification accuracy. The produced model is the *Markov blanket* of the class variable, which consists of the class node's parents, the class node's children, and the parents of class node's children. Algorithm 1 shows the outline of ABC-Miner+.

Algorithm 1. Pseudo-code of ABC-Miner+.

Begin
$BNC_{final} = \phi; STR_{bsf} = \phi;$
$sets = trainingSet.Split();$ /* split training set into learning and validation sets */
$learningSet = sets[0];$ $validationSet = sets[1];$
$Initialize();$ $t = 1;$
repeat
    $STR_{tbest} = \phi;$ /* an empty network structure */
    for $i = 1 \rightarrow$ colony_size **do**
        $STR_i = FindRelationshipTypes(ant_i);$ /* create a candidate solution */
        $LearnParameters(STR_i, learningSet);$
        if $Quality(STR_i, validationSet) > Quality(STR_{tbest}, validationSet)$
        then
            $STR_{tbest} = STR_i;$
        end if
    end for
    $UpdatePheromone(STR_{tbest});$
    if $Quality(STR_{tbest}, validationSet) > Quality(STR_{bsf}, validationSet)$ **then**
        $STR_{bsf} = STR_{tbest};$
    end if $t = t + 1;$
until $t =$ max_iterations **or** $Convergence();$
$STR_{final} = PerformLocalSearch(STR_{bsf});$
$BNC_{final} = ExecuteABCMiner(STR_{final});$ /* extend the final structure */
return $BNC_{final};$
End

ABC-Miner+ executes in two sequential phases. First, it finds the dependency relationship type between the class variable and each of the input variables. Second, it finds the dependency relationships among the input variables. Each step is considered a different ACO procedure and has a different construction graph. In the first phase, the product is a BN structure STR that defines the edges only between the input variables and the class variable, if any exists, and does not contain edges between the input variables. The decision components in the construction graph of the first phase are a set of relationship types between pairs of variables (attributes). More precisely, there are three decision components for each variable, representing the various relationship types that the variable can have with the class node: 1) "cause", where the class is a parent of the variable, 2) "effect", where the class is a child of the variable, and 3) "none", where there is no relationship between the class and the variable in the network, so that the algorithm can perform variable (feature) selection.

The idea is to find the best edges between the class and the input variables. Each ant_i constructs a candidate solution (BN structure), via the FindRelation-shipTypes() method, as follows. For each input variable, ant_i probabilistically selects a relationship type, according to the pheromone amounts currently associated to the decision components in the construction graph, and adds its corresponding directed edge to the current candidate BN structure STR_i. Note that no edge is added between a variable and the class node in the case of selecting the "none" relationship type. The method returns a complete candidate solution (i.e. a BN structure where all the relationships between the class and the input variables are defined) before the BN parameters are learnt, and then the quality of the solution is evaluated. The algorithm learns the BN parameters using the learning set (containing 70% of the training cases), while the quality is evaluated on a validation set (containing the remaining 30% of the training cases), in order to try avoiding over-fitting to the training set. The quality of a candidate solution is evaluated as a BN classifier, using classification accuracy (Equation 3), before the iteration-best STR_{tbest} is used to perform pheromone update. The best-so-far STR_{bsf} structure undergoes local search, and the optimized STR_{final} structure is produced to be used in the next phase.

$$Accuracy = \frac{|Correctly\_Classified\_Cases|}{|Validation\_Set|} \tag{3}$$

In the second phase, the best constructed and optimized STR_{final} structure of the BN is extended to a complete class Markov blanket, by finding the dependency relationships among the input variables. Note that the BN structure discovered in the first step contains no edges between the input variables. To include this type of edge in the network structure, we execute the original ABC-Miner algorithm in this phase. However, in the context of ABC-Miner+, the solution creation procedure starts with the STR_{final} structure constructed in the previous phase, rather than a Naïve-Bayes structure as in the original ABC-Miner algorithm. The process of extending the BN structure to a candidate class Markov blanket, which takes place in the second phase of ABC-Miner+, is described in Algorithm 2. The algorithm shows just the process for each ant, for the

Algorithm 2. ABC-Miner+'s Second Phase: Ant Solution Creation Procedure.

Begin CreateSolution(ant) /* initialize the candidate Markov blanket solution
with the structure of STR_{final} discovered in phase 1 */
$MB \leftarrow STR_{final}$;
$k = ant.SelectMaxParents()$;
while $GetValidEdges() \neq \phi$ **do**
    $\{i \rightarrow j\} = ant.SelectEdgeProbablistically()$;
    $MB = MB \cup \{i \rightarrow j\}$;
    $RemoveInvalidEdges(MB, k)$;
end while
$MB.LearnParameters(learningSet)$;
return MB;
End

sake of simplicity, since the overall pseudo-code of ABC-Miner has been already
described in [19].

The execution of the procedure shown in Algorithm 2 is more efficient than its
corresponding solution creation procedure in the original ABC-Miner algorithm
in several ways. First, the search space of this procedure in the context of ABC-
Miner+ is smaller than the search space in context of the original ABC-Miner.
The reason is that, in ABC-Miner, the initial structure is the Naïve-Bayes' structure, where all the input variables are children of the class variable, so all the
candidate edges between the input variables are available for selection by an ant
(i.e. any variable can be a parent to any other variable). On the other hand, in
ABC-Miner+, the initial structure has some input variables as parents of the
class variable, and others are not even related to the class variable. In this case,
the candidate edges available for selection to be added to the network are only
the edges that satisfy two conditions, namely: the edge is connecting two input
variables (rather than connecting an input variable to the class), and the edge
is pointing to a child node of the class node. The algorithm does not consider
adding edges between the class variable's parents because these edges do not
affect the predictions (posterior probability calculation) of the BN classifier.

Second, in the Markov blanket produced by ABC-Miner+, the size of the
CPT for the variables that do not have the class variable as parent is relatively
smaller compared to the CPT of the BN classifiers produced by ABC-Miner,
where the class node has to be a parent to all the variables, besides their other
parents. Smaller CPT size means less computational time.

Note that in the case of a Markov blanket (MB) classifier, both causal (parent)
variables, and the effect (child) variables of the class variable are used to compute
the posterior probability $P(c|\mathbf{x})$ of class c given case $\mathbf{x}$, along with the parents
of the class node's children, according to the following formula:

$$P(c|\mathbf{x}) \propto P(c|\mathbf{Parents}(c)) \prod_{v \in m} P(x_v|\mathbf{Parents}(X_v), MB), \quad (4)$$

where m is the set of the input variables that have the class variable as parent.

5   Experiments and Computational Results

We compared the performance of our proposed ABC-Miner+ with two other BN classifier algorithms. The first one is basically the first phase of the ABC-Miner+ algorithm, where the BN classifiers produced have a structure only with the relationships between the class and the input variables, without discovering the dependency relationships among the input variables. The algorithm is denoted as ABC. The second algorithm is the original ABC-Miner, which produces BN classifiers with the structure of a BAN (where the class variable is a parent of all the input variables). The evaluation criteria consist of the following three types of performance measures: predictive accuracy (in general the most important criterion in the classification task of data mining), model size (measured by the total number of the edges in the network), and the running time.

The performance of the algorithms was evaluated using 18 public-domain datasets from the University of California at Irvine (UCI) dataset repository. The main characteristics of the datasets are shown in the URL in [1]. The experiments were carried out using the well-known *stratified* 10-fold cross validation procedure [20]. Since the ACO algorithms are stochastic, we run each 10 times – using a different random seed to initialize the search each time – for each of the 10 iterations of the cross-validation procedure. As for the parameter configurations, we set `colony_size` to 10 and `max_iterations` to 1000. Note than in the case of ABC-Miner+, each phase is allocated half of the total maximum number of iterations (i.e. 500 iterations in our experiments).

Table 1 shows the experimental results of the algorithms in three parts, one for each type of performance measure: predictive accuracy, model size, and running time. The entries in the table represent the mean values obtained by 10-fold cross validation. For each performance measure, the best result for each dataset is underlined.

In terms of predictive accuracy, the extended ABC-Miner+ algorithm obtained the best results in 14 out of 18 datasets, while ABC-Miner and ABC obtained the best results in 5 and 3 datasets, respectively. ABC-Miner+ outperformed ABC-Miner in 13 datasets plus 1 tie, while ABC outperformed ABC-Miner in 3 datasets. It is interesting to notice that ABC, which is only the first phase of ABC-Miner+ can find the best BN classification model in some datasets, and in those datasets the second phase of ABC-Miner+ does not improve its performance. This can be noticed in datasets `hayes`, `monk` and `pima`, where ABC and ABC-Miner+ have the same predictive performance and almost the same model (size).

We used the matched-pair samples Wilcoxon Signed-Rank statistical test [9] to compare the predictive accuracies of ABC-Miner+ and ABC-Miner, where the samples are the datasets. According to the Wilcoxon test, the Z-value is -2.2012, and the p-value is 0.0139. Therefore, the results of ABC-Miner+ are statistically significantly better at the 5% significance level.

In terms of model size, ABC obviously produces the smallest BN models, whose sizes are lower limits for the sizes of the models produced by ABC-Miner+,

Table 1. Results of predictive accuracy (%), BN size (number of edges), and running time (in seconds) for the three ACO-based Bayesian classification algorithms

Dataset	Predictive Accuracy			Model Size			Running Time		
	ABC-Miner	ABC	ABC-Miner+	ABC-Miner	ABC	ABC-Miner+	ABC-Miner	ABC	ABC-Miner+
balace	77.5	75.8	82.4	6.5	3.1	5.1	40	10	35
breast-w	92.7	90.4	95.8	23.6	8.3	14.8	540	130	280
car	97.2	86.7	98.1	14.5	6.0	11.2	480	240	310
contraceptive	66.5	75.0	76.7	18.7	7.2	10.4	220	150	550
credit-a	86.5	81.8	84.2	23.6	10.3	15.3	230	140	210
credit-g	71.7	70.2	73.8	28.2	11.2	18.5	520	120	500
dermatology	99.1	96.9	98.4	34.6	17.4	26.6	440	240	450
glass	93.3	87.6	91.3	11.5	3.0	4.1	140	90	110
hayes-roth	80.0	80.2	80.2	12.7	6.0	7.3	60	10	30
heart-c	83.4	74.6	86.9	18.4	10.2	21.3	510	250	480
inf	82.4	78.7	85.7	5.1	3.0	4.2	20	10	13
ionosphere	96.2	92.3	96.8	21.6	20.4	25.5	560	200	310
monk	65.2	74.2	74.2	17.4	6.3	6.3	30	10	50
nursery	98.2	94.7	96.7	22.6	6.0	14.6	430	290	320
pima	77.8	81.4	81.4	7.4	3.0	3.1	16	10	18
soybean	95.6	92.9	95.6	21.8	18.4	28.6	420	180	360
tic-tac-to	86.4	85.6	86.8	27.8	8.0	9.6	360	160	320
vote	94.8	94.3	95.6	34.6	12.4	15.6	400	280	280

since the second phase of that algorithm can only add (and not remove) more edges to the BNs learnt by ABC. Moreover, the maximum number of the edges in a BN produced by ABC equals to the number of the input attributes (if all the input variables have relationships to the class), which is also the minimum number of edges that a BAN produced by ABC-Miner may have (if the local search procedure removed all the edges between the input variables and reduced the BAN to a Naïve-Bayes structure).

Besides, in terms of model size ABC-Miner+ outperformed the original ABC-Miner in all the datasets, producing BN classification models with fewer edges. The feature selection process implicitly performed by our extended algorithm can be easily noticed in the results of ABC. In some datasets, such as `breast-w`, `credit-a`, `credit-g`, and `dermatology`, the number of edges in the model produced by ABC-Miner+ is less than the number of input attributes in the corresponding dataset. This means that the produced BN classification model does not have all the input variables related to the class variable.

In terms of running time, as expected, ABC took the least amount of time to finish its execution in all the datasets. On the other hand, the two-phase ABC-Miner+ algorithm achieved a shorter execution time than ABC-Miner in 14 datasets. The reason behind that, as explained in Section 4, is that the first phase of the ABC-Miner+ algorithm reduced the search space for the second phase, after producing a BN structure with different dependency relationship types defined between the input and the class variables, and the first phase (ABC) does not consume a large amount of time, as shown in the results.

6   Concluding Remarks

In this paper, we have introduced ABC-Miner+ an extended version of an ACO algorithm for learning BN classifiers. ABC-Miner+ builds class Markov blanket-based BN classification models, in which it is not necessary to have an edge between the class variable and each of the input variables, and the edges between the class and the input variables may have different directions; unlike ABC-Miner, which learns BAN models. Empirical results showed that, overall, the ABC-Miner+ algorithm has an improved performance over the original ABC-Miner in terms of predictive accuracy, model size, and running time.

As a future research direction, we would like to investigate a different approach to learn the Markov blankets in a single integrated phase, rather than in two sequential phases as in ABC-Miner+. Moreover, we would like to try techniques to avoid over-fitting on the learning set during the training phase, like using different random partitions of learning\validation sets each iteration.

References

[1] UCI Repository of Machine Learning Databases,
    http://www.ics.uci.edu/~mlearn/MLRepository.html
    (retrieved October 2011)

12   K.M. Salama and A.A. Freitas

[2] de Campos, L.M., Fernandez-Luna, J.M., Gamez, J.A., Puerta, J.M.: Ant colony optimization for learning Bayesian networks. International Journal of Approximate Reasoning 31(3), 291–311 (2002)

[3] Cheng, J., Greiner, R.: Learning bayesian belief network classifiers: Algorithms and system. In: Stroulia, E., Matwin, S. (eds.) Canadian AI 2001. LNCS (LNAI), vol. 2056, pp. 141–151. Springer, Heidelberg (2001)

[4] Chickering, D.M.: Learning Bayesian Networks is NP-Complete. Advanced Technologies Division, Microsoft Corporation, Redmond, WA. Technical Report (1996)

[5] Daly, R., Shen, Q., Aitken, S.: Learning bayesian networks: Approaches and issues. Knowledge Engineering Reviews 26(2), 99–157 (2011)

[6] Dorigo, M., Stützle, T.: Ant Colony Optimization. MIT Press (2004)

[7] Friedman, N., Geiger, D., Goldszmidt, M., Provan, G., Langley, P., Smyth, P.: Bayesian Network Classifiers. Machine Learning 29, 131–163 (1997)

[8] Heckerman, D.: A Tutorial on Learning with Bayesian Networks. In: Holmes, D.E., Jain, L.C. (eds.) Innovations in Bayesian Networks. SCI, vol. 156, pp. 33–82. Springer, Heidelberg (2008)

[9] Japkowicz, N., Shah, M.: Evaluating Learning Algorithms: A Classification Perspective. Cambridge University Press (2011)

[10] Jiang, L., Wang, D., Cai, Z., Yan, X.: Survey of improving naïve-bayes for classification. In: Alhajj, R., Gao, H., Li, X., Li, J., Zaïane, O.R. (eds.) ADMA 2007. LNCS (LNAI), vol. 4632, pp. 134–145. Springer, Heidelberg (2007)

[11] Kononenko, I.: Semi-naive bayesian classifier. In: Kodratoff, Y. (ed.) EWSL 1991. LNCS, vol. 482, pp. 206–219. Springer, Heidelberg (1991)

[12] Martens, D., Backer, M.D., Haesen, R., Vanthienen, J., Snoeck, M., Baesens, B.: Classification with ant colony optimization. IEEE Transactions on Evolutionary Computation 11, 651–665 (2007)

[13] Martens, D., Baesens, B., Fawcett, T.: Editorial survey: swarm intelligence for data mining. Machine Learning 82(1), 1–42 (2011)

[14] Parpinelli, R.S., Lopes, H.S., Freitas, A.A.: Data Mining with an Ant Colony Optimization Algorithm. IEEE Transactions on Evolutionary Computation 6(4), 321–332 (2002)

[15] Pearl, J.: Causality: Models, Reasoning and Inference. Cambridge University Press (2000)

[16] Pinto, P.C., Nägele, A., Dejori, M., Runkler, T.A.: Using a Local Discovery Ant Algorithm for Bayesian Network Structure Learning. IEEE Transactions on Evolutionary Computation 13(4), 767–779 (2009)

[17] Salama, K.M., Abdelbar, A.M., Otero, F.E., Freitas, A.A.: Utilizing Multiple Pheromones in an Ant-based Algorithm for Continuous-Attribute Classification Rule Discovery. Applied Soft Computing 13(1), 667–675 (2012)

[18] Salama, K.M., Abdelbar, A., Freitas, A.A.: Multiple Pheromone Types and Other Extensions to the Ant-Miner Classification Rule Discovery Algorithm. Swarm Intelligence 5(3-4), 149–182 (2011)

[19] Salama, K.M., Freitas, A.A.: ABC-Miner: an Ant-based Bayesian Classification Algorithm. In: Dorigo, M., Birattari, M., Blum, C., Christensen, A.L., Engelbrecht, A.P., Groß, R., Stützle, T. (eds.) ANTS 2012. LNCS, vol. 7461, pp. 13–24. Springer, Heidelberg (2012)

[20] Witten, I.H., Frank, E.: Data Mining: Practical Machine Learning Tools and Techniques, 3rd edn. Morgan Kaufmann (2010)

[21] Wu, Y., McCall, J., Corne, D.: Two novel Ant Colony Optimization approaches for Bayesian network structure learning. In: International Conference on Evolutionary Computation (CEC), pp. 1–7 (2010)

A Multiple Pheromone Ant Clustering Algorithm

Jan Chircop and Christopher D. Buckingham

Aston University,
Birmingham, United Kingdom
janchircop@gmail.com, C.D.Buckingham@aston.ac.uk

Abstract. Ant colony optimisation algorithms model the way ants use pheromones for marking paths to important locations in their environment. Pheromone traces are picked up, followed, and reinforced by other ants but also evaporate over time. Optimal paths attract more pheromone and less useful paths fade away. The main innovation of the proposed Multiple Pheromone Ant Clustering Algorithm (MPACA) is to mark objects using many pheromones, one for each value of each attribute describing the objects in multidimensional space. Every object has one or more ants assigned to each attribute value and the ants then try to find other objects with matching values, depositing pheromone traces that link them. Encounters between ants are used to determine when ants should combine their features to look for conjunctions and whether they should belong to the same colony. This paper explains the algorithm and explores its potential effectiveness for cluster analysis.

Keywords: Ant Colony Algorithms, Swarm Intelligence, Emergent Behaviour, Cluster Analysis, and Classification.

1 Introduction

Making sense of large data sets can be approached from two directions. One is to exploit existing knowledge, often using humans who are experts in the domain. The alternative is to have no preconceptions and tackle the data from the bottom up, which is how computational models of ants and other social insects do it. Social insects do not have the sophisticated brain power of humans but compensate by pooling their resources in very large numbers. The results are impressive enough to inform human knowledge and understanding of the world with, for example, architects using ideas from termite nests to incorporate natural air conditioning into multi-storey buildings [1]. This paper investigates how computer models of ant colony behaviours can help humans sort data into meaningful classes.

The paper begins by reviewing current computational models of ants that are applied to cluster analysis and classification models. Conclusions from the literature review are used to introduce the main innovations of the proposed model. The algorithm is then tested on a number of classical data sets [2]. The

G. Terrazas et al. (eds.), *Nature Inspired Cooperative Strategies for Optimization*
(NICSO 2013), Studies in Computational Intelligence 512,
DOI: 10.1007/978-3-319-01692-4_2, © Springer International Publishing Switzerland 2014

paper concludes with a discussion of the model, its effectiveness, and how it can be applied to more varied data.

2   Background

Swarm intelligence (SI) describes how the interactions of multiple separate entities with limited individual cognitive capabilities can lead to more sophisticated intelligent behaviour [3], [4], [5]. Many examples exist in the insect world including the aforementioned termites and our particular focus on ant colonies. There are several types of ant behaviour that have been used as metaphors for computer science [6] but the two most relevant to the proposed model are how ants sort larvae and corpses and how they forage for food.

The larvae or corpse-sorting metaphor is known as the Basic Model (BM) [7], because it is purely about clustering ant bodies into piles, referenced as the Standard Ant Clustering Algorithm (SACA). The key operator is a similarity function, whereby the ants attempt to move items into areas where other similar items are located. In contrast, the ant foraging metaphor simulates how ants lay scent trails or pheromones to create paths that other ants follow. It has led to Ant Colony Optimisation (ACO) algorithms where the build-up of pheromones is used to optimise the shortest path on a fully connected graph [8]. The proposed Multiple Pheromone Ant Clustering Algorithm, MPACA, is derived from foraging behaviour, which has many computational variants described by researchers. The following section explores those most relevant to the MPACA.

2.1   Algorithms Inspired by the Foraging Behaviour of Ants

The use of scent or pheromone to form paths is a form of stigmergy, where information is placed in the environment for communication purposes [6] [23]. During foraging, real ants deposit a pheromone trace to mark the paths to food. The pheromone is laid down at full strength but then evaporates over time. The shorter the distance between the nest and food, the less time for evaporation to occur and the faster the ants are able to reinforce the path. Hence, the shorter paths tend to have higher levels of pheromone, which in turn attracts more ants, and eventually the colony learns the best path to the food source. Stigmergy thus works as a function of distance and evaporation rate. Whether or not an ant follows a path is probabilistically related to the strength of the path; there will always be a few ants that go off the beaten track, which is important for learning new or changing situations. A coherent mathematical description is given by [24], where they obtain encouraging results from applying ACO to the real world domain of road safety at intersections.

Ant Clustering Algorithms Using Pheromones. Many ant colony clustering algorithms such as ACLUSTER [20] and ANTCLUST [14], [15] use pheromones to operationalise the process (see [13] for a full review). ANTCLUST constructs a colonial odour for determining an ant's nest membership so that ants can discriminate between nest mates and intruders. It accords with natural ant societies and variants have been shown to separate noise from data [16], conduct credit evaluation of small enterprises [17], and mine information on the web [18].

Induction of multiple colonies within a graph space is proposed by [19], where each colony is represented by a distinct colour. Ants travel along the graph and lay pheromones corresponding to their colony colour. Colonies compete to colonise separate sub-sections of the graph space, being attracted by the pheromone representing their colour, and repulsed by pheromones of other colours (colonies). Organisational information emerges from the global behaviour of ants in their colonies.

Some clustering approaches widen the horizon within which ants can see or sense the world such as the adaptive ant-based clustering algorithm (AACA) [21] and the Aggregation Pheromone Density Based Clustering (APC) algorithm [22]. They allows ants to see further than the single step ahead, overcoming shortsightedness of the BM and its derivatives.

The Adaptive Graphic Clustering (AGCM) algorithm [25] is based on a digraph where the similarity between objects in the space of object attributes is calculated as weights on the directed edges of a pheromone map. The weight on each edge is adaptively updated by the ants during the search process, in relation to the pheromone being deposited.

The Ant-Miner algorithm [26] and its derivatives are also clustering algorithms using pheromones but their output is a set of association rules defining the clusters. This has the advantage that data representation is more comprehensible to the user. The main loop of the algorithm consists of three key steps, namely rule construction, rule pruning, and pheromone updating. Results show that Ant-Miner has good classification performance on test data sets and the ability to constrain the number of rules required [26], [27].

Multi-colony and Multi-pheromone Approaches. In real ant colonies, there are subsets of ants within the same colony that have different objectives and they may lay down distinctive pheromones accordingly [9]. However, computational models of ACO usually exploit multiple pheromones to distinguish between colonies, not ants within colonies. [10] provide an exception, but still only use two pheromone types: a trailing pheromone for leading ants towards a nest or clusters of other ants and a foraging pheromone for locating new food sources.

In multi-colony models, ants are sub-divided into several colonies, each evolving independently of each other. ACO models use colonies to subdivide the problem into separate elements with the various results integrated at the end to provide the overall solution [11],[12].

2.2   Conclusions from the Literature

ACO algorithms are powerful methods for analysing data without having any prior knowledge of their inherent structure. They rely on independent entities with minimal intelligence whose actions are governed by specific information available within their immediate environment. Despite these limitations, the collective behaviour of the ant population or colony provides emergent properties that represent pattern-recognition information in the data.

The literature review shows many potential candidates that can enhance cluster analysis. Some depend on ants picking up and dropping objects to generate the clusters; others require ants to move themselves into positions reflecting the structure. This paper focuses on the latter and, more specifically, those algorithms where ants follow pheromone trails that keep them within a localised area of the hyperdimensional space, which is often represented by a linked graph.

Although some previous ant models have multiple pheromones, none of them have a different pheromone associated with each distinguishing feature of the objects to be analysed. This is a key innovation for our proposed model, which is why we have called it the Multiple Pheromone Ant Clustering Algorithm, MPACA. It differs from alternative models by being both a multi-pheromone and multi-colony ACO approach. It incorporates the ability of individual ants to detect feature combinations and to form colonies based on the number of encounters they have with other ants. Ants don't stop when they reach somewhere they 'like', as in the ASM, and they don't have nests to which they need to return. Instead, their movements maintain a dynamic equilibrium where ants that are interested in similar objects end up clustered together. The next section will introduce the model and describe its operation in detail.

3   Method for MPACA

The MPACA behaviour depends on a number of system elements and their parameters. These include the domain architecture, ant movements, pheromone deposition, ant encounters, and the colony and feature merging processes.

3.1   Domain Architecture

The MPACA is applicable to multiple dimensions and can accommodate any type of data, whether it is continuous or discrete, ordinal or nominal. However, only ordinal dimensions are used to set up the hyperdimensional problem space. Non-ordinal variables are still part of the feature set detected by ants but influence clustering behaviour through their pheromone trails rather then as part of the domain architecture.

Ordinal dimensions are normalised to help prevent bias due to types of distributions and ranges. The values are converted into the number of standard deviations (SDs) from the mean, z, where $z = (x - \mu)/SD$, x, is the original value and μ, is the mean. This provides all dimensions in the hyperdimensional space with equitable units based on their distributions. Object values are likewise normalised so that objects can be appropriately placed in the graph space, after which they are joined by edges. The resulting graph, G, has a vertex, v, for every object. In theory, ants should be able to move from any object to any other object but the graph space is kept more manageable by connecting objects with an edge, e, that is within a distance parameter, d: if two objects are beyond that distance apart, then they are considered to be outliers and not joined.

Ants are placed on every vertex or object with at least one ant for each attribute of the object. This means the value for each dimension of the object's graph location (e.g. length, width) and the value for each of its nominal features (e.g. colour, shape) has one or more ants assigned to it. The ant's own attribute value becomes the distinctive pheromone for the ant, which it lays down on the edges as it travels from objects with matching values. Edges therefore have multiple layers of pheromone placed on them corresponding to the different features of objects and ants respond to the pheromones which are of interest to it. There will be as many pheromone scents in the domain as there are distinct attribute values, including nominal features and ordinal dimensions.

The Step Size. The granularity of differences between objects along a dimension is determined by how far the ant can travel in each time step, which is the step size. A reasonable assumption is that four SDs from the mean (plus or minus 2 SDs) covers most population values on that dimension except the outliers. A step size of 0.1 SDs gives 40 steps along each dimension, which is enough to detect meaningful differences between objects. Values are converted into positive equivalents, which means 0, is the lowest negative deviation from the mean and all other values are positive, measured from this point.

When joining objects by edges, the edge distance is the Euclidean function of the individual dimension lengths. As the number of dimensions in the hyperdimensional space increases, the edges between objects become longer by default. To counteract this, the step length of the edge, $EdgeStep$, is calculated as the Euclidean distance between points that are one step away along every dimension.

$$EdgeStep = \sqrt{\sum_{i=1}^{D} DimensionStep_i^2} \qquad (1)$$

where D is the number of dimensions and $DimensionStep$ is the step size along each dimension in SDs from the mean. The length of each edge is thus its Euclidean distance divided by $EdgeStep$, which is now normalised to ensure dimensionality does not increase the edge length excessively compared to dimensions.

3.2 Ant Movement

Ants move one step at a time and each movement is recorded as one timestep for the whole system. Ants on a vertex choose which edge to visit next using a stochastic approach that distinguishes it from [6]. This mechanism does not require any foresight about the potential vertices that can be visited, and has the single restriction that ants cannot go back along an edge they have just traversed. The path or edge to follow is chosen as a probabilistic function of the strength of matching pheromone on the first step of each edge leading from the vertex: the higher the strength, the more likely the path will be chosen.

Whenever ants reach a vertex that has a feature set matching their own, they will deposit a trail when leaving it but will go into non-deposit mode otherwise. Vertices with similar feature values will tend to have connecting paths with higher levels of pheromone traces than objects connecting different feature values. Ants will lay trails in both directions for similar objects but only in one direction for dissimilar ones. Pheromone trails are diminished by a parameterised amount due to evaporation and must be reinforced if they are to be maintained. The longer the path between objects, the weaker the connection, all things being equal.

3.3   The Multi-pheromone and Multi-colony Approach

The MPACA makes use of multiple pheromones and multiple colonies. Initially, each ant belongs to its own distinct colony and carries only one feature value for detection. Over time, colonies begin to merge but this does not change the movements of the ants. Feature merging does, though, because ants then respond only to objects that match their multiple feature set and lay down trails for all of their features. The deposited pheromones are independent of the colony to which the ant belongs but obviously ants looking for similar features are likely to be in the same vicinity and thus become part of the same colony. Both colony and feature merging are governed by meetings or encounters between ants.

3.4   Ant Encounters

Each ant records meetings or encounters with other ants in a list structure that is only accessible by the individual ant. These encounters are used to determine whether ants should share the same features or join the same colony.

On each encounter, the ant records the following information of the other ant: the ant identifier (id), the colony id, the carried feature id, the timestep, and a boolean flag holding the deposit mode of the encountered ant at that time stamp. This is put into the *AntSeenRecord*, within the *AntSeenList*. The size of the list structure is kept in check by the time stamp which is placed on it. On exceeding the time-window parameter, this encounter is removed.

Encounters are only recorded for an ant when it is at a vertex. This ant will be called the focus ant to prevent referential ambiguity; it is the only one that has its AntSeenList updated when the encounters are recorded. Other ants are recognised as an encounter if they are within the focus ant's visibility range, v, which is a parameterised number of steps from the vertex along its edges. The focus ant's AntSeenList is only updated if it is put in pheromone deposition mode at the vertex. Ants in deposit mode and travelling on a path away from the object are recorded as having encountered the ant as well as ants coming towards the vertex the ant is on, irrespective of their deposit mode (both types of ant must be within the visibility range).

3.5   Colony and Feature Merging

Feature and colony merging are both triggered asynchronously at the ant level on arrival at a vertex. The ant's deposit mode is updated, ant encounters recorded as described earlier, and the updated AntSeenRecord used to compute whether to merge colonies and/or features. The colony memberships of encountered ants in the focus ant's AntSeenList are counted if those ants were in deposit mode. If the number for the largest colony is more than the threshold parameter, the ant joins this colony. A similar process applies to the feature merging except that the encountered ants do not have to be in deposit mode. If the number of ants with the same feature is above the feature-merging threshold, the focus ant adds the feature to the one or more features it is already detecting and drops any that are now below the threshold.

3.6   The MPACA Algorithm

Require: Graph space with connecting edges and ants assigned to each feature.

    while (*Termination not met*) **do**
        for (*Each ant in antlist*) **do**
            Increment *StepNumber* against all encounters in AntSeenList by one
            if (*StepNumber > threshold*) **then**
                Remove encounter from AntSeenList
            end if
            if (*Ant at vertex*) **then**
                Update *AntSeenList* counts;
                if (*Ant features match object*) **then**
                    Activate pheromone deposition mode;
                    Process *AntSeenList* for colony and feature merging
                else
                    Deactivate pheromone deposition mode;
                end if
                Choose next edge stochastically taking pheromone values into account;
            end if
            EdgeTraversal ← *EdgeTraversal* − 1;
            if (*Ant in deposition mode*) **then**
                deposit pheromone for each feature;
            end if
        end for
        if (*Stopping criterion reached*) **then**
            Output cluster definitions;
        else
            Perform system wide evaporation;
        end if
    end while

In the MPACA, each step of the ants is a single time interval so edges which are n steps long will take n timesteps to traverse. The MPACA terminates when ants reach a stable dynamic equilibrium in the colonies they form. This is indicated by a consistent number of colonies and a stable population number in each one.

Clustering begins when the stigmergy of pheromone trails draws ants towards areas of the search space that have objects matching their features of interest. When the density of ants goes over the colony threshold, ant colonies will merge. If there are enough ants searching for a different feature in the same space, because the vertices also match their feature values, an ant can take on this feature in addition to its own. This enables ants to pick up feature combinations and provides colonies with the ability to detect non-linear relationships. For example, if an object of colour blue is always found next to a large object, then features *colour=blue*, *size=large* can be merged into one; the ants are no longer looking for just blue or large, but for the combined features blue AND large. In theory, this means the algorithm can solve the XOR problem (see [29]), which depends on knowing how values of two separate features occur together.

3.7   MPACA Parameters

The philosophy, properties, rationale, and operation of the MPACA model have been introduced but its overall emergent behaviour is dependent on the parameters controlling colony and feature combination. The following list discusses parameter functions in more detail.

Distance parameters
    The distance parameters help ants sense the environment beyond their immediate vicinity, similarly to AACA described earlier. The edge-length parameter, d, removes objects that are too far away to be worth exploring. The visibility parameter, v, controls the range within which encounters between ants are recorded. Together, they improve computational tractability and the influence of social interactions.

Pheromone deposition and evaporation
    If all the features being detected by an ant match those of an object, the ant will lay a pheromone trail down for each matching feature on the path leading away from the object. The same fixed amount is deposited for each feature for all ants, as prescribed by a parameter, ph. In line with the natural ant model, as soon as pheromone deposition takes place, evaporation occurs. A maximum pheromone level, $ph.max$, is set to prevent the amount on a path increasing indefinitely and overwhelming the influence of low-scent paths.

Ant path choice and pheromone parameters
    A residual parameter r, determines the percentage of total matching pheromone on all edges that is placed on each of them by default. It enables ants to go down paths with little or no scent and explore new areas, which helps prevent local minima and allows the system to evolve if the domain knowledge changes. Given N potential paths from a vertex with pheromone scent s on the first step of each path, where s is the pheromone matching

the features of the ant, the probability of selecting a particular path, p, is given by

$$P(p) = \frac{(s+r)}{\sum\limits_{i=1}^{N} s_i + (r \times N)}.$$

Detection range for continuous dimensions
The ant responds to (i.e. lays down a pheromone for) all values of a dimension that are within a parameterised range of its own object's value. The range is based on the step size for the dimension.

Ant complement
The ant complement, ac, is the number of ants associated with each feature of an object. This enables more sensitive detection of clusters and merging decisions because of an increased population density but the trade-off is greater computational load for each step. The optimal balance depends on both the density of objects within the given domain space as well as it's dimensionality.

Merging thresholds
There are two merging thresholds, one for colonies and one for features. The colony threshold, ct, determines when the population density of ants is high enough to trigger the ant joining a colony. The feature threshold, ft, is linked to the number of times a particular feature has been seen in other ants. When features do merge, the number of objects matching the ant's detectors will reduce and this will concomitantly reduce the number of paths laid by the ant. The consequence is a natural check on the combination process: specialising detection reduces the number of matching objects and the probability of an ant being in deposit mode, which is when it records other ant encounters.

Time-window
All ant encounters are recorded within a time window, defined by the maximum number of steps that can be remembered. The time-window, tw, parameter ensures two things: firstly, that the chances of over-fitting are reduced and secondly, the system can evolve if the knowledge domain changes structure.

3.8   Emergent Properties of MPACA

Emergent properties of the system as a whole come from analysing the state of the ants at the end of the learning process. The two influential phenomena are feature merging and colony formation. Most ants will have joined a colony, apart from the outliers, and the specification of the colony's membership criteria is given by the relative frequencies of ants with different feature detectors. Some will have single detectors, others will have multiple ones; together they provide a precise description of the cluster. This description can be converted into a classification algorithm for assigning unknown objects to classes. In this paper, a simpler method was used. The centre of each colony was taken to be the multivariate mean and each object was assigned to the colony with the centre that had the shortest Euclidean distance.

4   Evaluation and Results

The MPACA evaluation consists of a two-step process. First, the ant colonies and their location clusters are formed. Then each object is assigned to the nearest colony as described earlier. Table 1 shows the results of applying the MPACA to three data sets from the Machine Learning Repository [2] that have commonly been used to compare ant algorithms. The Iris data have two non-linearly separable classes, 'versicolor' and 'virginica', and one linearly separable class, 'setosa'. The Wine data is a 13 dimension set having all three classes linearly separable from each other. The reduced Wisconsin Breast Cancer (WBC) data set has 9 dimensions in two overlapping classes of malignant and benign cases.

Accuracy of cluster formation was evaluated using the F-Measure and Rand Index [31] standard metrics, which enabled results to be compared with other cluster and classification techniques found in the literature. Table 1 shows the mean and standard deviations of these indices for a variety of algorithms including the MPACA, which also has the measures for its best fit at the bottom.

Table 1. Application of classifiers to the Iris, Wine, and WBC data sets, with mean and standard deviations (in brackets) of runs. N/A means the results are not available; ABC is a derivation of the Standard Ant Clustering Algorithm, SACA; Average link is a hierarchical classifier; C4.5 is a decision-tree algorithm; RIPPER is an inductive rule learner; 1NN is Nearest Neighbour analysis; Logit is based on regression; SVM is a Support Vector Machine; and PB-ACA is a pheromone-based hybrid with SACA, density based DBSCAN and distribution-based EM-Clustering.

	Iris		Wine		WBC	
Classifier	**F**	**Rand**	**F**	**Rand**	**F**	**Rand**
ABC [30]	.82(.015)	.83(.009)	N/A	N/A	.97(.002)	.94(.003)
K-Means[30] [34]	.83(.085)	.82(.101)	.82(.034)	.84(.090)	.97(.000)	.93(.065)
Average link [30],[34]	.81(.000)	.83(.000)	.84(.000)	.81(.000)	.97(.000)	.93(.000)
PACE [34]	.82(.014)	.82(.008)	.88(.003)	.82(.090)	N/A	N/A
Ant-Miner [28]	N/A	.77(.039)	N/A	.85(.057)	N/A	.91(.024)
Ant-Miner+ [28]	N/A	.95(.008)	N/A	.95(.017)	N/A	.96(.005)
Ant-Miner2 [28]	N/A	.82(.032)	N/A	.85(.064)	N/A	.92(.027)
Ant-Miner3 [28]	N/A	.77(.038)	N/A	.84(.050)	N/A	.91(.030)
RIPPER [28]	N/A	.93(.019)	N/A	.91(.049)	N/A	.95(.025)
C4.5 [28]	N/A	.94(.022)	N/A	.90(.050)	N/A	.95(.027)
1NN [28]	N/A	.91(.022)	N/A	.95(.018)	N/A	.96(.015)
Logit [28]	N/A	.94(.029)	N/A	.94(.026)	N/A	.97(.010)
SVM [28]	N/A	.95(.026)	N/A	.95(.035)	N/A	.93(.058)
PB-ACA [32]	N/A	.79	N/A	N/A	N/A	N/A
DBSCAN [37]	N/A	.76(.000)	N/A	.73(.000)	N/A	.83(.000)
EM-Clustering [37],[38]	N/A	.91(.000)	N/A	.97(.000)	N/A	.96(.000)
MPACA	.82(.036)	.88(.023)	.87(.045)	.91(.029)	.93(.037)	.93(.036)
MPACA best run	.88	.92	.95	.97	.97	.97

Twenty MPACA runs were generated over 2000 time intervals. In this exploratory phase of testing, parameters were initialised with intuitively sensible settings as explained for the domain architecture, where step size and dimension ranges were based on covering the main population with enough granularity to distinguish individuals appropriately (see Section 3.1). Other parameters were manually adjusted to evaluate their impact and to approximate an optimal setting. For each of the 20 runs, the same settings were applied to all three data sets apart from the time window. The following list shows the mean parameter values with standard deviations in brackets (see Section 3.7 for details of the parameter operations).

- *Detection range* for continuous dimensions: 1.5 steps (0.5).
- *Ant Complement*: 3 (2.1).
- *Pheromone evaporation* on each timestep: 7.9 units (4.5).
- *Pheromone deposition* on each step of ants in deposit mode: 100 units (31.4).
- *Residual percentage* of pheromone placed on all steps: 3.5 (1.1).
- *Max pheromone* on a step, which is operationalised as a multiplier of the pheromone deposition parameter (i.e. if deposition per step is 100 units, the maximum allowed on a step is 150 if the multiplier is 1.5): 1.5 (0.5).
- *Feature merging threshold*: 5.9 (1).
- *Colony merging threshold*: 4 (1.0).
- *Time-Window*: Iris, 100 (32.4); Wine, 112.5 (26.3); WBC, 75 (16.2).

Preliminary results show that the MPACA generates sensible results that are equitable with alternative approaches, both in general and compared to ant colony optimisation models. At this stage, the results have provided support for the MPACA principles but it is clear that a better method for testing model parameters is required, along with more sophisticated assignation of objects to colonies. These are undoubtedly reasons why, for example, the rule-extraction approach of Ant Miner is giving it a better performance across all data sets. The best fit for the MPACA suggests improvements can be made and that the multiple pheromone approach is worth further investigation.

4.1   How the MPACA Compares to Other Clustering Techniques

It is not easy determining the "active ingredients" that distinguish one model from another when they all have the same classification goal. This paper has tried to show the high-level distinctions of the MPACA, which is around its use of multiple pheromone deposition linked to separate object features. However, much of the output performance depends on operational details of parameters and their settings; it is unclear at this early stage how influential these are compared to the overall mechanisms. For example, the MPACA is a bottom-up approach that first puts neighbouring objects into the same cluster and gradually extends the objects belonging to that cluster as colonies merge. This is not entirely disimilar to hierarchical clustering in principle but the distance measures are very different, being based on ant movements for the MPACA. Table 1 shows that the average-link hierarchical clustering approach [30] does have similar output to the MPACA and further work would be needed to determine which operational elements are distinguishing their behaviours.

The MPACA can be viewed both as a distribution-based (e.g. EM-clustering [35]) and a density-based algorithm (e.g. DBSCAN [36]). Ants merge into clusters depending on the density of other ants on surrounding objects. Like DBSCAN, outliers in the MPACA will fail to form bigger clusters and can be ignored. Similarities exist with EM-clustering when assigning objects to learned classes on the basis of ant colonies and their demographics. For this paper, much more work is needed on how to optimise this for the MPACA but the same caveats for over-fitting will apply.

The MPACA ant density is controlled by pheromones which are in turn controlled by the feature cues being sought. Unlike most other clustering algorithms, where clusters form over dimensions, the MPACA allows for mixed types of data: although nominal features do not contribute to the hyperdimensional space, they are included in the pheromone deposition. The advantage is better exploitation of all available clustering information.

5   Conclusion

This paper has described a new Ant Colony Clustering model called the Multi-pheromone Ant Clustering Algorithm, MPACA. It is related to other ant models by using pheromones within a graph domain space and, to some extent, by its use of more than one pheromone for laying down information along a path. However, no other algorithm has a pheromone for every attribute value of the objects in the domain space, which is the core innovation of the MPACA. The ants are able to link similar features of objects, to combine the features they detect, and to form colonies based on local ant population densities. Together, these enable ants to learn the feature profile for different clusters and for mapping colony membership onto those clusters.

The remit of this paper was to describe the MPACA and compare it with other ant models and standard classifiers to demonstrate the potential for its new properties. However, much more work needs to be done on how to adapt the parameter settings for optimising the final cluster analyses. This is non-trivial because it is computationally time consuming for the ants to stabilise their colonies and this process would have to be repeated many times if the parameter settings are to be adapted to reduce errors by, for example, simulated annealing. On the other hand, the algorithm has been deliberately prevented from using any global operators, which means it is ideally suited to parallel processing. At each timestep, for example, the data structure for recording ant encounters is updated for an individual ant independently of the data structures for any other ant and only using information that is in the local "visibility range" of the ant.

The method for assigning objects to classes based on ant colonies can certainly be improved because distance to single centroid points is not a very subtle measure. The next steps will be to investigate alternative classification methods, to optimise the program so that it can learn more quickly, and to apply the algorithm to real-world, much larger, data sets. The idea is to see how the bottom-up learning of ants compares to the top-down classification algorithms used in cognitive modelling of human experts. Two domains where the authors already have a cognitive model driving decision making are in mental-health risk assessment [39] and hub-and-spoke logistics [40]. The domains have extremely high dimensions (over 200 for the mental-health one) and extremely high numbers of cases (many millions for the logistics domain). These present serious challenges for the tractability of the MPACA but the rewards are high. If the MPACA can form accurate clusters, these will have ant populations that represent a detailed analysis of the relative importance of features and feature combinations required for cluster membership. They can be used to output sophisticated classification rules that complement decision making used in the cognitive models and may even be able to improve those models.

Acknowledgement. Work presented in this paper was partly supported by the EU FP7 under Grant No. 257398, "Advanced predictive-analysis-based decision-support engine for logistics" `www.advance-logistics.eu`.

References

[1] French, J.R.J., Ahmed, B.M.: The challenge of biomimetic design for carbon-neutral buildings using termite engineering. InsectScience 17(2), 154–162 (2010)
[2] Bache, K., Lichman, M.: UCI Machine Learning Repository. University of California, School of Information and Computer Science, Irvine (2013), `http://archive.ics.uci.edu/ml`
[3] Guerona, S., Levin, S.A., Rubenstein, D.I.: The dynamics of herds: From Individuals to Aggregations. Journal of Theoretical Biology 182, 85–89 (1996)
[4] Parrish, J.K., Hamner, W.M.: Animal Groups in Three Dimensions, How Species Aggregate. Cambridge University Press (1997)
[5] Murray, J.D.: Mathematical Biology. Springer, New York (1989)
[6] Dorigo, M., Birattari, M., Stutzle, T.: Ant colony optimization, vol. 1, pp. 28–39 (November 2006)
[7] Deneubourg, J.L., Gross, S., Franks, N., Sendova-Franks, A., Detrain, C., Chrétien, L.: The dynamics of collective sorting robot-like ants and ant-like robots. In: Proceedings of the First International Conference on Simulation of Adaptive Behavior on From Animals to Animats, pp. 356–363 (1990)
[8] Dorigo, M.: Optimisation, Learning, and Natural Algorithms. Ph.D. Thesis. Dipartimento Di Elettronica, Politecnico Di Milano, Milan, Italy (1992)
[9] Dussutour, A., Nicolis, S.C., Shephard, G., Beekman, M., Sumpter, D.J.T.: The role of multiple pheromones in food recruitment by ants. The Journal of Experimental Biology 212(4), 2337–2348 (2009)
[10] Ngenkaew, W., Ono, S., Nakayama, S.: Pheromone-based concept in Ant Clustering. In: 3rd International Conference on Intelligent System and Knowledge Engineering, ISKE 2008, Xiamen, November 17-19, vol. 1, pp. 308–312 (2008)
[11] Middendorf, M., Reischle, F., Schmeck, H.: Multi Colony Ant Algorithms. Journal of Heuristics 8(3), 305–320 (2002), `http://dx.doi.org/10.1023/A:1015057701750`, doi:10.1023/A:1015057701750

[12] Guntsch, M.: Ant Algorithms in Stochastic and Multi-Criteria Environments (2004)

[13] Jafar, O.A.M., Sivakumar, R.: Ant-based Clustering Algorithms: A Brief Survey. International Journal of Computer Theory and Engineering 2(5), 1793–8201 (2010), http://www.ijcte.org/papers/242-G730.pdf

[14] Labroche, N., Monmarché, N., Venturini, G.: A New Clustering Algorithm Based on the Chemical Recognition System of Ants. In: Proc. of 15th European Conference on Artificial Intelligence (ECAI 2002), Lyon, France, pp. 345–349 (2002)

[15] Labroche, N., Richard, F.J., Monmarché, N., Lenoir, A., Venturini, G.: Modelling of the Chemical Recognition System of Ants

[16] Zaharie, D., Zamfirache, F.: Dealing with noise in ant-based clustering. In: The 2005 IEEE Congress on Evolutionary Computation, September 2-5, vol. 3, pp. 2395–2401 (2005)

[17] Liang, X.-C., Chen, S.-F., Liu, Y.: The study of small enterprises credit evaluation based on incremental AntClust. In: IEEE International Conference on Grey Systems and Intelligent Services, GSIS 2007, Nanjing, November 18-20, pp. 294–298 (2007)

[18] Inbarani, H.H., Thangavel, K.: Clickstream Intelligent Clustering using Accelerated Ant Colony Algorithm. In: International Conference on Advanced Computing and Communications, ADCOM 2006, December 20-23, pp. 129–134 (2006)

[19] Bertelle, C., Dutot, A., Guinand, F., Olivier, D.: Organization Detection Using Emergent Computing. International Transactions on Systems Science and Applications (ITSSA) 2(1), 61–69 (2006)

[20] Ramos, V., Muge, F., Pina, P.: Self-Organized Data and Image Retrieval as a Consequence of Inter-DynamicSynergistic Relationships in Artificial Ant Colonies. In: Hybrid Intelligent Systems, Frontiers of Artificial Intelligence and Applications, AEB 2002, vol. 87, pp. 500–509 (December 2002)

[21] El-Feghi, I., Errateeb, M., Ahmadi, M., Sid-Ahmed, M.A.: An adaptive ant-based clustering algorithm with improved environment perception. In: IEEE International Conference on Systems Man and Cybernetics Systems, SMC 2009, San Antonio, TX, October 11-14, pp. 1431–1438 (2009)

[22] Kothari, M., Ghosh, S., Ghosh, A.: Aggregation Pheromone Density Based Clustering. In: 9th International Conference on Information Technology, ICIT 2006, Bhubaneswar, December 18-21, pp. 259–264 (2006)

[23] Shelokar, P.S., Jayaraman, V.K., Kulkarni, B.D.: An ant colony approach for clustering. Analytica Chimica Acta 509(2), 187–195 (2004)

[24] Jiang, H., Chen, S.: A new ant colony algorithm for a general clustering. In: IEEE International Conference on Grey Systems and Intelligent Services, GSIS 2007, Nanjing, November 18-20, pp. 1158–1162 (2007)

[25] Yang, H., Li, X., Bo, C., Shao, X.: A Graphic Clustering Algorithm Based on MMAS. In: IEEE Congress on Evolutionary Computation, CEC 2006, Vancouver, BC, September 11, pp. 1592–1597 (2006)

[26] Parpinelli, R.S., Lopes, H.S., Freitas, A.A.: Data mining with an ant colony optimization algorithm. IEEE Transactions on Evolutionary Computation 6(4), 321–332 (2002)

[27] Parpinelli, R.S., Lopes, H.S., Freitas, A.A.: An Ant Colony Based System for Data Mining: Applications To Medical Data. In: Proceedings of the Genetic and Evolutionary Computation Conference, GECCO 2001, pp. 791–797 (2001)

[28] Martens, D., De Backer, M., Haesen, R., Vanthienen, J., Snoeck, M., Baesens, B.: Classification With Ant Colony Optimization. IEEE Transactions on Evolutionary Computation 11(5), 651–665 (2007); Sponsored by : IEEE Computational Intelligence Society

[29] Elizondo, D.: The Linear Separability Problem: Some Testing Methods. IEEE Transactions on Neural Networks 17(2), 330–344 (2006)

[30] Handl, J., Knowles, J., Dorigo, M.: On the performance of ant-based clustering. In: Proceedings of the Third International Conference on Hybrid Intelligent Systems Frontiers in Artificial Intelligence and Appliations, vol. 104, pp. 204–213 (2003)

[31] Sasaki, Y.: The truth of the F-measure,
http://www.toyota-ti.ac.jp/Lab/Denshi/COIN/people/yutaka.sasaki/
index-e.html (accessed June 30, 2011)

[32] Li, L., Wu, W.-C., Rong, Q.-M.: Research on Hybrid Clustering Based on Density and Ant Colony Algorithm. In: 2010 Second International Workshop on Education Technology and Computer Science (ETCS), Wuhan, March 6-7, vol. 2, pp. 222–225 (2010)

[33] Mahmoodi, M.S., Bigham, B.S., Khan Rostam, A.N.-A., Mahmoodi, S.A.: Using Fuzzy Classification Sysstem for Diagnosis of Breast Cancer. In: CICIS 2012, IASBS, Zanjan, Iran, May 29-31, pp. 412–417 (2012)

[34] Chandrasekar, R., Vijaykumar, V., Srinivasan, T.: Probabilistic Ant based Clustering for Distributed Databases. In: 3rd International IEEE Conference Intelligent Systems, pp. 538–545 (September 2006)

[35] Dempster, A.P., Laird, N.M., Rubin, D.B.: Maximum Likelihood from Incomplete Data via the EM Algorithm. Journal of the Royal Statistical Society, Series B 39(1), 1–38 (1977)

[36] Ester, M., Kriegel, H.-P., Sander, J., Xu, X.: A density-based algorithm for discovering clusters in large spatial databases with noise. In: Simoudis, E., Han, J., Fayyad, U.M. (eds.) Proceedings of the Second International Conference on Knowledge Discovery and Data Mining (KDD 1996), pp. 226–231. AAAI Press (1996) ISBN 1-57735-004-9

[37] Xiong, Z., Chen, R., Zhang, Y., Zhang, X.: Multi-density DBSCAN Algorithm Based on Density Levels Partitioning. Journal of Information and Computational Science 9(10), 2739–2749 (2012)

[38] Hall, M., Frank, E., Holmes, G., Pfahringer, B., Reutemann, P., Witten, I.H.: The WEKA Data Mining Software: An Update. SIGKDD Explorations 11(1) (2009)

[39] Buckingham, C.D., Ahmed, A., Adams, A.E.: Using XML and XSLT for flexible elicitation of mental-health risk knowledge. Medical Informatics and the Internet in Medicine 32(1), 65–81 (2007)

[40] Buckingham, C.D., Buijs, P., Welch, P.G., Kumar, A., Ahmed, A.: Developing a cognitive model of decision-making to support members of hub-and-spoke logistics networks. In: Ilie-Zudor, E., Kemény, Z., Monostori, L. (eds.) Proceedings of the 14th International Conference on Modern Information Technology in the Innovation Processes of the Industrial Enterprises. Hungarian Academy of Sciences, Computer and Automation Research Institute, pp. 14–30 (2012), igor.xen.emi.sztaki.hu/mitip/media/MITIP2012proceedings.pdf

An Island Memetic Differential Evolution Algorithm for the Feature Selection Problem

Magdalene Marinaki and Yannis Marinakis

Department of Production Engineering and Management,
Technical University of Crete, Chania, Greece
magda@dssl.tuc.gr, marinakis@ergasya.tuc.gr

Abstract. The Feature Selection Problem is an interesting and important topic which is relevant for a variety of database applications. This paper applies a hybridized version of the Differential Evolution algorithm, the Island Memetic Differential Evolution algorithm, for solving the feature subset selection problem while the Nearest Neighbor Classification method is used for the classification task. The performance of the proposed algorithm is tested using various benchmark datasets from the UCI Machine Learning Repository. The algorithm is compared with variants of the differential evolution algorithm, a particle swarm optimization algorithm, an ant colony optimization algorithm and a genetic algorithm and with a number of algorithms from the literature.

Keywords: Feature Selection Problem, Differential Evolution, Island Models, Memetic Algorithms, Variable Neighborhood Search.

1 Introduction

Recently, there has been an increasing need for novel data-mining methodologies that can analyze and interpret large volumes of data. The proper selection of the right set of features for classification is one of the most important problems in designing a good classifier. The basic feature selection problem is to search through the space of feature subsets to identify the optimal or near-optimal subset with respect to the performance measure. However, as finding the optimum feature subset has been proved to be NP-hard [27], many algorithms are, thus, proposed to find suboptimum solutions in comparably smaller amount of time [25].

Differential Evolution (DE) is a stochastic, population-based algorithm that was proposed by Storn and Price [18]. Recent books for the DE can be found in [18, 39]. DE has the basic characteristics of the evolutionary algorithms as it is an evolutionary algorithm. It focuses in the distance and the direction information of the other solutions. In the differential evolution algorithms [17], initially, a mutation is applied to generate a trial vector and, afterwards, a crossover operator is used to produce one offspring. The mutation step sizes are not sampled from an a priori known probability distribution function as in

G. Terrazas et al. (eds.), *Nature Inspired Cooperative Strategies for Optimization (NICSO 2013)*, Studies in Computational Intelligence 512,
DOI: 10.1007/978-3-319-01692-4_3, © Springer International Publishing Switzerland 2014

other evolutionary algorithms but they are influenced by differences between individuals of the current population.

In this paper, a hybridized version of the Differential Evolution, the **Island Memetic Differential Evolution (IMDE)** algorithm is applied, analyzed and used for solving the feature selection problem. In order to give more exploration abilities in the proposed algorithm, instead of a whole population, a number of subpopulations are used. The interaction between the populations is realized with a migration policy. These kinds of subpopulations are called Island models and they are initially applied to genetic algorithms (Island Genetic Algorithms [17]). Also, a local search phase is used in each individual in order to effectively explore the solution space. It should be noted that a memetic strategy usually improves the performance of the algorithm [34]. A number of Memetic Differential Evolution algorithms has be presented in the literature [6, 33, 35, 36, 46]. Also, island models have been incorporated in the past in a Differential Evolution algorithm [3, 14, 24, 51, 52]. But, at least to our knowledge, there is no algorithm that combines both of these characteristics and is applied to the feature selection problem.

The proposed algorithm is compared with variants of the DE algorithm, with a Genetic Algorithm [20], an Ant Colony Optimization [13] algorithm and a Particle Swarm Optimization [26] algorithm and with the number of algorithms from the literature. The rest of the paper is organized as follows. In the next section, the Feature Selection Problem is presented while in the third section, the Island Memetic Differential Evolution algorithm is presented and analyzed in details. Afterwards, in the fourth section, computational results are given and analyzed while the last section gives the conclusions.

2   Feature Selection Problem

Recently, there has been an increasing need for novel data-mining methodologies that can analyze and interpret large volumes of data. The proper selection of the right set of features for classification is one of the most important problems in designing a good classifier. The basic feature selection problem is an optimization problem with a performance measure for each subset of features to measure its ability to classify the samples.

A formulation of the problem is the following [8]:

- V is the original set of features with cardinality m.
- d represents the desired number of features in the selected subset, X, where $X \subseteq V$.
- $F(X)$ is the feature selection criterion function for the set X.

Let us consider a high value of F to indicate a better feature subset. Formally, the problem of feature selection is to find a subset $X \subseteq V$ such that $\mid X \mid = d$ and

$$F(X) = \max_{Z \subseteq V, |Z| = d} F(Z) \qquad (1)$$

In the literature, many successful feature selection algorithms have been proposed. These algorithms can be classified into two categories. If feature selection depends on learning algorithm, the approach is referred to as a *wrapper model*. Otherwise, it is said to be a *filter model*. Filters, such as mutual information (MI), are based on the statistical tools. Wrappers assess subsets of features according to their usefulness to a given classifier [47]. Unfortunately, finding the optimum feature subset has been proved to be NP-hard [27]. Many algorithms are, thus, proposed to find suboptimal solutions in comparably smaller amount of time [25].

Stochastic algorithms, including Simulated Annealing (SA) [29, 42], Scatter Search algorithms [10, 19], Ant Colony Optimization [1, 2, 11, 23, 37, 41, 53], Genetic Algorithms (GA) [4, 5, 7–9, 16, 21, 22, 40, 43, 44], Particle Swarm Optimization [12, 28, 30, 38, 48–50], Greedy Randomized Adaptive Search Procedure [8] and Tabu Search [8] are of great interest because they often yield high accuracy and are much faster.

3 The Proposed Island Memetic Differential Evolution Algorithm for Feature Selection Problem

Initially, in the Island Memetic Differential Evolution (IMDE) algorithm, a population is created at random as in the classic DE. Then, contrary to the classic DE, the initial population in the IMDE algorithm is divided into a number of islands, depending to the selection of the user (the number of islands is one of the parameters of the algorithm). Each island represents a different population, where the DE's operators are applied independently to each other island. The basic difference of the IMDE algorithm from the classic DE is the **migration policy** of the population. The migration policy is used in order to exchange information between different islands. There is a number of different ways to realize the migration strategy [17]. In this paper, the user gives the percentage of the population that will migrate in another island and the times that a migration policy will be applied. A ring topology is used for the migration of the population, which means that the migrants will migrate to the neighborhood island. The migration occurs a number of times depending on the selection of the user. In IMDE, a randomly selected migrant replaces a randomly selected individual and the individual that will be replaced is not eliminated but it is used as a migrant for another island.

For each island, each member of the population is randomly placed in the d-dimensional space as a candidate solution (in the feature selection problem, d corresponds to the number of activated features). Every candidate feature in IMDE is mapped into a binary individual where the bit 1 denotes that the corresponding feature is selected and the bit 0 denotes that the feature is not selected. As the initial random values for every member of the population are continuous values in the (0,1) interval, they are transformed into discrete values using the following equations:

$$sig(s_i(t)) = \frac{1}{1 + exp(-s_i(t))} \tag{2}$$

$$x_i(t) = \begin{cases} 1, & \text{if } rand1 < sig(s_i(t)) \\ 0, & \text{if } rand1 \geq sig(s_i(t)) \end{cases} \tag{3}$$

where $s_i(t)$ and $x_i(t)$ are the continuous and discrete values of each member of the population, $rand1$ is a random number in (0,1) interval and t is the iteration number. We use in each iteration both continuous and discrete values for each individual as the whole procedure of IMDE is realized in the continuous space and the Feature Selection Problem needs discrete values. Afterwards, the fitness of each individual is calculated using the following equation:

$$OCA = 100 \frac{\sum\limits_{i=1}^{C} c_{ii}}{\sum\limits_{i=1}^{C} \sum\limits_{j=1}^{C} c_{ij}}. \tag{4}$$

The fitness function measures the quality of the produced members of the population. In this problem, the quality is measured with the overall classification accuracy. Thus, for each individual the classifiers (1-Nearest Neighbor, k-Nearest Neighbor or wk- Nearest Neighbor) are called and the produced overall classification accuracy (OCA) gives the fitness function. In the fitness function, we would like to maximize the OCA. The previously mentioned formula for OCA (Eq. 4) is defined taking into account that the accuracy of a C class problem can be described using a $C \times C$ confusion matrix. The element c_{ij} in row i and column j describes the number of samples of true class j classified as class i, i.e., all correctly classified samples are placed in the diagonal and the remaining misclassified cases in the upper and lower triangular parts.

It should be noted that Nearest Neighbor methods [15] are among the most popular for classification. The classic **1 - Nearest Neighbor (1-nn)** method works as follows: In each iteration of the feature selection algorithm, a number of features are activated. For each sample of the test set, the Euclidean Distance from each sample in the training set is calculated. With this procedure the nearest sample from the training set is calculated. Thus, each test sample is classified in the same class to which its nearest sample from the training set belongs. The previous approach may be extended to the **k-Nearest Neighbor (k-nn)** method, where we examine the k-nearest samples from the training set and, then, classify the test sample by using a voting scheme. More weight is attached to those members that are closer to the test samples in the **Weighted k Nearest Neighbor (wk-nn)**.

Then, in each island, separately, the operators of a classic DE are used as described in the following. The mutation operator produces a trial vector for each individual of the current population by mutating a target vector with a weighted difference. This trial vector will, then, be used by the crossover operator to produce offspring. In the mutation phase, for each parent, $s_i(t)$, the trial vector, $u_i(t)$, is generated as follows: a target vector, the best vector of each island, $s_{opt_j}(t)$, is selected from the population, where j is the number of island. Then, two individuals, s_{i_1} and s_{i_2}, are selected randomly from the population of the island such that i, i_1 and i_2 are all different. Using these individuals, the trial vector is calculated by perturbing the target vector as follows:

$$u_i(t) = s_{opt_j}(t) + \beta(s_{i_1}(t) - s_{i_2}(t)) \tag{5}$$

where $\beta \in (0, \infty)$ is the scale factor. The upper bound of β is usually the value 1 because as it has been proved if the $\beta > 1$ there is no improvement in the solutions [17, 39] and the most usually utilized value is $\beta = 0.5$.

After the completion of the mutation phase of the algorithm a **uniform crossover operator** [39] is applied. In this crossover operator, the points are selected randomly from the trial vector and from the parent. Initially, a crossover operator number (Cr) is selected [39] that controls the fraction of parameters that are selected from the trial vector. The offspring $s_i'(t)$ is given by:

$$s_i'(t) = \begin{cases} u_i(t), & \text{if } rand_i(0, 1) \leq Cr \\ s_i(t), & \text{otherwise.} \end{cases} \tag{6}$$

where $rand_i(0, 1)$ is the output of a random number generator.

Then, the equations (2) and (3) are used in order to transform the continuous values calculated by equations (5) and (6) into discrete values. After the crossover operator, the fitness function of the offspring $s_i'(t)$ is calculated. For each individual, a local search phase is used. Initially, a local search number (Lr) is selected that controls the fraction of bits that will be changed with the local search. Thus, for all bits of the solution the value of Lr is compared with the values of a random number generator $rand_i(0, 1)$ that is calculated for each bit of the solution. If the random number is less or equal to the Lr, then, the corresponding bit will be changed, otherwise the corresponding bit will be the same as before the local search. Thus, the choice of the Lr is very significant because if the value is close or equal to 1, then, most of the bits will change in the local search phase but if the value is close to 0, then, almost none of the bits will change. Finally, the solution of the offspring $s_i'(t)$ and of the parent $s_i(t)$ are compared and the fittest survives in the next generation.

In the following, a pseudocode of the Island Memetic Differential Evolution is presented.

Initialization
Initialize the control parameters β, Cr and Lr
Select the mutation operator
Select the number of islands
Select the number of generations
Select the number of migrations in the generations
Select the percentage of the population that migrates in another island

Generate the initial population for each island
Calculate the initial cost function value
    (fitness function) of each member of the population
Main Phase
Do while maximum number of generations has not been reached
    For each island **do**
        Select the parent vector $s_i(t)$
        Create the trial vector $u_i(t)$ by applying the mutation operator
        Create the offspring $s_i'(t)$ by applying the crossover operator
        Perform local search in each individual
        Calculate the cost function (*fitness*) of the offspring
        if $fitness(s_i'(t)) \geq fitness(s_i(t))$
            Replace the parent with the offspring for the next generation
        else
            Add the parent in the next generation
        endif
    endfor
    if conditions for migration holds **then**
        Perform migration strategy
    endif
Enddo
Return the best individual (the best solution).

4   Computational Results

The performance of the proposed methodology is tested on 10 benchmark instances taken from the UCI Machine Learning Repository. The datasets were
chosen to include a wide range of domains and their characteristics are given in
Table 1. In two cases (Breast Cancer Wisconsin, Hepatitis) the data sets are appeared with different size of observations as in these datasets there is a number
of missing values. The problem of missing values was faced with two different
ways. In the first way where all the observations are used, we took the mean values of all the observations in the corresponding feature while in the second way
where we have less values in the observations, we did not take into account the
observations that they had missing values. Some data sets involve only numerical
features, and the remaining include both numerical and categorical features. For
each data set, Table 1 reports the total number of features and the number of
categorical features in parentheses. All the data sets involve 2-class problems and
they are analyzed with 10-fold cross validation. The algorithm was implemented
in Fortran 90 and was compiled using the Lahey f95 compiler on a Intel Core 2
DUO CPU T9550 at 2.66 GHz, running Suse Linux 9.1.

Table 1. Data Sets Characteristics

Data Sets	Observations	Features
Australian Credit (AC)	690	14(8)
Breast Cancer Wisconsin 1 (BCW1)	699	9
Breast Cancer Wisconsin 2 (BCW2)	683	9
German Credit (GC)	1000	24 (13)
Heart Disease (HD)	270	13(7)
Hepatitis 1 (Hep1)	155	19 (13)
Hepatitis 2 (Hep2)	80	19 (13)
Ionosphere (Ion)	351	34
Spambase (spam)	4601	57
Pima Indian Diabetes (PID)	768	8

As it has already been mentioned, three approaches that use different classifiers, the 1-nn, k-nn and the wk-nn, are used. In all algorithms, the value of k is changed dynamically depending on the number of iterations. Each generation uses different k. The reason why k does not have a constant value is that we would like to ensure the diversity of solutions in each iteration of the algorithms. The determination of k is done by using a random number generator with a uniform distribution $(0, 1)$ in each iteration. Then, the produced number is converted to an integer k (e.g., if the produced number is in the interval $0.2 - 0.3$, then $k = 3$).

The parameters of the proposed algorithms were selected after thorough testing. A number of different alternative values were tested and the ones selected are those that gave the best computational results concerning both the quality of the solution and the computational time needed to achieve this solution. Thus, the selected parameters for the IMDE are: The number of individuals is set equal to 200, the number of generations is set equal to 1000, $\beta = 0.5$ and $Cr = 0.8$. Ten different islands are used (10 different subpopulations). The times of migration was set equal to 10, meaning that we use 100 generations before a percentage of the population migrates to another population. Finally, the percentage of the population that migrates in another population is set equal to 20% of random chosen individuals. After the selection of the final parameters, 20 different runs with the selected parameters were performed for each of the instances.

In the comparisons, three different variants of the DE algorithm are used, the classic differential evolution with random target vector (DEr), the classic differential evolution with the optimum individual as the target vector (DEo) and the Island Differential Evolution (IDE) without the use of the memetic operator (the local search phase). The parameters are the same with the ones used in the proposed algorithm.

For comparison purposes, three other metaheuristic algorithms, a genetic based metaheuristic (GA), an ant colony optimization (ACO) algorithm and a particle swarm optimization (PSO) algorithm, are used. For analytical description of these algorithms and how they are designed and applied for the solution of the Feature Selection Problem please see ([31, 32]). We used two different sets of parameters. In the first one we selected the parameters of these three algorithms based on the fact that we would like to have the same function evaluations for all the algorithms. Thus, when we ensured that the function evaluations will be the same we optimized the other parameters of the algorithms separately. In the second set of parameters we tried to find the best parameters for each algorithm independently from the parameters of the other algorithms. The first set of parameters for the genetic based metaheuristic is: The population size is set equal to 200, the number of generations is set equal to 1000, the probability of crossover is set equal to 0.8 and the probability of mutation is set equal to 0.25. The second set of parameters for the genetic based metaheuristic is: The population size is set equal to 100, the number of generations is set equal to 200, the probability of crossover is set equal to 0.7 and the probability of mutation is set equal to 0.1. The first set of parameters for the ACO based metaheuristic is: The number of ants used is set equal to 200. The number of iterations that each ant constructs a different solution, based on the pheromone trails, is set equal to 1000, and $q = 0.5$. The second set of parameters for the ACO based metaheuristic is: The number of ants used is set equal to 50. The number of iterations that each ant constructs a different solution, based on the pheromone trails, is set equal to 500, and $q = 0.9$. Finally, the first set of parameters for the PSO based metaheuristic is: The number of swarms is set equal to 1, the number of particles is set equal to 200, the number of generations is set equal to 1000, and $c_1 = 2, c_2 = 2$, $w_{max} = 0.9$, $w_{min} = 0.01$, while the second set of parameters for the PSO based metaheuristic is: The number of swarms is set equal to 1, the number of particles is set equal to 80, the number of generations is set equal to 200, and $c_1 = 1.8, c_2 = 1.7$, $w_{max} = 0.8$, $w_{min} = 0.01$.

Table 2 shows the Overall Classification Accuracy (OCA and OCA1) and the average selected features (SNF) for the proposed algorithm and for all algorithms used in the comparisons. The difference between OCA and OCA1 is that in the OCA1 the results with the second set of parameters for the ACO, GA and PSO are presented. In this Table the best method for each dataset is denoted with bold letters and the best method and the best classifier for each dataset are denoted with bold and italic letters. The proposed algorithm performs better than all the other algorithms used in the comparisons. The results of the two classic versions of DE (DEr and DEo) and of the PSO and ACO are almost identical. The results of the GA are slightly inferior from the results of all other algorithms. The results of the PSO, ACO and GA using the second set of parameters are improved, however, they are still inferior than the results of the proposed method. The use of the Island model (IDE) improves the results of the classic differential evolution. Finally, the use of the local search phase in each solution of the algorithm improves even more the results (IMDE).

Table 2. Classification Results (OCA(%)) for all algorithms

Clas-sifier	IMDE		IDE		DEr		DEo		GA			ACO			PSO		
	OCA	SNF	OCA	SNF	OCA	SNF	OCA	SNF	OCA	OCA1	SNF	OCA	OCA1	SNF	OCA	OCA1	SNF
Data Set: AC																	
1nn	90.15	7.17	89.28	7.35	87.12	7.45	86.35	7.28	86.42	87.17	8.12	87.88	88.58	8.21	86.95	87.35	8.17
knn	92.31	7.28	91.49	7.41	89.32	7.52	89.44	7.92	87.12	88.24	8.11	89.37	90.15	8.05	90.01	90.28	7.98
wknn	91.57	7.41	90.37	7.52	88.32	8.01	87.35	8.18	86.12	87.31	8.07	88.18	89.24	8.23	89.39	90.28	8.12
Data Set: BCW1																	
1nn	99.35	4.42	99.21	4.57	99.08	4.68	99.17	4.61	98.01	98.57	4.57	98.65	98.89	5.01	99.15	99.24	4.98
knn	99.28	4.51	99.15	4.72	98.96	4.57	99.01	4.76	97.37	97.58	5.12	98.41	98.85	5.07	98.85	99.01	4.95
wknn	99.45	4.58	99.08	4.71	98.23	4.88	97.85	4.79	97.48	98.15	5.01	98.08	98.47	5.18	99.05	99.18	4.84
Data Set: BCW2																	
1nn	99.52	4.93	99.12	5.17	99.15	5.08	98.77	5.23	98.55	98.87	5.34	99.17	99.24	5.21	99.23	99.35	5.15
knn	99.21	5.01	99.02	5.21	98.77	5.18	98.56	5.35	97.88	98.12	5.66	98.79	98.95	5.73	99.09	99.17	5.43
wknn	99.15	4.87	98.23	5.12	97.76	5.21	98.01	5.34	97.45	98.35	5.87	98.03	98.58	5.49	97.65	98.27	5.53
Data Set: GC																	
1nn	80.17	13.01	78.13	13.88	75.34	14.19	76.19	14.28	70.96	72.95	15.31	75.37	76.49	15.52	76.23	77.38	15.28
knn	82.81	13.17	80.28	13.76	80.19	14.52	79.24	14.47	76.01	76.57	14.87	77.57	78.95	15.01	78.01	78.92	15.31
wknn	80.98	13.21	80.27	13.98	79.34	14.42	78.29	14.78	77.35	78.95	14.32	78.27	79.18	15.12	77.11	78.37	15.31
Data Set: HD																	
1nn	96.18	6.51	93.21	6.72	91.29	6.67	90.88	6.84	88.45	89.74	6.91	90.38	91.28	6.92	90.49	91.82	6.88
knn	94.81	6.47	91.23	6.82	90.89	6.79	92.01	6.65	87.45	89.27	6.87	91.78	92.49	6.98	91.24	92.47	6.91
wknn	92.17	6.31	90.89	6.72	91.23	6.88	90.01	6.57	89.39	89.94	6.94	91.22	91.57	6.88	90.88	91.47	6.77
Data Set: Hep1																	
1nn	98.24	10.37	97.11	10.58	96.88	10.69	97.01	10.81	94.37	95.18	10.78	97.08	97.35	11.02	98.15	98.21	11.17
knn	97.56	10.42	96.49	10.88	96.35	10.79	96.11	10.95	94.32	95.14	11.12	96.23	96.74	10.88	96.92	97.15	10.79
wknn	97.55	10.23	96.89	10.87	96.11	10.88	96.08	11.01	95.31	95.89	11.23	96.23	96.47	11.09	97.12	97.21	10.98
Data Set: Hep2																	
1nn	100	9.75	100	9.88	100	9.95	100	10.01	98.85	99.35	10.88	100	100	9.85	100	100	9.88
knn	99.37	10.11	98.98	10.76	98.44	10.95	99.01	10.65	95.23	96.19	10.87	97.15	98.21	10.85	98.23	98.87	10.95
wknn	99.68	10.05	97.14	10.34	98.01	10.57	98.17	10.42	94.29	95.17	10.74	96.75	97.35	10.82	98.18	98.91	10.95
Data Set: Ion																	
1nn	98.54	15.12	98.32	15.88	97.88	15.75	96.14	16.12	95.08	96.15	16.05	97.12	97.31	16.34	98.17	98.35	16.21
knn	98.33	15.34	97.12	15.47	97.21	15.98	95.18	16.14	95.11	95.95	16.45	96.37	96.74	16.38	96.14	96.87	16.21
wknn	98.79	15.07	98.03	15.96	97.72	15.74	97.49	15.38	94.21	95.17	15.91	96.44	96.71	15.85	97.01	97.58	16.02
Data Set: spam																	
1nn	89.04	21.88	87.32	22.57	85.14	22.64	84.27	22.49	82.23	83.52	23.12	83.01	84.18	22.85	84.28	85.71	22.98
knn	86.59	21.57	85.11	22.01	83.28	23.55	84.01	22.47	81.32	82.51	22.76	83.44	84.47	24.65	83.79	84.95	24.32
wknn	84.57	21.40	83.12	21.98	82.11	22.05	81.23	22.88	81.18	81.87	25.01	82.11	82.57	24.97	83.18	83.87	23.88
Data Set: PID																	
1nn	79.55	4.02	74.14	4.11	73.18	4.15	72.21	4.13	71.57	72.81	4.21	74.11	75.49	4.37	75.18	76.27	4.28
knn	79.37	4.07	78.23	4.21	77.14	4.12	76.33	4.19	74.19	74.88	4.23	76.37	77.57	4.26	78.21	79.35	4.37
wknn	80.18	4.09	79.21	4.29	78.14	4.31	77.35	4.21	74.24	75.47	4.28	76.47	77.32	4.37	75.39	78.24	4.25

The purpose of feature variable selection is to find the smallest set of features that can result in satisfactory predictive performance. Because of the curse of dimensionality, it is often necessary and beneficial to limit the number of input features in order to have a good predictive and less computationally intensive model. In general, there are $2^{number\ of\ features} - 1$ possible feature combinations. The average selected features for all algorithms in all runs of the algorithms are presented in Table 2. The significance of the solution of the feature selection problem using the proposed method is demonstrated by the fact that with this algorithm the best solutions were found by using less features than the other algorithms used in the comparisons. More precisely, in the most difficult instance, the Spambase instance, the proposed algorithm needed between 21.40 to 21.88 average number of features in order to find their best solutions while the other algorithms needed between 21.98 - 25.01 average number of features to find their best solutions.

A statistical analysis based on the Mann-Whitney U-test for all algorithms is presented in Table 3. In this Table, a value equal to 1 indicates a rejection of the null hypothesis at the 5% significance level, which means that the method is statistically significant different from the other methods. On the other hand, a value equal to 0 indicates a failure to reject the null hypothesis at the 5% significance level, meaning that no statistical significant difference exists between the two methods. As it can be seen from this Table, at the 5% significance level the results with the IMDE are statistically significant different from the results with the GA and the ACO.

Table 3. Results of Mann - Whitney test for all algorithms

	5% significance level						
	IMDE	IDE	DEr	DEo	GA	ACO	PSO
IMDE	-	0	0	0	1	1	0
IDE	0	-	0	0	0	0	0
DEr	0	0	-	0	0	0	0
DEo	0	0	0	-	0	0	0
GA	1	0	0	0	-	0	0
ACO	1	0	0	0	0	-	0
PSO	0	0	0	0	0	0	-

The results of the algorithm are, also, compared (Table 4) with the results of a number of metaheuristic approaches from the literature. In these implementations, the same databases are used as the ones we use in this paper and, thus, comparisons of the results can be performed. More precisely, in Table 4 the results of the proposed algorithm are compared with the results of the following algorithms:

1. The Parallel Scatter Search algorithm proposed by Garcia Lopez et al. [19]. In this paper, three different versions of Scatter Search are proposed, named Sequential Scatter Search Greedy Combination (SSSGC), Sequential Scatter Search Reduced Greedy Combination (SSSRGC) and Parallel Scatter Search (PSS).
2. The Particle Swarm Optimization Linear Discriminant Analysis (PSOLDA) proposed by Lin and Chen [28].
3. The Particle Swarm Optimization Support Vector Machines (PSOSVM1, PSOSVM2) proposed by Lin et al. [30].
4. The Simulated Annealing Support Vector Machines (SASVM1, SASVM2) proposed by Lin et al. [29].
5. A Particle Swarm Optimization algorithm with a Nearest Neighbour (PSONN) classifier proposed by Pedrycz et al. [38].
6. A Genetic Algorithm using an adjacency matrix-encoding, GWC operator, and fitness function based on the VC dimension of multiple ODTs combined with naive Bayes (GOV - genetic algorithm for ODTs using VC dimension upper bound) proposed by Rokach [40].
7. A Scatter Search (SS-ensemble) with different classifiers like Support Vector Machines (SVM), Decision Trees (DT) and Back Propagation Networks (BPN) proposed by Chen et al. [10].
8. Three different metaheuristics (GRASP, Tabu Search and a Memetic algorithm) proposed by Casado Yusta [8].

Table 4. Comparison of the proposed algorithm with other metaheuristic approaches

Method	Data Set							
	AC	BCW	GC	HD	Hep	Ion	Spam	PID
IMDE-1nn	90.15	99.52	80.17	96.18	100	98.54	89.04	79.55
IMDE-knn	92.31	99.28	82.81	94.81	99.37	98.33	86.59	79.37
IMDE-wknn	91.57	99.45	80.98	92.17	99.68	98.79	84.57	80.18
SSSGC	-	95.22	-	74.99	-	87.75	-	67.92
SSSRGC	-	94.88	-	74.99	-	87.12	-	67.66
PSS	-	95.11	-	74.91	-	87.35	-	68.10
PSOLDA	84.5	96.5	75.6	84.7	-	92.2	-	76.7
PSOSVM1	91.03	99.18	81.62	92.83	-	99.01	-	82.68
PSOSVM2	88.09	97.95	79.00	88.17	-	97.50	-	80.19
SASVM1	92.19	99.38	-	93.33	-	99.07	-	82.22
SASVM2	88.34	97.95	-	87.97	-	97.50	-	80.19
PSONN	-	-	74.7	83.9	-	94.6	-	-
GOV	85.35	97.13	-	-	81.29	-	-	-
SS-ensemble	91.74	99.46	85.49	96.24	97.46	-	-	83.92
GRASP	-	93	92.7	-	-	90.4	84.6	-
Tabu Search	-	92.6	92.7	-	-	90.6	82.64	-
Memetic	-	91.8	92.6	-	-	89	79.76	-

As it can be seen from Table 4, the proposed algorithm gives better results in four instances, the Australian Credit (AC), the Breast Cancer Wisconsin (BCW), the spambase (Spam) and the Hepatitis. For the other four instances the algorithms that perform better are: for the German Credit, the GRASP and the Tabu Search proposed by Casado Yusta [8], for the Heart Disease and for the Pima Indian Diabetes (PID), the Scatter Search ensemble proposed by Chen et al. [10] and for the ionosphere, the Simulated Annealing Support Vector Machines (SASVM1) proposed by Lin et al. [29].

5 Conclusions

In this paper, a hybridized version of Differential Evolution, the Island Memetic Differential Evolution, is applied for solving the Feature Subset Selection Problem. Three different classifiers are used for the classification problem, based on the Nearest Neighbor classification rule (the 1-Nearest Neighbor, the k-Nearest Neighbor and the wk-Nearest Neighbor). The performance of the proposed algorithm is tested using various benchmark datasets from UCI Machine Learning Repository. The objective of the computational experiments, the desire to show the high performance of the proposed algorithm in searching for a reduced set of features with high accuracy, was achieved as the algorithm gave very efficient results.

References

[1] Al-Ani, A.: Feature subset selection using ant colony optimization. International Journal of Computational Intelligence 2(1), 53–58 (2005)

[2] Al-Ani, A.: Ant colony optimization for feature subset selection. Transactions on Engineering, Computing and Technology 4, 35–38 (2005)

[3] Apolloni, J., Leguizam, G., Garcia-Nieto, J., Alba, E.: Island Based Distributed Differential Evolution: An Experimental Study on Hybrid Testbeds. In: International Conference on Hybrid Intelligent Systems, pp. 696–701 (2008)

[4] Cantú-Paz, E.: Feature subset selection, class separability, and genetic Algorithms. In: Deb, K., Tari, Z. (eds.) GECCO 2004. LNCS, vol. 3102, pp. 959–970. Springer, Heidelberg (2004)

[5] Cantu-Paz, E., Newsam, S., Kamath, C.: Feature selection in scientific application. In: Proceedings of the 2004 ACM SIGKDD International Conference on Knowledge Discovery and Data Mining, pp. 788–793 (2004)

[6] Caponio, A., Neri, F., Tirronen, V.: Super-fit control adaptation in memetic differential evolution frameworks. Soft Computing 13, 811–831 (2009)

[7] Carvalho, D.R., Freitas, A.A.: A hybrid decision tree/genetic algorithm method for data mining. Information Sciences 163(1-3), 13–35 (2004)

[8] Casado Yusta, S.: Different metaheuristic strategies to solve the feature selection problem. Pattern Recognition Letters 30, 525–534 (2009)

[9] Casillas, J., Cordon, O., Del Jesus, M.J., Herrera, F.: Genetic feature selection in a fuzzy rule-based classification system learning process for high-dimensional problems. Information Sciences 136(1-4), 135–157 (2001)

[10] Chen, S.C., Lin, S.W., Chou, S.Y.: Enhancing the classification accuracy by scatter-search-based ensemble approach. Applied Soft Computing 11(1), 1021–1028 (2011)

[11] Chen, Y., Miao, D., Wang, R.: A rough set approach to feature selection based on ant colony optimization. Pattern Recognition Letters 31, 226–233 (2010)

[12] Chuang, L.Y., Yang, C.H., Li, J.C.: Chaotic maps based on binary particle swarm optimization for feature selection. Applied Soft Computing (2009), doi:10.1016/j.asoc.2009.11.014

[13] Dorigo, M., Stutzle, T.: Ant Colony Optimizationm. A Bradford Book. The MIT Press Cambridge, Massachusetts (2004)

[14] Dorronsoro, B., Bouvry, P.: Improving Classical and Decentralized Differential Evolution with New Mutation Operator and Population Topologies. IEEE Transactions on Evolutionary Computation 15(1), 67–98 (2011)

[15] Duda, R.O., Hart, P.E., Stork, D.G.: Pattern Classification and Scene Analysis, 2nd edn. John Wiley and Sons, New York (2001)

[16] ElAlami, M.E.: A filter model for feature subset selection based on genetic algorithm. Knowledge-Based Systems 22, 356–362 (2009)

[17] Engelbrecht, A.P.: Computational Intelligence: An Introduction. John Wiley and Sons (2007)

[18] Feoktistov, V.: Differential Evolution - In Search of Solutions. Springer, NY (2006)

[19] Garcia Lopez, F., Garcia Torres, M., Melian Batista, B., Moreno Perez, J.A., Moreno Vega, J.M.: Solving feature subset selection problem by a parallel scatter search. European Journal of Operational Research 169, 477–489 (2006)

[20] Holland, J.H.: Adaptation in Natural and Artificial Systems. University of Michigan Press, Ann Arbor (1975)

[21] Hsu, W.H.: Genetic wrappers for feature selection in decision tree induction and variable ordering in Bayesian network structure learning. Information Sciences 163(1-3), 103–122 (2004)

[22] Huang, J., Cai, Y., Xu, X.: A hybrid genetic algorithm for feature selection wrapper based on mutual information. Pattern Recognition Letters 28, 1825–1844 (2007)

[23] Huang, C.L.: ACO-based hybrid classification system with feature subset selection and model parameters optimization. Neurocomputing 73, 438–448 (2009)
[24] Izzo, D., Rucinski, M., Ampatzis, C.: Parallel global optimisation meta-heuristics using an asynchronous island-model. In: IEEE Congress on Evolutionary Computation (CEC 2009), pp. 2301–2308 (2009)
[25] Jain, A., Zongker, D.: Feature Selection: Evaluation, Application, and Small Sample Performance. IEEE Transactions on Pattern Analysis and Machine Intelligence 19, 153–158 (1997)
[26] Kennedy, J., Eberhart, R.: Particle Swarm Optimization. In: Proceedings of 1995 IEEE International Conference on Neural Networks, vol. 4, pp. 1942–1948 (1995)
[27] Kohavi, R., John, G.: Wrappers for Feature Subset Selection. Artificial Intelligence 97, 273–324 (1997)
[28] Lin, S.W., Chen, S.C.: PSOLDA: A Particle swarm optimization approach for enhancing classification accurate rate of linear discriminant analysis. Applied Soft Computing 9, 1008–1015 (2009)
[29] Lin, S.W., Lee, Z.J., Chen, S.C., Tseng, T.Y.: Parameter determination of support vector machine and feature selection using simulated annealing approach. Applied Soft Computing 8, 1505–1512 (2008)
[30] Lin, S.W., Ying, K.C., Chen, S.C., Lee, Z.J.: Particle swarm optimization for parameter determination and feature selection of support vector machines. Expert Systems with Applications 35, 1817–1824 (2008)
[31] Marinakis, Y., Marinaki, M., Doumpos, M., Matsatsinis, N., Zopounidis, C.: Optimization of Nearest Neighbor Classifiers via Metaheuristic Algorithms for Credit Risk Assessment. Journal of Global Optimization 42, 279–293 (2008)
[32] Marinakis, Y., Marinaki, M., Doumpos, M., Zopounidis, C.: Ant Colony and Particle Swarm Optimization for Financial Classification Problems. Expert Systems with Applications 36(7), 10604–10611 (2009c)
[33] Mininno, E., Neri, F.: A memetic Differential Evolution approach in noisy optimization. Memetic Computing 2, 111–135 (2010)
[34] Moscato, P., Cotta, C.: A Gentle Introduction to Memetic Algorithms. In: Glover, F., Kochenberger, G.A. (eds.) Handbooks of Metaheuristics, pp. 105–144. Kluwer Academic Publishers, Dordrecht (2003)
[35] Muelas, S., La Torre, A., Pena, J.M.: A Memetic Differential Evolution Algorithm for Continuous Optimization. In: Proceedings of the 2009 Ninth International Conference on Intelligent Systems Design and Applications, pp. 1080–1084 (2009)
[36] Neri, F., Tirronen, V.: On memetic Differential Evolution frameworks: A study of advantages and limitations in hybridization. In: IEEE Congress on Evolutionary Computation (CEC 2008), pp. 2135–2142 (2008)
[37] Parpinelli, R.S., Lopes, H.S., Freitas, A.A.: An ant colony algorithm for classification rule discovery. In: Abbas, H., Sarker, R., Newton, C. (eds.) Data Mining: A Heuristic Approach, pp. 191–208. Idea Group Publishing, London (2002)
[38] Pedrycz, W., Park, B.J., Pizzi, N.J.: Identifying core sets of discriminatory features using particle swarm optimization. Expert Systems with Applications 36, 4610–4616 (2009)
[39] Price, K.V., Storn, R.M., Lampinen, J.A.: Differential Evolution: A Practical Approach to Global Optimization. Springer, Berlin (2005)
[40] Rokach, L.: Genetic algorithm-based feature set partitioning for classification problems. Pattern Recognition Letters 41, 1676–1700 (2008)
[41] Shelokar, P.S., Jayaraman, V.K., Kulkarni, B.D.: An ant colony classifier system: Application to some process engineering problems. Computers and Chemical Engineering 28, 1577–1584 (2004)

[42] Siedlecki, W., Sklansky, J.: On automatic feature selection. International Journal of Pattern Recognition and Artificial Intelligence 2(2), 197–220 (1988)
[43] Siedlecki, W., Sklansky, J.: A note on genetic algorithms for large-scale feature selection. Pattern Recognition Letters 10, 335–347 (1989)
[44] Srinivasa, K.G., Venugopal, K.R., Patnaik, L.M.: A self-adaptive migration model genetic algorithm for data mining applications. Information Sciences 177(20), 4295–4313 (2007)
[45] Storn, R., Price, K.: Differential Evolution - A Simple and Efficient Heuristic for Global Optimization over Continuous Spaces. Journal of Global Optimization 11(4), 341–359 (1997)
[46] Tirronen, V., Neri, F., Karkkainen, T., Majava, K., Rossi, T.: An Enhanced Memetic Differential Evolution in Filter Design for Defect Detection in Paper Production. Evolutionary Computation 16(4), 529–555 (2008)
[47] Uncu, O., Turksen, I.B.: A novel feature selection approach: Combining feature wrappers and filters. Information Sciences 177(2), 449–466 (2007)
[48] Unler, A., Murat, A.: A discrete particle swarm optimization method for feature selection in binary classification problems. European Journal of Operational Research 206, 528–539 (2010)
[49] Wang, Y., Feng, X.Y., Huang, Y.X., Pu, D.B., Zhou, W.G., Liang, Y.C., Zhou, C.G.: A novel quantum swarm evolutionary algorithm and its applications. Neurocomputing 70(4-6), 633–640 (2007)
[50] Wang, X., Yang, J., Teng, X., Xia, W., Jensen, R.: Feature selection based on rough sets and particle swarm optimization. Pattern Recognition Letters 28, 459–471 (2007)
[51] Weber, M., Neri, F., Tirronen, V.: Distributed differential evolution with explorative exploitative population families. Genetic Programming Evolvable Machines 10, 343–371 (2009)
[52] Weber, M., Neri, F., Tirronen, V.: A study on scale factor/crossover interaction in distributed differential evolution. Artificial Intelligence Reviews (2011), doi:10.1007/s10462-011-9267-1
[53] Zhang, C., Hu, H.: Ant colony optimization combining with mutual information for feature selection in support vector machines. In: Zhang, S., Jarvis, R.A. (eds.) AI 2005. LNCS (LNAI), vol. 3809, pp. 918–921. Springer, Heidelberg (2005)

Using a Scouting Predator-Prey Optimizer to Train Support Vector Machines with non PSD Kernels

Arlindo Silva[1] and Teresa Gonçalves[2]

[1] Escola Superior de Tecnologia do Instituto Politécnico de Castelo Branco, Portugal
arlindo@ipcb.pt
[2] Universidade de Évora, Portugal
tcg@uevora.pt

Abstract. In this paper, we investigate the use of an heterogeneous particle swarm optimizer, the scouting predator-prey optimizer, to train support vector machines with non positive definite kernels, including distance substitution based kernels. These kernels can arise in practical applications, resulting in multi-modal optimization problems where traditional algorithms can struggle to find the global optimum. We compare the scouting predator-prey algorithm with the previous best evolutionary approach to this problem and a standard quadratic programming based algorithm, on a large set of benchmark problems, using various non positive definite kernels. The use of cooperating scout particles allows the proposed algorithm to be more efficient than the other evolutionary approach, which is based on an evolution strategy. Both are shown to perform better than the standard algorithm in several dataset/kernel instances, a result that underlines the usefulness of evolutionary training algorithms for support vector machines.

Keywords: particle swarm optimization, heterogeneous particle swarms, support vector machines, non PSD kernels.

1 Introduction

Support vector machines (SVMs) are one of the most popular kernel based methods [1]. These use a problem dependent similarity measure between objects - the kernel function or simply kernel - to implicitly map the original data onto a feature space where simple linear relations can be found. In the case of SVMs, a hyper-plane that maximizes a margin between data points of different classes is learned and then used to classify new objects [2]. The low computational cost, strong theoretical foundation (from statistical learning) and general applicability have made SVMs the state-of-the-art approach in many domains.

The choice of an appropriate kernel function is essential to the application of SVMs to a specific problem. In one hand, the generality of the approach results from the possibly of using kernels to measure similarities in different domains, which can range from simple vectorial data to general discrete structures. In the

G. Terrazas et al. (eds.), *Nature Inspired Cooperative Strategies for Optimization (NICSO 2013)*, Studies in Computational Intelligence 512,
DOI: 10.1007/978-3-319-01692-4_4, © Springer International Publishing Switzerland 2014

other hand, the kernel should embody as much problem dependent information as possible, in order to guarantee the quality of the final classifier. Finally, the algorithms used to train the SVM are quadratic programming based approaches, which depend on the kernel function to be positive semi-definite (PSD) to find the optimum solution, so this condition must be satisfied by every new kernel.

For many practical applications, specially in non vectorial domains, the requirement for the kernel function to be PSD can be very strict. If a domain specialized kernel doesn't exist, it must be developed by the user, often based on preexistent similarity or distance measures. One of the possibilities for kernel construction is to imbed a distance measure on a previously existent PSD kernel, e.g. the Gaussian radial basis function (RBF). Unfortunately, depending on the distance function, the resulting kernel may not be PSD [3]. While with non PSD kernels the SVM standard geometrical interpretation is lost, there are theoretical results that suggest alternative interpretations [4]. Traditional training algorithms can still be used to train this SVMs, frequently with good empirical results [5], but they may fail to find the optimal solution, since the training problem may become multi-modal as a result of using a non PSD kernel.

Since evolutionary algorithms are (population based) global optimization algorithms, they are natural candidates to tackle the non-convex problem resulting from a non PSD kernel. In this paper, we investigate the usefulness of evolutionary computation, when applied to the training of SVMs with non PSD kernels, by comparing the best evolutionary approach found in the literature and a specially tailored heterogeneous particle swarm optimizer (PSO), called scouting predator-prey optimizer (SPPO), with a popular standard SVM training algorithm (MySVM). The empirical comparison is done using a set of 10 classification benchmark problems and three different non PSD kernels: the Epanechnikov kernel, the Sigmoid kernel and a distance substitution kernel.

2   Support Vector Machines

Support vector machines are most frequently used in classification tasks [1, 2, 6]. As input, they receive a training set of n examples with m real attributes, $T = \{(\mathbf{x}_1, y_1), ..., (\mathbf{x}_n, y_n)\}$, where $\mathbf{x}_i \in \mathbb{R}^m$ and $y_i \in \{\pm 1\}$. The algorithm learns a hyperplane $\langle \mathbf{w}, \mathbf{x} \rangle + b = 0$, with $\mathbf{w} \in \mathbb{R}^m$ and $b \in \mathbb{R}$, which separates the positive from negative examples. Following this separation, new examples can be classified using $f(\mathbf{x}) = \text{sgn}(\langle \mathbf{w}, \mathbf{x} \rangle + b)$.

One of the ways in which support vector machines differ from other classification methods, is in the fact that their training doesn't try to minimize just the empirical risk. It is intuitive to consider that, of all hyperplanes that correctly classify the training examples (thus minimizing empirical risk), the one most distant from the closest examples, is also the one that will best generalize for new data. By maximizing the distance from the hyperplane to the regions occupied by each class, a greater tolerance is ensured when classifying new instances that are between these regions and the hyperplane. As a consequence, it can be said that SVMs minimize, not only the empirical risk, but also the structural risk of the classifier. Computing the hyperplane is typically done by finding the α values that maximize equation (1), subject to restrictions (2) and (3).

$$\text{maximize } \sum_i \alpha_i - \frac{1}{2} \sum_i \sum_j \alpha_i \alpha_j y_i y_j k(\mathbf{x}_i, \mathbf{x}_j), \tag{1}$$

$$\text{subject to } \sum_i \alpha_i y_i = 0 \tag{2}$$

$$\text{and } \forall i : 0 \leqslant \alpha_i \leqslant C \tag{3}$$

The use of a kernel function $k(\mathbf{x}_i, \mathbf{x}_j)$ in equation (1), instead of the inner product $\langle \mathbf{x}_i, \mathbf{x}_j \rangle$, extends the SVM approach to problems where non-linear separation is needed. The kernel function performs an implicit mapping of the data to a space with higher dimensionally, where the linear separation becomes possible. Kernels are problem dependent, and many different are used, with the radial basis function (RBF) kernel being the most common for applications with real valued data.

The optimization problem is usually solved using quadratic programming based approaches, with mySVM [7] and LIBSVM [8] being probably the most widely known. There approaches are successful (and very efficient) provided the kernel function is positive definite. In this case, the objective function is concave and possesses a single global optimum, which considerably simplifies the optimization problem. The use of a non positive definite kernel can, however, result in a multi-modal objective function, with several local optima, where traditional algorithms can stagnate, producing sub-optimal classifiers. In this case, the clear geometric interpretation presented is not valid and an alternative explanation for the experimental success of such approaches is needed.

Recently, theoretical work by Haasdonk suggests such an alternative interpretation for SVMs based on non PSD kernels [4]. In this work, instead of maximizing the margin in an induced Euclidean space, an optimal hyperplane is computed by minimizing the distance between convex hulls in pseudo-Euclidean spaces. One of the conclusions of this formulation is that traditional methods (like MySVM and LIBSVM) can still find good solutions for optimization problems with non-PSD kernels, but there is no guarantee of the solution being optimal, since, while those methods will converge to a stationary point, since the problem is non concave, that point may be only a local optimum. This conclusion confirms previous work by other authors [5].

Since SVMs based on non PSD kernels can be of practical interest (see [9] for a detailed discussion about learning with non PSD kernels and [4, 5, 10] on the use of these kernels with support vector machines), the use of global optimization algorithms, like the swarm based optimizer proposed in this work, constitutes an interesting alternative to standard training methods.

3   Evolutionary Computation and Support Vector Machines

Most applications of evolutionary computation to SVMs have been centered on the optimization of C and the kernels' parameters. Genetic programming has also been used to evolve new kernels for specific problems, often finding kernel functions with promising performance. Recent examples of both approaches can be found, respectively, in [11] and [12]. In this work, we are mainly interested in the problem of training SVMs using evolutionary based approaches and, more specifically, swarm optimization algorithms, since these approaches can facilitate the training of SVMs based on non PSD kernels [10].

Training SVMs with non PSD kernels is not only an interesting practical problem for evolutionary computation, with a high-dimensional, multi-modal and non-separable objective function. It also constitutes an important research area for several reasons: as we have already stated, traditional methods are not guaranteed to find the global optimum on the resulting optimization problem; proving a new kernel to be PSD can be a difficult task; some kernels that are proven non PSD, e.g. the sigmoid kernel, can be of practical interest and show promising empirical results [5]; some kernel learning approaches (including the GP based methods) return kernels that are not guaranteed to be PSD [9].

Interestingly, the first evolutionary SVM training approach we found in the literature was PSO based [13]. It combined a linear particle swarm optimizer with a traditional decomposition based method to train a SVM with a PSD kernel. It had some severe problems and experimental results were very limited. Genetic algorithms were used to optimize the primal formulation of this problem [14], which has several drawbacks when compared with the more common dual version. The most interesting work in the area, however, has been done by Mierswa using evolution strategies (ES).

Mierswa initially compared several ES based algorithms with a standard PSO optimizer on 6 benchmark problems and found that the ES methods were competitive with the traditional quadratic programming based approaches in terms of accuracy [15]. On the other hand, the PSO algorithm performed significantly poorer than the other algorithms. The same author proposed a new multi-objective evolutionary SVM formulation [16], which allows the simultaneous independent optimization of the classification error and model complexity. This formulation was based on the best previously found ES approach, which was then used to train a SVM with the Epanechnikov kernel, in the first reported use of an evolutionary algorithm to train a SVM with a non PSD kernel [10].

4   The Scouting Predator-Prey Optimiser

While the past use of particle swarm optimizers in SVM training hasn't been particularly successful, we believe that the general characteristics of these algorithms, when allied with problem specific customizations, should allow the development of efficient swarm based SVM training algorithms. The standard PSO, described in [17], presented several limitations, mainly in balancing exploration and exploitation, maintaining diversity in the swarm after local convergence and solution fine-tuning. Since its introduction, a substantial amount of research has been done to overcome those drawbacks [18], producing variants with successful application to hard optimization problems in many domains [19].

The original PSO was based on the flocking behavior of birds. Individuals (particles) were simultaneously attracted to the best solutions found by themselves and the group (swarm). In previous work [20], we presented a swarm optimizer that extended the metaphor by adding two other flock/swarm inspired behaviors: predation and scouting. The scouting predator-prey optimizer is an heterogeneous particle swarm algorithm, where the overall search behavior is the result of the cooperation of various groups of particles with distinct roles and update rules. A predator particle introduces an adaptive level of disturbance in the swarm, allowing better control of diversity and exploration/exploitation balance. Scout particles are used to add both general improvements and problem specific search strategies, while keeping the computational costs low. The version of the SPPO described here was specifically tailored to the training of SVMs. A more detailed description, together with comparative results for many benchmark functions can be found in [20].

4.1   The Swarm

To optimize a function $f(\mathbf{x})$ in $\mathbb{R}^m$, each particle i in the swarm is represented by three m-size vectors: the current position, $\mathbf{x}_i$; the best position food so far, $\mathbf{p}_i$; and the the particle's velocity $\mathbf{v}_i$. The position and velocity are updated iteratively using equations (4) and (5), where the distance between a particle and its previous best position is represented by $(\mathbf{p}_i^t - \mathbf{x}_i^t)$, while $(\mathbf{p}_g^t - \mathbf{x}_i^t)$ represents the distance to the the swarm best position, $\mathbf{p}_g$. Vectors of uniformly distributed random numbers, $\mathbf{u}(0, \phi_1)$ and $\mathbf{u}(0, \phi_2)$, control the strength of attraction in each dimension ($\otimes$ is a vector component-wise multiplication).

$$\mathbf{v}_i^{t+1} = w\mathbf{v}_i^t + \mathbf{u}(0, \phi_1) \otimes (\mathbf{p}_i^t - \mathbf{x}_i^t) + \mathbf{u}(0, \phi_2) \otimes (\mathbf{p}_g^t - \mathbf{x}_i^t) \qquad (4)$$

$$\mathbf{x}_i^{t+1} = \mathbf{x}_i^t + \mathbf{v}_i^{t+1} \qquad (5)$$

After a particle has moved, $f(\mathbf{x}_i)$ is computed and $\mathbf{x}_i$ is saved in $\mathbf{p}_i$ if $f(\mathbf{x}_i) > f(\mathbf{p}_i)$. Collaboration between particles is ensured by the updating of g, the index of the particle with the best solution (in all of the swarm or some neighborhood), which allows for all other particles to orient their search to more promising areas of the search space. In the SPPO all particles use the described representation, but predators and scouts use different update rules.

4.2   The Predator

The role of the predator particle is similar to that of an adaptive mutation operator and is used to maintain some diversity in the swarm, even when it has already converged, thus allowing it to escape possible local optima. It is an useful tool to control the balance between global exploration of the search space and the exploitation of promising areas, which was difficult to do in the original PSO. The predator movement is controlled by equation (6), which effectively makes it pursue the best particle in the search space.

$$\mathbf{v}_p^{t+1} = w\mathbf{v}_p^t + \mathbf{u}(0, \phi_1) \otimes (\mathbf{x}_g^t - \mathbf{x}_p^t) + \mathbf{u}(0, \phi_2) \otimes (\mathbf{p}_g^t - \mathbf{x}_p^t) \tag{6}$$

The predator introduces a disturbance element in the swarm, which increases with proximity, as described by equation (7), where $u(-1, 1)$ and $u(0, 1)$ are uniformly distributed random numbers, x_{max} and x_{min} are the upper and lower limit to the search space and r is the user defined perturbation probability. Equation (7) states that the velocity of particle i in dimension j can suffer a perturbation with a probability that depends on the particles's distance to the predator in that dimension. When the distance is large, this effect is negligenciable, but, as the the swarm converges and distance approximates 0, disturbance probability becomes maximum (r). This mechanism allows for particles to escape local optima even in the last phases of convergence and naturally adapts its intensity to the current state of exploration/exploitation.

$$v_{ij}^{t+1} = v_{ij}^t + u(-1, 1)|x_{max} - x_{min}|, \text{ if } u(0, 1) < r\exp^{-|x_{ij} - x_{pj}|} \tag{7}$$

4.3   Scout Particles

Scout particles are the second type of heterogeneous particles in the SPPO. Many successful variants of the PSO algorithm are based in its hybridization with other evolutionary or local search algorithms. Generally, however, the behavior of all particles is changed by the hybridization, which makes it hard to understand an control the effect of every algorithm component. Scouts are an inexpensive way of introducing new behaviors in the swarm, as only a few particles are affected and can easily be turned on and off to study their influence in the overall algorithm. These particles are typically updated before the swarm main update cycle, where they can cumulatively be updated using equations (4) and (7).

We chose two scout particles to tailor the SPPO to this specific problem. The first is a local search scout which, from previous work [20], we know can be used to increase the convergence speed without compromising the final results. For this scout we choose the best particle at each iteration and perform a random mutation on one of its dimensions j using the equation $p'_{gj} = p_{gj} + N(0, \sigma)$, where $N(0, \sigma)$ is a a random number drawn from a normal distribution with average 0 and standard deviation $\sigma = x_{max}/10$. The new $\mathbf{p}'_g$ will substitute $\mathbf{p}_g$ if $f(\mathbf{p}'_g) > f(\mathbf{p}_g)$. This allows for a local search to be made around $\mathbf{p}_g$ over time.

Problem specific knowledge is incorporated into the second scout particle. In a typical solution to the SVM training problem the majority of the α_i will be 0. From the remaining, which correspond to the support vectors, most will have the maximum value C. To explore this knowledge we change, at each iteration, a random dimension of this particle to 0, with an 80% probability, or to C, with

20% probability. As a consequence, this scout particle will, during the search process, iteratively explore the limits of the search space, where we expect of find the solution in most dimensions.

4.4   Application to SVM Training

The application of the previously described algorithm to the SVM training problem was straightforward. For a problem with m training examples, each particle was represented by three m sized vectors, corresponding the particle's current position, best position and velocity in the m search dimensions. The search space was limited, in each dimension, to the interval $[0..C]$, where C is the SVM regularization parameter. The position of a particle represents the α values in equation (1). The particle swarm optimizer was used to search for the α vector that maximizes the objective function (1). To to this, we embedded the algorithm in a SVM training operator for the well known data mining software package Rapidminer [21]. The software already included an evolutionary training operator for SVMs, which facilitated our implementation, since we could share some functionalities, including SVM model management and evaluation.

Contrary to many other applications of evolutionary optimization, the individual evaluation during the evolutionary process differs from the final quality measure reported for the best individual found. The evaluation is done using a process called cross-validation, where the training set is divided in k sets of equal cardinality, called folds. The evolutionary algorithm is ran k times, each time using the reunion of $k-1$ folds to train the SVM. During this training, a particle is simply evaluated using equation (1). When the training ends, the α values corresponding to the best particle are used to build the final SVM. It is then applied to the classification of the examples in the extra fold that was not used during the training process. The classification error on this fold is the final measure of the classifier quality. In the experimental work described in the following section, we use the two different performance measures to compare different elements of our approach. The final (averaged over the k folds) value of the objective function is mainly used to compare the efficiency of the different evolutionary approaches, when using similar computational resources. The final classification accuracy is used to compare the quality of the classifiers generated by the different algorithms (evolutionary and classical).

As a final implementation issue, it must be stressed that the evolutionary implementations are significantly slower than the classical algorithms. This is expected and was already noticed in previous approaches to the evolutionary training of SVMs [15]. The main reason for this results from the the nature of the problem: quadratic programming based approaches were developed to be particularly efficient in the concave optimization function. However, this is also their major weakness, when applied to problems with several optima. It is in this class of training problems, resulting from the use of non PSD kernels, that evolutionary approaches are useful, as we illustrate in the next section.

While the evolutionary training algorithms are slower, it should be stressed that the temporal complexity is still in the same order of magnitude for both classes of approaches, and that there is still vast space for improvement in those algorithms. E.g., while the standard SVM training algorithms use efficient stop criteria, we always ran the evolutionary algorithms for a fixed number of generations. While this allows for an easier comparison between the evolutionary algorithms, it also means that many iterations (and time) are wasted by the faster algorithms.

5   Experimental Results

In previous work [22], we have shown that evolutionary approaches, more specifically the evolution strategy based algorithm proposed by [10] and the SPPO, are competitive with traditional methods when training SMVs with PSD kernels. In fact, in a set of 10 benchmark problems, both algorithms achieved results similar, both in terms of classification accuracy and robustness, to the ones from two very popular quadratic programming based methods, MySVM and LIBSVM. These results demonstrate that the SPPO can efficiently tackle the uni-modal SVM training problem, in opposition to past PSO based approaches, which were found not to be competitive with the ES based approaches [15].

Table 1. Dataset and kernel parameters

Dataset	Source	n	m	err	σ_E	d	a	b	σ_D
Checkerboard	Synthetic	1000	2	48.60	0.92	6.54	1.258	1.452	0.006
Spirals	Synthetic	500	2	50.00	0.12	3.84	1.119	-1.737	9.684
Threenorm	Synthetic	500	2	50.00	61.60	9.38	0.110	1.746	0.018
Credit	UCI MLR	690	14	44.49	564.62	0.65	0.378	-1.894	0.002
Diabetes	UCI MLR	768	8	34.90	220.51	4.87	0.616	-0.065	0.001
Ionosphere	UCI MLR	351	34	35.90	2.44	7.48	0.132	-0.318	0.247
Liver	UCI MLR	345	6	42.03	61.59	6.90	0.116	0.770	0.052
Lupus	StatLib	87	3	40.23	241.63	7.42	0.120	0.786	0.002
Musk	UCI MLR	476	166	43.49	63.12	6.93	0.006	0.197	0.187
Sonar	UCI MLR	208	60	46.63	61.63	6.90	0.025	0.752	0.063

As we already stated, there is no particular reason to use, in practical applications, the evolutionary training algorithms in the concave training problem, where the quadratic programming based approaches are much faster. Consequently, in this work, we further investigate the use of these algorithms with several non PSD kernels. We are also interested in comparing the proposed PSO based training algorithm with the best previous ES based approach [10], in terms of both optimization efficiency and accuracy.

All the experiments described in this section were performed on a set of 10 benchmark problems, where the first three were synthetically generated and the remaining seven are real world benchmark problems. Name, source, number of attributes n and number of instances m are listed, for each dataset, in Table 1. In the same table, err represents the expected error of a 0-R classifier, i.e. an algorithm that always returns the most frequent class in the training set. Also listed are the parameters of the kernels presented in the following subsections. Those were obtained using a short run of the evolutionary parameter optimizer in the Rapidminer data mining software [21], which was used to run all experiments, with additional operators written for that purpose.

We used $C = 1$ in all experiments. While we understand that this may not be the ideal choice for some problem/kernel combinations, we chose to do this because it eased comparison with previous work in the area (which already used $C = 1$) and we were not interested in finding the best possible classifier for each problem, but only in comparing the different algorithms in similar experimental conditions. This choice also allowed thus to simplify the experimental setup by saving us the optimization of an extra parameter.

While the goal of this paper is mainly to evaluate the usefulness of evolutionary SVM training algorithms on different non PSD kernels, while also comparing the efficiency of ES and particle swarm based methods, the use of standard benchmark problems like the ones described also allows for some level of comparison with previous approaches using diverse machine learning techniques. E.g., Meyer presents an extensive empirical comparison of SVMs and 16 other classification algorithms using many of this datasets [23]. While the results cannot be directly compared, as the experimental setups were not the same, they still allow an assessment of the validity of the approaches described here.

We compare a classic SVM training approach (MySVM) with the SPPO and the best previous ES based approach, which were implemented to search for the vector α that maximizes equation (1). We report the error percentage averaged over a 20-fold cross validation, as well as the standard deviation. For the evolutionary approaches, we also present the average and standard deviation of the highest objective function value found. Statistically significance tests were performed for the classification accuracy (ANOVA and pairwise t-tests). The evolutionary algorithms were ran for a fixed number of generations, different for each kernel. The evolution strategy used 20 individuals, while the scouting predator-prey algorithm was limited to 18, to compensate for the extra function evaluations of the scout particles.

5.1   Learning with Non PSD Kernels

In the first two sets of experiments, we used common non PSD kernels, the Epanechnikov kernel and the Sigmoid kernel. The Epanechnikov kernel is a non PSD kernel which has already been used in previous work in the area of evolutionary SVM training [10]. The Sigmoid kernel was popular in SVMs due to its origin in neural networks, but can also be non PSD for a subset of its parameters values [5]. Table 2 presents the results obtained for the Epanechnikov kernel in the benchmark datasets, with the evolutionary algorithms being ran for 150 iterations for the synthetic problems and 100 iterations for the real world problems. The iteration limit was 30 for the sigmoid kernel and the results obtained are listed in Table 3.

Table 2. Experimental results (error percentage) using the Epanechnikov kernel

Dataset	MySVM	ES ($f(\alpha^*)$)	ES	SPPO ($f(\alpha^*)$)	SPPO
Checkerboard	6.5 (4.6)	-303.6 (32.6)	8.5 (4.3)	51.5 (9.2)	7.8 (4.4)
Spirals	7.2 (3.9)	163.5 (3.0)	13.6 (7.6)	188.4 (2.8)	7.8 (5.3)
Threenorm	14.0 (7.5)	24.5 (17.0)	15.2 (8.2)	133.7 (4.0)	14.0 (5.4)
Credit	14.2 (6.4)	253.4 (5.7)	13.3 (6.8)	300.8 (4.3)	13.8 (7.5)
Diabetes	24.2 (3.5)	-114.4 (158.8)	26.9 (7.6)	296.1 (8.1)	25.3 (5.0)
Ionosphere	26.3 (10.3)	81.1 (3.7)	25.9 (10.2)	100.1 (1.9)	**15.7 (5.9)***
Liver	42.1 (2.8)	180.1 (6.0)	36.8 (10.8)	228.9 (5.0)	**35.4 (10.6)***
Lupus	28.2 (14.6)	49.0 (2.0)	22.5 (16.1)	58.5 (0.8)	**19.5 (12.3)**
Musk	7.8 (3.8)	101.8 (4.2)	11.3 (5.2)	117.4 (3.2)	8.7 (6.5)
Sonar	12.4 (11.4)	50.5 (1.9)	12.9 (10.2)	61.8 (1.56)	12.5 (11.0)

The first conclusion we can draw from the presented results is that all algorithms were generally able to learn with the non PSD kernels. Classification accuracy for the Lupus and Sonar datasets, using the Epanechnikov kernel and the SPPO algorithms, was even superior to the accuracy obtained in previous work using the PSD RBF kernel [22]. With the sigmoid kernel, there were three problems for which the algorithms were not able to learn useful (i.e. superior to 0-R) classifiers (Checkerboard, Spirals and Liver). However, looking at previous work using this kernel [5], this can possibly be attributed to the fact that we didn't optimize the C parameter.

Table 3. Experimental results (error percentage) using the sigmoid kernel

Dataset	MySVM	ES ($f(\alpha^*)$)	ES	SPPO ($f(\alpha^*)$)	SPPO
Checkerboard	48.3 (0.8)	1086.6 (79.2)	48.5 (6.2)	1736.8 (115.2)	48.0 (7.1)
Spirals	51.6 (10.7)	1262.9 (353.5)	50.0 (8.0)	4334.9 (367.2)	49.8 (10.4)
Threenorm	16.6 (9.1)	231.8 (3.8)	16.8 (8.7)	261.7 (4.1)	17.0 (6.0)
Credit	22.2 (7.4)	-7170.3 (593.6)	14.5 (5.5)	-4843.0 (310.6)	**14.1 (5.9)***
Diabetes	38.7 (5.3)	-2119.7 (492.5)	30.5 (7.4)	-682.0 (218.5)	**27.2 (7.7)***
Ionosphere	19.9 (6.8)	-557.3 (97.4)	25.9 (8.5)	-175.7 (43.1)	**12.8 (8.6)***
Liver	42.1 (2.8)	168.4 (7.1)	41.8 (5.8)	187.4 (4.9)	40.9 (6.3)
Lupus	26.7 (19.1)	42.1 (1.8)	28.0 (19.3)	49.4 (1.8)	**21.8 (21.5)**
Musk	24.5 (8.7)	181.3 (6.4)	29.9 (9.2)	204.5 (5.3)	25.6 (7.3)
Sonar	33.0 (14.6)	98.4 (3.12)	28.5 (14.1)	113.0 (2.7)	26.4 (15.8)

We can also observe that the SPPO algorithm behaved substantially better than the ES based approach. Classification error was similar or lower in all problems, while the optimization performance of the SPPO, measured by the best value found for the objective function, was substantially higher for all problems. Looking at the convergence plots, it is possible to conclude that the swarm based optimizer needed from $3\times$ to $5\times$ less iterations to achieve the optimization values (and consequent classification accuracy) that the ES based approach reached at the iteration limit. We present plots for the Cherkerboard and Liver datasets (using the Epanechnikov kernel) in Figure 1 as typical examples of the algorithms' convergence behaviors.

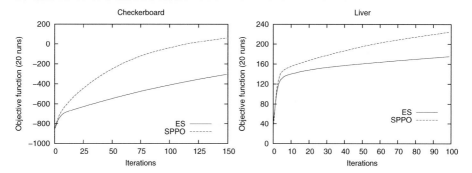

Fig. 1. Convergence plots for the SPPO and ES based algorithms, applied to the Cherkerboard and Liver datasets and using the Epanechnikov kernel

Finally, over the two experiments sets, we can see seven dataset/kernel pairs for which the classification accuracy is substantially better for the SPPO, when compared with the MySVM results. These values are presented in bold in the results tables. Since the differences are not very large (and, in some cases, the standard deviation is high), we performed a statistical analysis of the average error results, having found that the differences were statistically significant ($p < 0.05$) for six of the seven cases (marked with an asterisk in the results tables). For the remaining pairs the classification error is similar. It should be noted that the ES algorithm frequently produces lower quality classifiers, when compared with the other algorithms, but this fact can be attributed to the strict iteration limit. If this limit was higher, we could expect both evolutionary algorithms to achieve similar results.

In previous experiments [22], classical SVM training approaches and evolutionary algorithms consistently produced comparable classifiers in terms of classification accuracy, when the same kernels and parameters were used. We assume the differences found in these latest results are a consequence of the non PSD nature of the used kernels and the resulting multimodal optimization problems. The classical training approaches may then be caught in local optima, failing to find a competitive solution. These results confirm our initial proposition that evolutionary algorithms could be useful tools in the training of SVMs when using non PSD kernels.

5.2   Learning with Distance Substitution Kernels

In this last set of experiments, we investigate a more practical issue, which can lead to the necessity of training a SVM with a non PSD kernel. In many real world problems, where specific distance data is available (e.g. as a result of a physical measuring process), one of the possible approaches to using this information in a kernel method is to use the RBF kernel and substituting the Euclidian distance by the problem specific distance data. This process has been formalized and generalized to other kernels by Haasdonk [3]. The resulting kernels are PSD if the used distance is isometric with the $\mathcal{L}^2$-norm. In practice, however, non PSD kernels can easily result from specific distances, e.g. distances which are non-metric or even $\mathcal{L}^p$ metrics with $p \neq 2$ [3]. To simulate this situation, we repeated the previous experiments using a distance substitution RBF kernel. We replaced the Euclidian distance by the $\mathcal{L}^1$ distance (frequently called the Manhattan distance), which can result in a non PSD kernel matrix [4]. The iteration limit was set at 300 for the synthetic datasets and 200 for the remaining. Results obtained for the distance substitution kernel are presented in Table 4.

Table 4. Experimental results (error percentage) using the distance based kernel

Dataset	MySVM	ES $(f(\alpha^*))$	ES	SPPO $(f(\alpha^*))$	SPPO
Checkerboard	4.5 (2.4)	45.6 (11.1)	5.8 (4.1)	135.9 (2.8)	4.7 (3.5)
Spirals	48.6 (0.9)	411.5 (3.1)	49.2 (1.5)	443.9 (1.9)	48.8 (1.2)
Threenorm	20.4 (6.7)	202.3 (1.3)	15.0 (6.6)	209.2 (0.7)	**14.0 (8.9)***
Credit	30.3 (5.9)	235.0 (5.0)	30.0 (5.3)	269.6 (1.8)	29.5 (7.7)
Diabetes	29.3 (6.4)	211.7 (5.3)	29.7 (6.5)	269.8 (3.8)	27.6 (8.0)
Ionosphere	32.8 (5.6)	105.5 (1.7)	15.7 (8.6)	117.8 (0.9)	**12.4 (9.4)***
Liver	40.6 (4.5)	105.5 (1.7)	37.7 (5.1)	117.8 (0.9)	**36.3 (6.5)***
Lupus	30.2 (20.3)	40.9 (1.1)	27.0 (18.1)	43.3 (1.4)	28.5 (17.7)
Musk	45.6 (5.4)	199.6 (2.4)	43.5 (2.4)	218.5 (1.1)	43.5 (2.4)
Sonar	12.0 (8.8)	57.4 (1.3)	11.0 (9.0)	59.8 (0.8)	11.5 (9.6)

These results mirror the ones obtained for the previous kernels, with the SPPO achieving better classification accuracy in three of the datasets, when compared with the MySVM algorithm. Differences in the remaining datasets are not statistically significant. Evolutionary approaches obtain similar results in terms of classification error, but the SPPO consistently achieves higher values for the objective function, which again implies that less iterations are needed by this algorithm to find same quality classifiers. As we already found for other non PSD kernels, evolutionary SVM training algorithms can be advantageous over traditional approaches, when new kernels are created using distance substitution.

6   Conclusions

The experimental results presented in this paper allow us to draw several conclusions. First, all algorithms were able to build successful classifiers for most problems using the different kernels. This fact confirms previous theoretical [4] and practical [10] results on the feasibility of learning with non PSD kernels and SVMs. This conclusion can be of particular practical interest when using problem specific distance substitutions kernels. Second, while traditional approaches (in this case MySVM) seem perfectly able to find good classifiers for most problems, there can be situations when they get stuck on sub-optima occurring in the objective function. In these cases, evolutionary algorithms, with their global optimization abilities, can prove to be useful and robust SVM training tools. Third, of the two evolutionary approaches in comparison, the newly introduced swarm intelligence based approach consistently achieved better optimization values for the objective function in the allotted iterations, which suggests that it this the most computationally efficient of the two, requiring significantly less function evaluations to produce similar or better classification accuracy.

References

[1] Shawe-Taylor, J., Cristianini, N.: Kernel methods for pattern analysis. Cambridge Univ. Press, Cambridge (2004)
[2] Cristianini, N., Shawe-Taylor, J.: An Introduction to Support Vector Machines. Cambridge University Press (2000)
[3] Haasdonk, B., Bahlmann, C.: Learning with distance substitution kernels. In: Rasmussen, C.E., Bülthoff, H.H., Schölkopf, B., Giese, M.A. (eds.) DAGM 2004. LNCS, vol. 3175, pp. 220–227. Springer, Heidelberg (2004)
[4] Haasdonk, B.: Feature space interpretation of svms with indefinite kernels. IEEE Transactions on Pattern Analysis and Machine Intelligence 27, 482–492 (2005)
[5] Lin, H.T., Lin, C.J.: A study on sigmoid kernel for svm and the training of non-psd kernels by smo-type methods. Technical report, National Taiwan University, Taipei, Department of Computer Science and Information Engineering (2003)
[6] Burges, C.J.: A tutorial on support vector machines for pattern recognition. Data Mining and Knowledge Discovery 2, 121–167 (1998)
[7] Rüping, S.: mySVM-Manual. University of Dortmund, Lehrstuhl Informatik 8 (2000)
[8] Chang, C.C., Lin, C.J.: Libsvm: a library for support vector machines (2001)
[9] Ong, C.S., Mary, X., Canu, S., Smola, A.J.: Learning with non-positive kernels. In: Proceedings of the Twenty-First International Conference on Machine learning, ICML 2004. ACM, New York (2004)
[10] Mierswa, I., Morik, K.: About the non-convex optimization problem induced by non-positive semidefinite kernel learning. Advances in Data Analysis and Classification 2, 241–258 (2008)
[11] Samanta, B., Nataraj, C.: Application of particle swarm optimization and proximal support vector machines for fault detection. Swarm Intelligence 3, 303–325 (2009)
[12] Gilsberts, A., Metta, G., Rothkrantz, L.: Evolutionary optimization of least-squares support vector machines. In: Data Mining. Annals of Information Systems, vol. 8, pp. 277–297. Springer, US (2010)

[13] Paquet, U., Engelbrecht, A.: Training support vector machines with particle swarms. In: Proceedings of the International Joint Conference on Neural Networks, vol. 2, pp. 1593–1598 (2003)

[14] Stoean, R., Preuss, M., Stoean, C., Dumitrescu, D.: Concerning the potential of evolutionary support vector machines. In: IEEE Congress on Evolutionary Computation, CEC 2007, pp. 1436–1443 (2007)

[15] Mierswa, I.: Evolutionary learning with kernels: a generic solution for large margin problems. In: GECCO 2006: Proceedings of the 8th Annual Conference on Genetic and Evolutionary Computation, pp. 1553–1560. ACM, New York (2006)

[16] Mierswa, I.: Controlling overfitting with multi-objective support vector machines. In: Proceedings of the 9th Annual Conference on Genetic and Evolutionary Computation, GECCO 2007, pp. 1830–1837. ACM, New York (2007)

[17] Kennedy, J., Eberhart, R.: Particle swarm optimization. In: IEEE International Conference on Neural Networks, vol. 4, pp. 1942–1948 (1995)

[18] Poli, R., Kennedy, J., Blackwell, T.: Particle swarm optimization. Swarm Intelligence 1, 33–57 (2007)

[19] Poli, R.: Analysis of the publications on the applications of particle swarm optimisation. J. Artif. Evol. App. 2008, 4:1–4:10 (2008)

[20] Silva, A., Neves, A., Gonçalves, T.: An heterogeneous particle swarm optimizer with predator and scout particles. In: Kamel, M., Karray, F., Hagras, H. (eds.) AIS 2012. LNCS, vol. 7326, pp. 200–208. Springer, Heidelberg (2012)

[21] Mierswa, I., Wurst, M., Klinkenberg, R., Scholz, M., Euler, T.: Yale: Rapid prototyping for complex data mining tasks. In: Proceedings of the 12th International Conference on Knowledge Discovery and Data Mining, KDD 2006 (2006)

[22] Silva, A., Gonçalves, T.: Training support vector machines with an heterogeneous particle swarm optimizer. In: Tomassini, M., Antonioni, A., Daolio, F., Buesser, P. (eds.) ICANNGA 2013. LNCS, vol. 7824, pp. 100–109. Springer, Heidelberg (2013)

[23] Meyer, D., Leisch, F., Hornik, K.: Benchmarking support vector machines. In: SFB Adaptive Information Systems and Modelling in Economics and Management Science, vol. 78. WU Vienna University of Economics and Business, Vienna (2002)

Response Surfaces with Discounted Information for Global Optima Tracking in Dynamic Environments

Sergio Morales-Enciso[1] and Juergen Branke[2]

[1] Centre for Complexity Science, The University of Warwick, CV4 7AL, UK
S.Morales-Enciso@warwick.ac.uk
[2] Warwick Business School, The University of Warwick, CV4 7AL, UK
Juergen.Branke@wbs.ac.uk

Abstract. Two new methods for incorporating old and recent information into a surrogate model in order to improve the tracking of the global optima of expensive black boxes are presented in this paper. The response surfaces are built using Gaussian processes fitted to data which is obtained through sequential sampling. The efficient global optimization (EGO) algorithm applied to the generated response surface is used to determine the next most promising sample (where the expected improvement is maximized). The goal is to find the global maxima of an expensive to evaluate objective function which changes after a given number of function evaluations with as few samples as possible. Exploiting old information in a discounted manner significantly improves the search, which is shown through numerical experiments performed using the moving peaks benchmark (MPB).

1 Introduction

In dynamic environments, tracking global optima of expensive black box functions has mainly been approached using evolutionary algorithms (EA) and particle swarm optimization (PSO), while in the static case response surfaces have also been used widely. In this paper we introduce two methods to build response surfaces for the dynamic version using Gaussian processes (GP). These techniques build on the efficient global optimization (EGO) algorithm proposed in [10], and aim to use old information efficiently either by resampling at previously known good regions or by introducing sampling noise to discount it.

Dynamic optimization naturally arises when trying to optimize a problem in which the environment constantly changes or new information constantly arrives. For instance, in the vehicle routing problem it might be optimal to incorporate a recently arrived delivery order into an already defined route. Another example is dynamic pricing, where the overall revenue is to be maximized, but price sensitivity has to be learned by quoting prices to the customers, taking into account seasonality, market saturation, and trends [11].

The main contribution of this paper is a mathematical model for incorporating discounted old information with new information to create a response surface

G. Terrazas et al. (eds.), *Nature Inspired Cooperative Strategies for Optimization (NICSO 2013)*, Studies in Computational Intelligence 512,
DOI: 10.1007/978-3-319-01692-4_5, © Springer International Publishing Switzerland 2014

using a Gaussian process. In Sect. 2, a literature review is presented first on the static version of the problem and then on the dynamic one. Later, in Sect. 3, the concepts and techniques of Gaussian processes and EGO are explained, which are the basis for the proposed sequential sampling strategy detailed in Sect. 4. Finally, some numerical experiments and results are provided and analysed in Sect. 5, and the conclusions are presented in Sect. 6.

2 Related Work

2.1 Global Optimization of Black Box Static Functions

Global optimization tackles the problem of finding the best solution –over their whole domain– to many different classes of problems [6]. This paper considers global optimization in the context of expensive evaluations of black box dynamic functions. Black box refers to the lack of analytic expression of the objective function so methods requiring an analytic expression of the objective function or of its gradient can not be applied. Expensive evaluation means that every sample or observation taken from the objective function requires relatively large amount of resources as compared to the additional cost of creating a model to aid the search. This situation is likely to arise for instance when optimizing parameters for engineering products or dealing with complex simulations [5].

The static version of this problem has been vastly studied. In [16] it is revealed that current research does not focus on trying to directly model and understand black boxes, but focuses instead on sampling strategies and finding clever uses of the scarce observed data in order to determine promising areas to sample.

Response surfaces (or surrogate models) are approximations of a function of interest created using available data, and are the output of some sort of regression model. These models are used when a direct measurement of the function is not practical, for instance if the outcome of interest is not easy to measure or if each measurement is expensive to obtain in time, money, or any other cost unit. Some of the most widely used response surface building techniques include, but are not limited to, radial basis functions (RBF), support vector machines (SVM), artificial neural networks (ANN), and Gaussian processes (GP) [9].

In global optimization the use of surrogates as a replacement for expensive objective functions is a common practice. There are two standard approaches for using surrogate models which might seem similar but are conceptually different. The first one is to generate a set of candidate solutions to be evaluated and, instead of directly evaluating the objective function at each of these candidate solutions, the previously generated response surface is used to estimate the fitness of each of the candidates. Then, only the most promising ones according to the response surface in use are accurately evaluated using the objective function. The most commonly used candidate generating techniques are evolutionary algorithms (EA) (see [8] for a survey) and more recently particle swarm optimization (PSO) [18]. The second approach is to explore and analyse the generated response surface to decide where to sample next, i.e., use the generated model directly to propose the best candidate solution.

One option to choose the next best (or most promising) sample, although not the best technique, is to find the global optimum of the surrogate model and choose it as the next sample to be taken. A far better use of the surrogate, as shown in [10], is to sample where the expected improvement is maximized. This technique is called efficient global optimization (EGO) and, due to its simplicity in concept and good performance, has become a popular choice in literature with many variations and adaptations.

2.2   The Dynamic Problem

The dynamic version of the problem deals with finding the global optima of an expensive black box function changing over time, which calls for a more sophisticated exploration strategy capable of keeping track of solutions close to the optimum that might become useful at later times. In the general case, the changes can happen each time the function is evaluated, after a given number of evaluations, or after a given period of time. The frequency of the changes depends on the nature of the problem to be solved, for instance, after a given number of performed experiments, or at the beginning of every season. Some studies focus on change detection [7],[15], but this paper assumes the frequency of changes to be known in advance in terms of function evaluations, emulating a fixed sampling budget.

A change in a function can be reflected in the number of modes (or peaks) that it has, their location, and the height and width of each peak. If the magnitude of the change is such that there are no similarities between the function before and after the change, it is of no use to transfer any knowledge and a restart of the optimizer is the best possible option since outdated information would just mislead the search. If the changes are less dramatic, old samples (i.e. samples taken before the change) can be reused so as to guide and accelerate the search. Nevertheless, some considerations have to be taken into account when introducing outdated information, which is the problem addressed in Sect. 4.

Like in the static case, most of the techniques used to track global optima in dynamic environments are based on EA [8],[4],[17], and PSO [13],[2].

To the best knowledge of the authors, models which build a response surface using old samples updated with new coming information are not found in the literature. So, in the remainder of this paper, five techniques to track the global optima of a dynamic expensive black box function based on a response surface are described and compared.

3   Efficient Global Optimization (EGO) on Gaussian Process (GP) Generated Response Surfaces

3.1   GP as a Surrogate Model

The advantage of GP as a technique to build response surfaces over other methods such as RBF, SVM, and ANN is the analytical tractability it provides not only for the predictions but also for the confidence on its predictions. Furthermore, it sets a natural framework to extend this technique to incorporate old information for the dynamic case as it is shown in Sect. 4.

Consider the observed dataset $\mathcal{D} = \{(\boldsymbol{x}_i, y_i)_{i=1}^n\} = \{\boldsymbol{X}, \boldsymbol{Y}\}$ of n D–dimensional samples taken at $\boldsymbol{x}_i = [x_{i,1}, ..., x_{i,D}]$ with corresponding response value y_i. The idea is to use a GP and the observed samples $\mathcal{D}$ to create a response surface in order to have a cheap way to provide an estimate of the objective function (make predictions) at any desired test point $\boldsymbol{x}_p$.

A GP is fully defined by a mean function which allows to introduce any prior information available into the model, and a covariance function which expresses the scaled correlation between the data points [14]. As a result of applying GP for regression to a dataset, we obtain a distribution on the function that generated the data, also called latent function f which is defined by a mean and a covariance function. Throughout this paper, a zero mean prior function (1) and the squared exponential covariance function (2) are used.

$$m(\boldsymbol{x}) = 0 \tag{1}$$

$$k(\boldsymbol{x}, \boldsymbol{x}') = \sigma_f^2 exp\left(-\sum_{d=1}^{D} \frac{(x_d - x_d')^2}{2\ell_d^2}\right) + \sigma_n^2 \delta(\boldsymbol{x}, \boldsymbol{x}') \tag{2}$$

Let $\boldsymbol{K}$ denote the matrix containing the covariances evaluated at all training points, while $\boldsymbol{K}_p$ is the augmented covariance matrix containing $\boldsymbol{K}$ plus the covariances between the testing point $\boldsymbol{x}_p$ and all the other points.

Then, the best estimate for our prediction $\hat{y}_p$ is given by (3), and the confidence about that estimate by (4), where $k(\boldsymbol{x}_p, \boldsymbol{X})$ denotes the last row of $\boldsymbol{K}_p$. This allows us to characterize the prediction on the outcome y_p at the test point $\boldsymbol{x}_p$ with a normal distribution (5).

$$\mu := \hat{y}_p = m(\boldsymbol{x}) + k(\boldsymbol{x}_p, \boldsymbol{X})\boldsymbol{K}^{-1}\boldsymbol{Y} \tag{3}$$

$$\sigma := Var[y_p] = k(\boldsymbol{x}_p, \boldsymbol{x}_p) - k(\boldsymbol{x}_p, \boldsymbol{X})\boldsymbol{K}^{-1}k(\boldsymbol{x}_p, \boldsymbol{X})^T \tag{4}$$

$$y_p \sim \mathcal{N}(y_p|\mu, \sigma) \tag{5}$$

In the general case, there are $D + 2$ parameters in total which are learnt from the available data $\mathcal{D}$ by using maximum likelihood estimation (MLE). σ_f^2 is the maximum possible covariance. The characteristic length-scales for each dimension $\boldsymbol{\ell} = [\ell_1, ..., \ell_D]$ represent how much each point influences one another –independently for each dimension– as a function of the distance. Finally, σ_n^2 is the noise associated to the sampling process. Since only deterministic objective functions are considered in this paper, σ_n^2 will be set to zero except for our proposed model as detailed in Sect. 4.5 where this parameter plays a major roll as a proxy to discount reliability of old samples. However, even in that case σ_n^2 will not be learnt from the data, so only $D + 1$ parameters are to be inferred. A complete and formal description on GP can be found in [14].

3.2   Using the Surrogate Model Efficiently: EGO

Once the surrogate model is available, a sampling strategy such as EGO [10] can be followed to determine where the next observation should be taken. EGO looks for the sample that maximizes the expectation of improvement over the currently known best sample, which is possible because the GP provides an analytic expression of the probability distribution for each predicted value (5).

In order to calculate the expected improvement $\mathbb{E}[I(\boldsymbol{x}_p)]$ (6) at the test point $\boldsymbol{x}_p$, the best observed value so far $y^* = max_{i=1}^{n}(y_i)$, is taken as a reference. Then, the expected improvement is given by the probability of the predicted value y_p (given by the GP prediction using 3 and 4) times the improvement, integrated over all possible values better than y^* which yields (6) [10].

$$\mathbb{E}[I(\boldsymbol{x}_p)] = (y_p - y^*)\Phi\Big(\frac{y_p - y^*}{\sigma}\Big) + \sigma\phi\Big(\frac{y_p - y^*}{\sigma}\Big) \tag{6}$$

The next sample $\boldsymbol{x}_{n+1}$ is finally taken where the expected improvement is maximized and, together with the observed response, is added to $\mathcal{D}$. This sampling strategy has proven to be successful in a variety of applications for static problems [1].

4   Proposed Surrogate Models for the Dynamic Case

The following sequential sampling strategies for parameter optimization of dynamic black boxes build on the principles of the static version of the problem. So, each time a new sample is obtained, the response surface is rebuilt by updating the GP with the new observation as described in Sect. 3.1. Once the response surface is built, the EGO algorithm is used to determine where to sample next.

The key difference when building response surfaces in dynamic environments is that the age of the available data samples must be tracked and old data samples should be considered less reliable than recent ones, however they must not be entirely discarded because they still contain information of previous times.

As stated in Sect. 2.2, we address the problem where the objective function changes after a known number of evaluations c_f (change frequency), and the periods in between changes are referred to as epochs (t) numbered in increasing order. Nevertheless, it is not the epoch at which each sample was obtained that is relevant to discount the reliability of the sample, but rather how long ago it was taken with respect to the current epoch (t_c). So instead of using the epoch number, it is the age of a sample with respect to the present ($\tau = t_c - t$) that is considered to reduce reliability of the samples.

Below, five sampling strategies are described. The first three are simple strategies used as benchmarks. First, a random sampling strategy is proposed to compare against the completely uninformed case. Second, two limiting cases are presented: the reset strategy as a memoryless model which starts solving the problem after every change, and the ignore strategy that dismisses all the changes so that all the information is considered equally reliable. Finally, two strategies are proposed as different ways of reusing and transferring information from old epochs to the new ones. These five methods are compared through numerical experiments in Sect. 5.

In order to build a response surface using a GP, it is necessary to start with at least $\lambda = 2$ data samples to be able to estimate the length-scale parameters of the process. So, for the first epoch it is assumed that there are at least $\lambda \geq 2$ observations previously obtained.

Let $\mathcal{D}$ be the set of all the samples collected throughout the history of the experiment, and $\mathcal{D}_\tau \subset \mathcal{D}$ the set of data points of age τ.

4.1   Random Sampling Strategy

The random sampling strategy explores the parameter space $x \in \mathbb{R}^D$ by independently drawing a random number from a uniform distribution for each dimension. This technique serves only as a benchmark in order to set a reference to asses the improvements of the other techniques, and there is no response surface built.

4.2   Reset Sampling Strategy

This strategy dismisses all the previously obtained samples every time a change on the objective function happens. This is equivalent to dismissing all the information already gathered in previous epochs and starting to sample again as if this were a new problem. So, at the current epoch ($\tau = 0$), the response surface will be estimated using only current information in $\mathcal{D}_0$ (see Sect. 3.1). Since previous samples are not considered, at the beginning of each epoch λ observations need to be sampled in order to start building the response surface one more time. The reset strategy also serves as a reference to measure the improvement obtained by other sampling strategies. Besides, it is useful in the presence of very drastic changes where there is no similarity between the objective function before and after each change.

4.3   Ignore Sampling Strategy

As its name suggests, the ignore sampling strategy overlooks the fact that a change has happened, which means that all the available samples in $\mathcal{D}$ are used to fit the response surface. Not only is this a bad strategy to find the global optima of a changing function because old information is taken to be as valid as new one, which completely misguides the search, but also because it unnecessarily increases the computational cost of generating the GP. This is the opposite extreme to the reset strategy and serves as another benchmark. The ignore strategy is useful when the magnitude of the changes is negligible close to the static version of the problem.

4.4   Reset* Sampling Strategy

Reset* differs from reset (Sect. 4.2) only in the way the first samples of a new epoch (other than the first one) are taken. Instead of taking λ initial observations at the beginning of a new epoch ($\tau = 0$), reset* looks for the best response found in the immediate previous epoch ($\tau = 1$) and resamples at the same place where this previously best response was obtained. Furthermore, the length-scale parameters ($\boldsymbol{\ell}$) found at the end of the immediate previous epoch are reused in order to overcome the inability of fitting a GP with only one data point and allow to take a second sample. Once the second sample is obtained, the sampling process continues as the reset strategy (i.e. refitting the GP parameters from the available data ($\mathcal{D}_0$) every time a new sample becomes available) until the next function change.

4.5   Discounted Information through Noise Sampling Strategy (DIN)

The idea behind this strategy is to consider newly obtained samples as deterministic –as it has been done throughout this paper–, but to introduce some artificial measurement noise in order to discount the old samples. In this process, all the information obtained so far (from the current epoch and from previous epochs) is used to fit the GP, so no information is discarded.

   The recent observations, being treated as deterministic (no noise added), force the response surface to go exactly through the measured sample, while the old observations, treated as noisy observations, allow the response surface to pass within some distance of the actually observed response values (proportional to the magnitude of the introduced noise) but not necessarily through them. By considering old information but discounting its accuracy, the search is guided to the regions where there used to be good responses in order to explore if that is still the case, but it is acknowledged that the landscape might have changed. GP provide a natural way of introducing noise in different magnitudes for each data sample through the noise measurement term (σ_n^2) in (2). Furthermore, the introduced noise can be a function of the age of the observations. This modification gives raise to (7) which is to be used to calculate the covariance matrix used to generate the response surface for the DIN model. An illustration of this model is provided in Fig. 1.

$$k(\boldsymbol{x}, \boldsymbol{x}') = \sigma_f^2 exp\left(-\sum_{d=1}^{D} \frac{(x_d - x_d')^2}{2\ell_d^2} \right) + \sigma_n^2(\tau)\delta(\boldsymbol{x}, \boldsymbol{x}') \qquad (7)$$

$\sigma_n^2(\tau)$ is now a function of the age of the samples and no longer a constant as in (2), and can be any strictly increasing function in τ such that $\sigma_n^2(\tau_c) = 0$, for instance (8), where s is some constant noise level.

$$\sigma_n^2(\tau) = \tau s^2 \qquad (8)$$

The introduced noise $\sigma_n^2(\tau)$ increases as a function of the age of the samples following a predefined functional form which is user defined rather than learnt. Since DIN uses samples from previous epochs, it is not necessary to generate any random sampling nor to reuse the GP parameters from previous epochs. Nonetheless, the first sample of each epoch is taken where the best response was obtained at the previous epoch, following the same procedure as in reset* (Sect. 4.4).

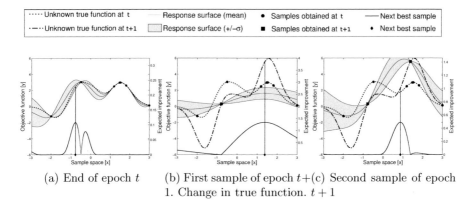

(a) End of epoch t (b) First sample of epoch $t+$(c) Second sample of epoch
1. Change in true function. $t+1$

Fig. 1. Sequential sampling using DIN with ($\sigma_n^2 = 0.1$). Figure 1(a) shows the end of an epoch, and the GP response surface ($\pm\sigma$) generated using samples obtained at t. The vertical line shows where the next best sample is. In 1(b) the new sample has been taken (square) after a change. The GP interpolates the new observation even if there are other old samples in the region. But in the absence of recent information, old data is used to guide the response surface. Figure 1(c) shows the next sample taken.

5   Experiments and Results

In order to compare the performance of the two proposed methods in Sect. 4 against the reference strategies, the MPB was implemented. In this Section, a brief description of such benchmark along with two performance measures are provided, followed by a description of the experiments performed to test the performance of each of the presented models.

5.1   The Moving Peaks Benchmark (MPB)

Even though there are many real examples of objective functions evolving over time, it is not easy to find cases which are both complex enough to present a challenge and simple enough to analyse and make an interpretation of the tuned parameters. The MPB provides a framework bridging this gap between very complex, hard to understand real-world problems and all too simple toy problems [3]. This benchmark consists of a D dimensional continuous function defined in a given interval with N peaks of different height and width. Each peak (p_i) is defined by its position $\boldsymbol{x}_i \in \mathbb{R}^D$, height h_i, and width w_i ($i \in [1, ..., N]$). At every change, each of the peaks suffers a slight change in its position, of fixed magnitude but in a random direction, and changes of a random magnitude in height, and width. An extensive survey on methods applied to this benchmark can be found in [12].

As described in [3], the average error ($\bar{\epsilon}$) measures the average deviation from the (unknown) optimum of each function evaluation performed so far. This is, the sum of the individual differences between the optimum value and each sample taken. The offline error (ϵ_o) is defined as the time average of the errors between the best currently known sample for a given epoch (period since last function change) and the optimum. Choosing which performance measure to use is problem specific. Figure 2(b) shows typical convergence curves for the offline error for the sampling strategies compared in this paper. The implementation of the MPB simulates the sequential sampling process applying the strategies described in Sect. 4 in the attempt of tracking the global optima of the moving peaks objective function.

The parameters governing the dynamics of the objective function are detailed in Table 1. All the simulations start with an initial number of $\lambda = 4$ samples, and when applicable the same number of initial samples is used at the beginning of later epochs. Then, one of the proposed strategies is followed to fit a GP to the available data together with the EGO method. A local hill climber with random multi-start is used to maximize the expected improvement.

Fitting a GP has computational complexity of $\mathrm{O}\left(|\mathcal{D}|^3\right)$, so the process slows down with each new sample incorporated to the data set. This has an indirect implication on the scalability of the presented technique to problems with a large number of dimensions given that the number of samples required would rapidly increase. For efficiency purposes, only data from the immediate previous epoch (i.e. $\tau = 1$) was considered.

Table 1. Parameters governing the dynamics of the MPB

Parameter	Value	Parameter	Value
Dimensions D	1	Min coordinate	0.0
Number of peaks	5	Max coordinate	100.0
Min peak height	30.0	Peak speed v_L	1.0
Max peak height	70.0	Peak height severity h_S	2.0
Min peak width	0.0001	Basis function used	false
Max peak width	1.0	Peak movement drift	0.0
Initial peak width	0.1	Change frequency c_f	25
Width change severity w_S	0.01	Epochs	80
Peak function	Inverse squared		

5.2   Numerical Results and Model Comparison

Since the DIN sampling strategy (Sect. 4.5) requires parameter tuning for the
noise level, each experiment has to be run in two steps. The first step is to
find out the optimal noise level s_n^* by running a first set of simulations of the
optimizer using the DIN sampling strategy with different noise discount values,
and empirically choosing the one with best performance. In this case, offline
error is chosen as the preferred measure of performance, so the remainder of the
experiments focus mainly on this performance measure, but the same procedure
would apply to the average error. Since the changes of the objective function are
stochastic, several replications are required to provide statistical significance to
the interpretation of the results. So, $R_{DIN} = 64$ replications were run in this
first part of the experiment. The results are shown in Fig. 2(a).

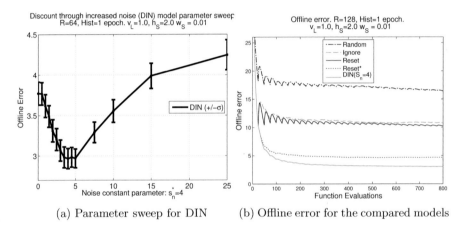

(a) Parameter sweep for DIN   (b) Offline error for the compared models

Fig. 2. A parameter sweep for different noise levels using the DIN sampling strategy,
considering the offline error (ϵ_o), is shown in 2(a). The offline error for the compared
models throughout the 80 epochs of the simulation is shown in 2(b), where the legend
follows the same order as the lines.

Once the DIN strategy has been tuned, the remaining strategies can be run
to asses their performance. For this part of the experiment $R = 128$ replications
were run using common random numbers across strategies. The offline error after 80 function changes, each happening after 25 function evaluations, are shown
in Fig. 3. Finally, in order to better understand how the offline error behaves
throughout the simulation and to verify that the comparison of the final values happens after convergence, the whole evolution across time is presented in
Fig. 2(b). These offline convergence curves are the mean across the 128 replications, and for sake of clarity their confidence intervals are omitted.

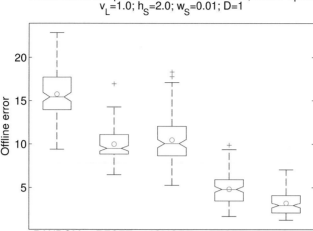

Fig. 3. Box plots showing the offline error (ϵ_o) of the five sampling strategies at the end of 80 function changes, each happening after 25 function evaluations.

6 Discussion, Conclusions, and Further Research

Two new methods to build surrogate models in order to track the global optima for dynamic expensive black box objective functions are presented in this paper. The first proposed method, reset*, transfers information from one epoch to the next one by resampling at the the previous best observation after each function change. DIN, the second proposed method, exploits the property of GPs that allows to control the noise level individually for each observation and discounts old information with such noise in order to construct a response surface using both old and new information. These new sampling strategies, together with three other sampling strategies that are not tailored for the dynamic environment and serve only as reference, are compared through numeric simulations implementing the MPB using the offline error as performance measure. Once the surrogate model is available, the selection of the next location to be sampled is determined by the well established EGO algorithm.

The poor performance of the random strategy confirms the advantages of using informed selection of the points to be sampled. These experiments show as well that sampling strategies using old information (reset*, and DIN) perform better than those which either discard it (reset) or treat it in the same way as recent information (ignore).

Because of how the offline error is defined, sampling at a good location at the beginning of an epoch highly increases the performance. So, in order to remove the component of the performance improvement coming from knowing where the previous best observation was made, and to get a better insight into how much the performance improves due only to the special treatment of old information, reset* was implemented.

The results presented in Fig. 3 show that discounting the information by introducing the right amount of noise, as the DIN sampling strategy does, clearly enhances the tracking of the global optima when considering the offline error as performance measure. However, a drawback of the DIN strategy is that it needs to be tuned, i.e. the noise discount parameter needs to be learnt through extensive simulations.

Future work in this area will focus on designing new response surface creation methods which do not require prior tuning. This might be achieved, for instance, by modifying the mean prior of the GP, by learning the magnitude of the changes through an auxiliary variable, or by tuning the noise parameter online. Furthermore, the scaling of these techniques to higher dimensions and the consistency of their performance when applied to other test functions are yet to be explored.

Acknowledgements. Sergio Morales-Enciso acknowledges the support from CONACyT.

References

[1] Biermann, D., Weinert, K., Wagner, T.: Model-based optimization revisited: Towards real-world processes. In: IEEE Congress on Evolutionary Computation 2008, pp. 2975–2982. IEEE (2008)

[2] Blackwell, T.M., Branke, J.: Multi-swarms, exclusion and anti-convergence in dynamic environments. IEEE Transactions on Evolutionary Computation 10(4), 459–472 (2004), http://eprints.gold.ac.uk/993/

[3] Branke, J.: Memory enhanced evolutionary algorithms for changing optimization problems. In: Proceedings of the Congress on Evolutionary Computation, vol. 3, pp. 1875–1882. IEEE Press (1999)

[4] Branke, J., Schmeck, H.: Designing evolutionary algorithms for dynamic optimization problems. In: Ghosh, A., Tsutsui, S. (eds.) Advances in Evolutionary Computing. Natural Computing Series, pp. 239–262. Springer, Heidelberg (2003)

[5] Burl, M.C., Wang, E.: Active learning for directed exploration of complex systems. In: Proceedings of the 26th Annual International Conference on Machine Learning, ICML 2009, pp. 89–96. ACM, New York (2009), doi:10.1145/1553374.1553386

[6] Floudas, C., Gounaris, C.: A review of recent advances in global optimization. Journal of Global Optimization 45, 3–38 (2009), doi:10.1007/s10898-008-9332-8

[7] Hu, X., Eberhart, R.: Adaptive particle swarm optimization: detection and response to dynamic systems. In: Proceedings of the 2002 Congress on Evolutionary Computation, CEC 2002, vol. 2, pp. 1666–1670. IEEE (2002), doi:10.1109/CEC.2002.1004492

[8] Jin, Y., Branke, J.: Evolutionary optimization in uncertain environments – a survey. IEEE Transactions on Evolutionary Computation 9(3), 303–317 (2005), doi:10.1109/TEVC.2005.846356

[9] Jones, D.R.: A taxonomy of global optimization methods based on response surfaces. Journal of Global Optimization 21, 345–383 (2001), doi:10.1023/A:1012771025575

[10] Jones, D.R., Schonlau, M., Welch, W.J.: Efficient global optimization of expensive black-box functions. Journal of Global Optimization 13(4), 455–492 (1998), http://dx.doi.org/10.1023/A:1008306431147, doi:10.1023/A:1008306431147

[11] Morales-Enciso, S., Branke, J.: Revenue maximization through dynamic pricing under unknown market behaviour. In: Ravizza, S., Holborn, P. (eds.) 3rd Student Conference on Operational Research, Schloss Dagstuhl–Leibniz-Zentrum fuer Informatik, Dagstuhl, Germany. OpenAccess Series in Informatics (OASIcs), vol. 22, pp. 11–20 (2012), http://drops.dagstuhl.de/opus/volltexte/2012/3542, doi:http://dx.doi.org/10.4230/OASIcs.SCOR.2012.11

[12] Moser, I.: All currently known publications on approaches which solve the moving peaks problem (2007)

[13] Parsopoulos, K., Vrahatis, M.: Unified particle swarm optimization in dynamic environments. In: Rothlauf, F., et al. (eds.) EvoWorkshops 2005. LNCS, vol. 3449, pp. 590–599. Springer, Heidelberg (2005)

[14] Rasmussen, C.E., Williams, C.: Gaussian Processes for Machine Learning. In: Adaptive Computation and Machine Learning. MIT Press, Cambridge (2006)

[15] Richter, H.: Detecting change in dynamic fitness landscapes. In: IEEE Congress on Evolutionary Computation, CEC 2009, pp. 1613–1620. IEEE (2009), doi:10.1109/CEC.2009.4983135

[16] Shan, S., Wang, G.: Survey of modeling and optimization strategies to solve high-dimensional design problems with computationally-expensive black-box functions. Structural and Multidisciplinary Optimization 41, 219–241 (2010), http://dx.doi.org/10.1007/s00158-009-0420-2, 10.1007, doi:10.1007/s00158-009-0420-2

[17] Simoes, A., Costa, E.: Improving memory's usage in evolutionary algorithms for changing environments. In: IEEE Congress on Evolutionary Computation, CEC 2007, pp. 276–283. IEEE (2007), doi:10.1109/CEC.2007.4424482

[18] Tang, Y., Chen, J., Wei, J.: A surrogate-based particle swarm optimization algorithm for solving optimization problems with expensive black box functions. Engineering Optimization, 1–20 (2012), doi:10.1080/0305215X.2012.690759

Fitness Based Self Adaptive Differential Evolution

Harish Sharma[1], Pragati Shrivastava[1],
Jagdish Chand Bansal[2], and Ritu Tiwari[1]

[1] ABV-Indian Institute of Information Technology and Management, Gwalior, [M.P.]
India
[2] South Asian University, New Delhi, India
{harish.sharma0107,pragatipiu,jcbansal,tiwariritu2}@gmail.com

Abstract. Differential Evolution (DE) is a well known optimization approach to solve nonlinear and complex problems. But, DE, like other probabilistic optimization algorithms, sometimes exhibits premature convergence and stagnation. DE exploration and exploitation capabilities depend on the two processes namely mutation process and crossover process. In these two processes exploration and exploitation are balanced using the fine tuning of scale factor F and crossover probability CR. In the solution search process of DE, there is a enough chance to skip the true solution due to large step size. Therefore, in this paper, to balance the diversity and convergence capability of DE, fitness based self adaptive F and CR are proposed. The proposed strategy is named as Fitness based Self Adaptive DE ($FSADE$). The experiments on 16 well known test problems of different complexities show that the proposed strategy outperforms the basic DE and recent variants of DE, namely Self-adaptive DE ($SaDE$) and Scale Factor Local Search DE ($SFLSDE$) in most of the experiments.

Keywords: Self adaptive scale factor, Self adaptive crossover, Self adaptive step size.

1 Introduction

Differential Evolution (DE) scheme, proposed by Storn and Price [17], is relatively a simple, fast and population based stochastic search technique. Researchers are continuously working to improve the performance of DE. Some of the recently developed versions of DE with appropriate applications can be found in [2, 3, 14–16]. Experiments over several numerical benchmarks [18] show that DE performs better than the genetic algorithm (GA) [5] and the particle swarm optimization (PSO) [6].

There are two fundamental processes which drive the evolution of DE population: the variation process, to explore different areas of the search space, and the selection process, for the exploitation of the previous experience. However, it has been shown that DE may occasionally stop proceeding towards the global optima even though the population has not converged to a local optima [7].

G. Terrazas et al. (eds.), *Nature Inspired Cooperative Strategies for Optimization*
(*NICSO 2013*), Studies in Computational Intelligence 512,
DOI: 10.1007/978-3-319-01692-4_6, © Springer International Publishing Switzerland 2014

Therefore, to maintain a proper balance between exploration and exploitation behavior of DE, a self adaptive position update strategy is proposed. In the proposed strategy, crossover probability CR and scale factor F are self adaptively changed for every individual based on the fitness of that individual. It is proposed that a better fit individual will be less perturbed and uses small step sizes in the position update process hence so exploits the already identified search area, while a low fit individual will be high perturb and uses large step sizes, which help in exploration of the solution search space. The proposed strategy is used for finding the global optima of a unimodal and/or multimodel functions using self adaptive CR and F for updating the candidate solution in the search space. Further, the proposed strategy is compared by experimenting on 16 well known test problems to the basic DE and its recent variants named, Self Adaptive DE ($SaDE$) [10] and Scale factor Local Search Differential Evolution ($SFLSDE$) [9].

Rest of the paper is organized as follows: Section 2 describes brief overview of the basic DE. Fitness based Self Adaptive DE (FSADE) is proposed and tested in Section 3. In Section 4, a comprehensive set of experimental results are provided. Finally, in Section 5, paper is concluded.

2   Brief Overview of Differential Evolution Algorithm

DE has several strategies based on method of selecting the target vector, number of difference vectors used and the type of crossover [11]. In this paper $DE/rand/1/bin$ scheme is used. Like other population based search algorithms, in DE a population of potential solutions (individuals) searches the solution. In a D-dimensional search space, an individual is represented by a D-dimensional vector $(x_{i1}, x_{i2}, ..., x_{iD})$, $i = 1, 2, ..., NP$ where NP is the population size.

In DE, there are three operators: mutation, crossover and selection. Initially, a population is generated randomly with uniform distribution, then the mutation, crossover and selection operators are applied to generate a new population. DE operators are explained briefly in following subsections.

2.1   Mutation

A trial vector is generated by the DE mutation operator for each individual of the current population. A target vector is mutated with a weighted differential to generate a new trial vector, which then produced an offspring in the crossover operation. If G is the index for generation counter, the mutation operator for generating a trial vector $u_i(G)$ is defined as follows:

- Select a target vector, $x_{i_1}(G)$, from the population, such that $i \neq i_1$.
- Again, randomly select two individuals, x_{i_2} and x_{i_3}, from the population such that $i \neq i_1 \neq i_2 \neq i_3$.

– Then the target vector is mutated for calculating the trial vector as follows:

$$u_i(G) = x_{i_1}(G) + F \times \overbrace{(x_{i_2}(G) - x_{i_3}(G))}^{\text{Variation Component}} \qquad (1)$$

$$\underbrace{\phantom{u_i(G) = x_{i_1}(G) + F \times (x_{i_2}(G) - x_{i_3}(G))}}_{\text{Step size}}$$

where $F \in [0, 1]$ is the mutation scale factor used to control the amplification of the differential variation [4].

2.2 Crossover

Offspring $x_i'(G)$ is generated using the crossover of parent vector, $x_i(G)$ and the trial vector, $u_i(G)$ as follows:

$$x_{ij}'(G) = \begin{cases} u_{ij}(G), & \text{if } j \in J \\ x_{ij}(G), & \text{otherwise.} \end{cases} \qquad (2)$$

where J is the set of crossover points or the points that will go under perturbation, $x_{ij}(G)$ is the jth element of the vector $x_i(G)$.

Different methods may be used to determine the set, J, of which binomial crossover and exponential crossover are the most frequently used [4]. In this paper, the DE and its variants are implemented using the binomial crossover. In this crossover, the crossover points are randomly selected from the set of possible crossover points, $\{1, 2, \ldots, D\}$, where D is the problem dimension. Algorithm 1 shows the steps of binomial crossover to generate crossover points [4]. In this algorithm, CR is the probability that the considered crossover point will be included, and $U(1, D)$ is a uniformly distributed random integer between 1 and D. The larger the value of CR, indicates that the more crossover points will be selected.

Algorithm 1. Binomial Crossover:

$J = \phi$
$j^* \sim U(1, D);$
$J \leftarrow J \cup j^*;$
  for each $j \in 1...D$ **do**
    if $U(0, 1) < CR$ and $j \neq j^*$ **then**
      $J \leftarrow J \cup j;$
    end if
  end for

2.3 Selection

The selection operator perform two tasks: First it selects an individual to generate the trial vector through mutation and then chooses the best between the parent and the offspring based on their fitness value for the next generation. If fitness of parent is greater than that of offspring the parent is selected otherwise offspring is considered. Therefore next generation component is decided by:

$$x_i(G+1) = \begin{cases} x_i'(G), & \text{if } f(x_i'(G)) > f(x_i(G)). \\ x_i(G), & \text{otherwise.} \end{cases} \tag{3}$$

This ensures that the population's average fitness does not deteriorate. The Differential Evolutionary strategy is described and represented by Algorithm 2 [4] given below in pseudo code:

Algorithm 2. Differential Evolution Algorithm

Initialize the control parameters, F and CR;
Generate and initialize the population, $P(0)$, of NP individuals;
while termination condition(s) **do**
  for each individual, $x_i(G) \in P(G)$ **do**
    Calculate the fitness, $f(x_i(G))$;
    Generate the trial vector, $u_i(G)$ by using the mutation operator;
    Generate an offspring, $x_i'(G)$, by using the crossover operator;
    if $f(x_i'(G))$ is better than $f(x_i(G))$ **then**
      Add $x_i'(G)$ to $P(G+1)$;
    else
      Add $x_i(G)$ to $P(G+1)$;
    end if
  end for
end while
Return the fittest individual as a solution;

Here, F, CR and P, represents the scale factor, crossover probability, and the population vector respectively. Here F and CR, are the control parameters and the choice of their values influences the performance of DE.

3   Fitness Based Self Adaptive DE (FSADE)

The inherent drawback with most of the population based stochastic algorithms is premature convergence. DE is not an exception. Any population based algorithm is regarded as an efficient algorithm if it is fast in convergence and able to explore the maximum area of the search space. In other words, if a population based algorithm is capable of balancing between exploration and exploitation of the search space, then the algorithm is regarded as an efficient algorithm. From this point of view, basic DE is not an efficient algorithm [7, 8]. Therefore, in this paper, to balance the diversity and convergence ability of DE, crossover probability (CR) and scale factor (F) are adaptively modified.

3.1 Why CR and F?

The CR and F are the crucial parameters, which are used to manage exploration and exploitation in DE. CR is the parameter which is responsible for the perturbation in the individual which is going to update for creating the offspring, while F is the parameter which decides a step size for an individual by which it can move in the search space.

It is obvious that the perturbation of an individual will be high for the high value of CR and so as the exploration, while a low value of CR may enhance the exploitation as for this case, the perturbation will be low (refer to Algorithm 1).

Further, F also plays an important role in balancing the exploration and exploitation. It is clear from equation (1), that a low value of F will force an individual to search a new solution in its neighborhood as the step size will be low proportionally and therefore, the individual exploits its neighborhood. Whereas a high value of F, makes an individual to take large size step and hence explore the search area to find new solutions.

It is evident from above discussion that CR and F have significant role to settle the exploration and exploitation capabilities. Therefore, these two (CR and F) are selected for modification in the view of their importance in managing the diversity in the population.

3.2 Self Adaptive Strategy

As explained in Section 3.1, in this paper, to balance the diversity and convergence ability of DE, CR and F are set adaptively and are different for every individual i, based on the fitness based probability $prob_i$ of the individual, which may be calculated as shown in equation (4):

$$prob_i = 0.9 \times \frac{fitness_i}{max\_fitness} + 0.1, \tag{4}$$

here $fitness_i$ is the fitness value of the ith solution and $max\_fitness$ is the maximum fitness in the population. It is clear from equation (4) that $prob_i \in [0.1, 1]$.

Further, based on the $prob_i$ of each individual i in the population, which is a function of fitness, the CR and F are adaptively changed as shown in equations (5) and (6):

$$CR_i = (C_1 - prob_i), \tag{5}$$

$$F_i = (2 \times C_1 - prob_i) \times U, \tag{6}$$

here, $C_1 = 1.1$ is a constant and U is a random number $\in (-0.5, 0.5)$. It is obvious from equation (5) and (6) that $CR_i \in [0.1, 1]$ and $F_i \in [-1.05, 1.05]$.

It is clear from equations (5) and (6) that for high value of $prob_i$ and that will be the case for high fitness solutions, the value of CR_i and F_i will be low and vice versa. Therefore, it could be expected that high fit individual will be less perturbs as well as the step sizes will be small in mutation and hence it could exploits the search area in its vicinity while the low fit individual explore the search area to find out new solutions due to high perturbation and large step sizes. The proposed population update process is shown in Algorithm 3:

Algorithm 3. Fitness based self adaptive position update process

---

Input: An individual x_i which is going to update its position, $prob_i$ and constant C_1;
G: Generation counter, $U(a, b)$: Uniform random number between a and b;
Select three random individuals (i_1, i_2, i_3) from population such that
$i \neq i_1 \neq i_2 \neq i_3$;
Calculate $CR_i(G) = (C_1 - prob_i(G))$;
for each $j \in 1...D$ **do**
    $F_i(G) = (2 \times C_1 - prob_i(G)) \times U(-0.5, 0.5)$;
    if $U(0, 1) < CR_i(G)$ **then**
        $u_{ij} = x_{i_1j}(G) + F_i(G) \times (x_{i_2j}(G) - x_{i_3j}(G))$;
    else
        $u_{ij} = x_{ij}(G)$
    end if
end for
if $f(u_i)$ is better than $f(x_i(G))$ **then**
    Add u_i to $P(G+1)$;
else
    Add $x_i(G)$ to $P(G+1)$;
end if

---

4   Experimental Results and Discussion

4.1   Test Problems under Consideration

In order to analyze the performance of $FSADE$, 16 different global optimization problems (f_1 to f_{16}) are selected (listed in Table 1) [1]. These problems are minimization problems and have different degree of complexity and multimodality.

4.2   Experimental Setting

To prove the efficiency of $FSADE$ algorithm, it is compared with the classical DE ($DE/rand/bin/1$) [11] and its variants namely, Self-adaptive DE ($SaDE$) [12] and Scale factor local search differential evolution ($SFLSDE$) [9].

Table 1. Test problems

Test Problem	Objective function	Search Range	Optimum Value	D	Acceptable Error
Griewank	$f_1(x) = 1 + \frac{1}{4000}\sum_{i=1}^{D}x_i^2 - \prod_{i=1}^{D}\cos\left(\frac{x_i}{\sqrt{i}}\right)$	[-600, 600]	$f(0) = 0$	30	$1.0E-05$
Rosenbrock	$f_2(x) = \sum_{i=1}^{D}(100(x_{i+1}-x_i^2)^2+(x_i-1)^2)$	[-30, 30]	$f(1) = 0$	30	$1.0E-02$
Rastrigin	$f_3(x) = 10D+\sum_{i=1}^{D}[x_i^2-10\cos(2\pi x_i)]$	[-5.12, 5.12]	$f(0) = 0$	30	$1.0E-05$
Michalewicz	$f_4(x) = -\sum_{i=1}^{D}\sin x_i\left(\sin\left(\frac{ix_i^2}{\pi}\right)^{20}\right)$	$[0, \pi]$	$f_{min} = -9.66015$	10	$1.0E-05$
Cosine Mixture	$f_5(x) = \sum_{i=1}^{D}x_i^2-0.1(\sum_{i=1}^{D}\cos 5\pi x_i)+0.1D$	[-1, 1]	$f(0) = -D \times 0.1$	30	$1.0E-05$
Exponential	$f_6(x) = -(exp(-0.5\sum_{i=1}^{D}x_i^2))+1$	[-1, 1]	$f(0) = -1$	30	$1.0E-05$
Cigar	$f_7(x) = x_0^2+100000\sum_{i=1}^{D}x_i^2$	[-10, 10]	$f(0) = 04$	30	$1.0E-05$
Step function	$f_8(x) = \sum_{i=1}^{D}(\lfloor |x_i+0.5| \rfloor)^2$	[-100, 100]	$f(-0.5 \leq x \leq 0.5) = 0$	30	$1.0E-05$
Inverted cosine wave	$f_9(x) = -\sum_{i=1}^{D-1}\left(\exp\left(\frac{-(x_i^2+x_{i+1}^2+0.5x_ix_{i+1})}{8}\right)\times \text{I}\right)$ where, $\text{I} = \cos\left(4\sqrt{x_i^2+x_{i+1}^2+0.5x_ix_{i+1}}\right)$	[-5, 5]	$f(0) = -D+1$	10	$1.0E-05$
Levy montalvo	$f_{10}(x) = 0.1(\sin^2(3\pi x_1)+\sum_{i=1}^{D-1}(x_i-1)^2\times(1+\sin^2(3\pi x_{i+1}))+(x_D-1)^2(1+\sin^2(2\pi x_D)))$	[-5, 5]	$f(1) = 0$	30	$1.0E-05$
Ellipsoidal	$f_{11}(x) = \sum_{i=1}^{B}(x_i-i)^2$	[-D, D]	$f(1,2,3,...,D) = 0$	30	$1.0E-05$
Branins's function	$f_{12}(x) = a(x_2-bx_1^2+cx_1-d)^2+e(1-f)\cos x_1+e$	$\begin{array}{l}-5 \leq x_1 \leq 10,\\ 0 \leq x_2 \leq 15\end{array}$	$f(-\pi,12.275) = 0.3979$	2	$1.0E-05$
Kowalik function	$f_{13}(x) = \sum_{i=1}^{11}\left[a_i - \frac{x_1(b_i^2+b_ix_2)}{b_i^2+b_ix_3+x_4}\right]^2$	[-5, 5]	$f(0.1928, 0.1908, 0.1231, 0.1357) = 3.07E-04$	4	$1.0E-05$
Shubert	$f_{14}(x) = -\sum_{i=1}^{5}i\cos((i+1)x_1+1)\sum_{i=1}^{5}i\cos((i+1)x_2+1)$	[-10, 10]	$f(7.0835, 4.8580) = -186.7309$	2	$1.0E-05$
Sinusoidal Problem	$f_{15}(x) = -[A\prod_{i=1}^{n}\sin(x_i-z)+\prod_{i=1}^{n}\sin(B(x_i-z))],$ $A=2.5, B=5, z=30$	[0, 180]	$f(90+z) = -(A+1)$	10	$1.0E-02$
Moved axis parallel hyper-ellipsoid	$f_{16}(x) = \sum_{i=1}^{D}5ix_i^2$	[-5.12, 5.12]	$f(x) = 0; x(i) = 5 \times i, i = 1:D$	30	$1.0E-15$

To test DE and DE variants over test problems, following experimental setting is adopted:

- The value of F and CR for $SaDE$ and $SFLSDE$ are kept same as suggested by their respective authors [9, 12].
- Population size NP=50.
- The stopping criteria is either maximum number of function evaluations (which is set to be 2.0×10^5) is reached or the corresponding acceptable error (mentioned in Table 1) has been achieved.
- The number of simulations/run =100.
- In order to investigate the effect of the parameter C_1, described by algorithm 3 on the performance of $FSADE$, its sensitivity with respect to different values of C_1 in the range $[1.1, 1.9]$, is examined in the Figure 1. It can be observed from Figure 1 that the FSADE is very sensitive towards C_1 and value 1.1 gives comparatively better results. Therefore $C_1 = 1.1$ is selected for the experiments in this paper.

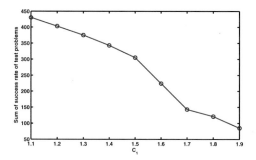

Fig. 1. Effect of parameter C_1 on sum of success rate for 30 runs of all the considered test problems

4.3 Results Comparison

Numerical results with experimental setting of subsection 4.2 are given in Table 2. In Table 2, standard deviation (SD), mean error (ME), average function evaluations (AFE) and, success rate (SR) are reported. Table 2 shows that most of the time $FSADE$ outperforms in terms of reliability, efficiency and accuracy as compare to the $DE/rand/bin/1$, $SFLSDE$ and $SaDE$. Some more intensive analyses based on acceleration rate (AR) [13], performance indices and boxplots have been carried out for results of DE and its variants.

$FSADE$, DE, $SFLSDE$ and $SaDE$ are compared through SR, ME and AFE in Table 2. First SR is compared for all these algorithms and if it is not possible to distinguish the algorithms based on SR then comparison is made on the basis of AFE. ME is used for comparison if it is not possible on the basis of SR and AFE both. Outcome of this comparison is summarized in Table 3. In Table 3, '+' indicates that the $FSADE$ is better than the considered algorithms and '-' indicates that the $FSADE$ is not better or the difference is very small. The last row of Table 3, establishes the superiority of $FSADE$ over DE, $SFLSDE$ and $SaDE$.

Table 2. Comparison of the results of test problems

Test Function	Algorithm	SD	ME	AFE	SR
f_1	FSADE	8.22E-07	9.01E-06	30294	100
	DE	4.52E-03	2.05E-03	64036.5	81
	SFLSDE	1.03E-03	1.57E-04	39382.59	98
	SaDE	8.85E-03	4.09E-03	45890.5	96
f_2	FSADE	1.69E+01	2.86E+01	200050	0
	DE	4.09E+01	4.24E+01	200050	0
	SFLSDE	2.82E+01	2.35E+01	189640.34	1
	SaDE	1.21E+01	8.46E+00	174741.5	13
f_3	FSADE	8.03E-07	9.16E-06	92932	99
	DE	5.71E+00	1.46E+01	200050	0
	SFLSDE	9.90E-02	9.96E-03	134474.92	99
	SaDE	1.25E-01	2.39E-01	142371.5	58
f_4	FSADE	1.18E-02	3.96E-03	43339.5	86
	DE	4.84E-02	4.90E-02	167536	23
	SFLSDE	4.48E-02	1.48E-02	66651.79	76
	SaDE	4.51E-02	1.72E-02	85810.5	67
f_5	FSADE	8.49E-07	9.02E-06	19217.5	100
	DE	2.90E-02	5.92E-03	30339	96
	SFLSDE	8.88E-04	5.88E-03	99779.94	55
	SaDE	1.47E-02	5.80E-01	109265	43
f_6	FSADE	6.88E-07	9.19E-06	16125.5	100
	DE	7.39E-07	8.99E-06	17018	100
	SFLSDE	9.82E-07	8.96E-06	23688.78	100
	SaDE	6.23E-07	9.12E-06	12503	100
f_7	FSADE	7.69E-07	9.01E-06	35806	100
	DE	8.77E-07	8.89E-06	39664.5	100
	SFLSDE	9.40E-07	8.99E-06	43179.52	100
	SaDE	7.29E-07	9.09E-06	23187	100
f_8	FSADE	1.10E-05	2.10E-05	13340	100
	DE	3.32E-01	1.00E-01	33846.5	91
	SFLSDE	2.30E-05	5.13E-05	16301.45	100
	SaDE	4.30E-02	1.80E-03	39404	92
f_9	FSADE	2.64E-01	9.95E-02	80435.5	85
	DE	6.30E-01	8.93E-01	176110.5	17
	SFLSDE	7.19E-01	5.63E-01	116706.29	52
	SaDE	8.12E-01	5.73E-01	117613	56
f_{10}	FSADE	7.45E-07	9.09E-06	18678	100
	DE	1.54E-03	2.29E-04	23981.5	98
	SFLSDE	7.48E-07	9.09E-06	24436.9	100
	SaDE	2.80E-03	7.77E-04	26025.5	93
f_{11}	FSADE	7.00E-07	9.13E-06	24994.5	100
	DE	7.39E-07	9.07E-06	27209	100
	SFLSDE	8.24E-07	9.01E-06	30359.56	100
	SaDE	6.78E-07	9.15E-06	16982.5	100
f_{12}	FSADE	3.28E-05	4.13E-05	1761	100
	DE	3.31E-05	4.11E-05	1790.5	100
	SFLSDE	3.26E-05	3.92E-05	3224.46	100
	SaDE	3.32E-05	4.14E-05	3355	100
f_{13}	FSADE	2.47E-04	1.76E-04	49103	90
	DE	3.63E-04	2.81E-04	63952.5	70
	SFLSDE	2.05E-04	5.55E-04	172173.74	16
	SaDE	3.00E-04	5.46E-04	155720	25
f_{14}	FSADE	5.36E-06	4.54E-06	7476	100
	DE	4.93E-06	4.06E-06	8348.5	100
	SFLSDE	5.79E-06	5.30E-06	26974.37	100
	SaDE	5.47E-06	4.80E-06	71139.5	100

Table 2. Comparison of the results of test problems (Cont.)

Test Function	Algorithm	SD	ME	AFE	SR
f_{15}	FSADE	2.60E-03	8.11E-03	109441.5	97
	DE	2.52E-01	5.02E-01	200050	0
	SFLSDE	4.83E-02	1.15E-01	199770.42	0
	SaDE	1.59E-01	6.31E-01	200000	0
f_{16}	FSADE	7.30E-17	9.11E-16	56257.5	100
	DE	8.16E-17	8.95E-16	59365	100
	SFLSDE	8.21E-17	8.99E-16	64566.58	100
	SaDE	7.58E-17	9.10E-16	35111.5	100

Table 3. Summary of Table 2 outcome

Test Problems	FSADE Vs DE	FSADE Vs SFLSDE	FSADE Vs SaDE
f_1	+	+	+
f_2	+	-	-
f_3	+	+	+
f_4	+	+	+
f_5	+	+	+
f_6	+	+	-
f_7	+	+	-
f_8	+	+	+
f_9	+	+	+
f_{10}	+	+	+
f_{11}	+	+	-
f_{12}	+	+	+
f_{13}	+	+	+
f_{14}	+	+	+
f_{15}	+	+	+
f_{16}	+	+	-
Total number of + sign	16	15	11

Further, we compare the convergence speed of the considered algorithms by measuring the average function evaluations (AFEs). A small value of AFEs means higher convergence speed. In order to minimize the effect of the stochastic nature of the algorithms, the reported function evaluations for each test problem is the average over 100 runs. we use the acceleration rate (AR) to compare convergence speeds, which is defined as follows, based on the AFEs for the two algorithms $ALGO$ and $FSADE$:

$$AR = \frac{AFE_{ALGO}}{AFE_{FSADE}}, \qquad (7)$$

where, $ALGO \in \{DE, SFLSDE, SaDE\}$ and $AR > 1$ means $FSADE$ converges faster. Table 4 shows a clear comparison between $FSADE$ and DE, $FSADE$ and $SFLSDE$, and $FSADE$ and $SaDE$ in terms of AR. It is clear from Table 4 that, for most of the test problems, convergence speed of $FSADE$ is faster among all the considered algorithms.

For the purpose of comparison in terms of consolidated performance, boxplot analyses have been carried out for all the considered algorithms. The boxplots for $FSADE$, DE, $SFLSDE$ and $SaDE$ are shown in Figure 2. It is clear from this figure that $FSADE$ is better than the considered algorithms as interquartile range and median are comparatively low.

Table 4. Acceleration Rate (AR) of $FSADE$ compare to the basic DE, $SFLSDE$ and $SaDE$

Test Prob-lems	DE	SFLSDE	SaDE
f_1	2.113834423	1.300012874	1.514837922
f_2	1	0.947964709	0.873489128
f_3	2.152649249	1.447024921	1.531996514
f_4	3.865665271	1.537899376	1.979960544
f_5	1.578717315	5.192139456	5.685703135
f_6	1.055347121	1.469026077	0.775355803
f_7	1.107761269	1.205929732	0.647573032
f_8	2.537218891	1.221997751	2.953823088
f_9	2.189462364	1.450930124	1.462202634
f_{10}	1.283943677	1.308325302	1.393377235
f_{11}	1.088599492	1.214649623	0.679449479
f_{12}	1.016751846	1.831039182	1.905167518
f_{13}	1.302415331	3.506379244	3.171292996
f_{14}	1.116706795	3.608128678	9.515716961
f_{15}	1.827917198	1.825362591	1.827460333
f_{16}	1.05523708	1.147697285	0.624121228

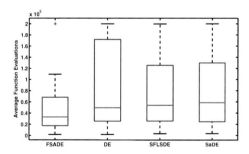

Fig. 2. Boxplots for average number of function evaluation

Further, to compare the considered algorithms, by giving weighted importance to the success rate, the standard deviation and the average number of function evaluations, performance indices (PI) are calculated [2]. The values of PI for the $FSADE$, DE, $SFLSDE$, and $SaDE$ are calculated by using following equations:

$$PI = \frac{1}{N_p} \sum_{i=1}^{N_p} (k_1 \alpha_1^i + k_2 \alpha_2^i + k_3 \alpha_3^i)$$

Where $\alpha_1^i = \frac{Sr^i}{Tr^i}$; $\alpha_2^i = \begin{cases} \frac{Mf^i}{Af^i}, & \text{if } Sr^i > 0. \\ 0, & \text{if } Sr^i = 0. \end{cases}$; and $\alpha_3^i = \frac{Mo^i}{Ao^i}$

$i = 1, 2, ..., N_p$

- Sr^i = Successful simulations/runs of ith problem.
- Tr^i = Total simulations of ith problem.
- Mf^i = Minimum of average number of function evaluations used for obtaining the required solution of ith problem.
- Af^i = Average number of function evaluations used for obtaining the required solution of ith problem.
- Mo^i = Minimum of standard deviation obtained for the ith problem.
- Ao^i = Standard deviation obtained by an algorithm for the ith problem.
- N_p = Total number of optimization problems evaluated.

The weights assigned to the success rate, the average number of function evaluations and the standard deviation are represented by k_1, k_2 and k_3 respectively where $k_1 + k_2 + k_3 = 1$ and $0 \leq k_1, k_2, k_3 \leq 1$. To calculate the PIs, equal weights are assigned to two variables while weight of the remaining variable vary from 0 to 1 as given in [2]. Following are the resultant cases:

1. $k_1 = W, k_2 = k_3 = \frac{1-W}{2}, 0 \leq W \leq 1$;
2. $k_2 = W, k_1 = k_3 = \frac{1-W}{2}, 0 \leq W \leq 1$;
3. $k_3 = W, k_1 = k_2 = \frac{1-W}{2}, 0 \leq W \leq 1$

The graphs corresponding to each of the cases (1), (2) and (3) for $FSADE$, DE, $SFLSDE$ and $SaDE$ are shown in Figures 3(a), 3(b), and 3(c) respectively. In these figures the weights k_1, k_2 and k_3 are represented by horizontal axis while the PI is represented by the vertical axis.

In case (1), average number of function evaluations and the standard deviation are given equal weights. PIs of the considered algorithms are superimposed in Fig. 3(a) for comparison of the performance. It is observed that PI of $FSADE$ are higher than the considered algorithms. In case (2), equal weights are assigned to the success rate and standard deviation and in case (3), equal weights are assigned to the success rate and average function evaluations. It is clear from Fig. 3(b) and Fig. 3(c) that, the algorithms perform same as in case (1).

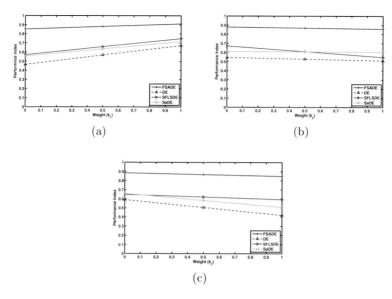

Fig. 3. Performance index for test problems; (a) for case (1), (b) for case (2) and (c) for case (3)

5 Conclusion

In this paper, a new poistion update strategy is proposed based on the fitness based CR and F. In the proposed strategy, CR and F are adaptively modified for every individual on the basis of its richness (in terms of fitness). The high fit individual is forced to exploit the search area in its vicinity by reducing perturbation as well as step sizes, while a low fit individual try to explore the search area for finding the new solutions. Further, the proposed algorithm is compared to the recent variants of DE, namely, Self Adaptive DE ($SaDE$) and Scale Factor Local Search DE ($SFLSDE$) and with the help of experiments over test problems, it is shown that the $FSADE$ outperforms to the considered algorithms in terms of reliability, efficiency and accuracy.

References

[1] Ali, M.M., Khompatraporn, C., Zabinsky, Z.B.: A numerical evaluation of several stochastic algorithms on selected continuous global optimization test problems. Journal of Global Optimization 31(4), 635–672 (2005)
[2] Bansal, J.C., Sharma, H.: Cognitive learning in differential evolution and its application to model order reduction problem for single-input single-output systems. Memetic Computing, 1–21 (2012)
[3] Chakraborty, U.K.: Advances in differential evolution. Springer (2008)
[4] Engelbrecht, A.P.: Computational intelligence: an introduction. Wiley (2007)

[5] Holland, J.H.: Adaptation in natural and artificial systems. The University of Michigan Press, Ann Arbor (1975)

[6] Kennedy, J., Eberhart, R.: Particle swarm optimization. In: Proceedings of the IEEE International Conference on Neural Networks, vol. 4, pp. 1942–1948. IEEE (1995)

[7] Lampinen, J., Zelinka, I.: On stagnation of the differential evolution algorithm. In: Proceedings of MENDEL, pp. 76–83 (2000)

[8] Mezura-Montes, E., Velázquez-Reyes, J., Coello Coello, C.A.: A comparative study of differential evolution variants for global optimization. In: Proceedings of the 8th Annual Conference on Genetic and Evolutionary Computation, pp. 485–492. ACM (2006)

[9] Neri, F., Tirronen, V.: Scale factor local search in differential evolution. Memetic Computing 1(2), 153–171 (2009)

[10] Omran, M.G.H., Salman, A., Engelbrecht, A.P.: Self-adaptive differential evolution. In: Hao, Y., Liu, J., Wang, Y.-P., Cheung, Y.-M., Yin, H., Jiao, L., Ma, J., Jiao, Y.-C. (eds.) CIS 2005. LNCS (LNAI), vol. 3801, pp. 192–199. Springer, Heidelberg (2005)

[11] Price, K.V.: Differential evolution: a fast and simple numerical optimizer. In: 1996 Biennial Conference of the North American Fuzzy Information Processing Society, NAFIPS 1996, pp. 524–527. IEEE (1996)

[12] Qin, A.K., Suganthan, P.N.: Self-adaptive differential evolution algorithm for numerical optimization. In: The IEEE Congress on Evolutionary Computation, vol. 2, pp. 1785–1791. IEEE (2005)

[13] Rahnamayan, S., Tizhoosh, H.R., Salama, M.M.A.: Opposition-based differential evolution. IEEE Transactions on Evolutionary Computation 12(1), 64–79 (2008)

[14] Sharma, H., Bansal, J.C., Arya, K.V.: Dynamic scaling factor based differential evolution algorithm. In: Deep, K., Nagar, A., Pant, M., Bansal, J.C. (eds.) Proceedings of the International Conference on SocProS 2011. AISC, vol. 130, pp. 73–86. Springer, Heidelberg (2012)

[15] Sharma, H., Bansal, J.C., Arya, K.: Fitness based differential evolution. Memetic Computing 4(4), 303–316 (2012)

[16] Sharma, H., Bansal, J.C., Arya, K.V.: Self balanced differential evolution. Journal of Computational Science (2012)

[17] Storn, R., Price, K.: Differential evolution-a simple and efficient adaptive scheme for global optimization over continuous spaces. International Computer Science Institute - Publications - TR (1995)

[18] Vesterstrom, J., Thomsen, R.: A comparative study of differential evolution, particle swarm optimization, and evolutionary algorithms on numerical benchmark problems. In: Congress on Evolutionary Computation, CEC 2004, vol. 2, pp. 1980–1987. IEEE (2004)

Adaptation Schemes and Dynamic Optimization Problems: A Basic Study on the Adaptive Hill Climbing Memetic Algorithm*

Jenny Fajardo Calderín[1], Antonio D. Masegosa[2],
Alejandro Rosete Suárez[1], and David A. Pelta[2]

[1] Department of Artificial Intelligence and Systems Infrastructure
Faculty of Computing, Polytechnic Institute José Antonio Echeverría, Havana, Cuba
{jfajardo,rosete}@ceis.cujae.edu.cu
[2] Models of Decision and Optimization Research Group
DECSAI, CITIC, University of Granada, Spain
{admase,dpelta}@decsai.ugr.es

Abstract. An open question that arises in the design of adaptive schemes for Dynamic Optimization Problems consists on deciding what to do with the knowledge acquired once a change in the environment is detected: forget it or use it in subsequent changes? In this work, the knowledge is associated with the selection probability of two local search operators in the Adaptive Hill Climbing Memetic Algorithm. When a problem change is detected, those probability values can be restarted or maintained. The experiments performed over five binary coded problems (for a total of 140 different scenarios) clearly show that keeping the information is better than forgetting it.

Keywords: dynamic optimization problems, metaheuristics, AHMA, adaptive schemes.

1 Introduction

Many real world situations present features, costs, objective functions or constraints that may vary over time. In order to understand how to better deal with such scenarios, researchers usually model them as Dynamic Optimization Problems (DOPs) [5].

In terms of optimization, the basic approach to address DOPs consists on restarting the search each time a change is detected, considering it as completely new problem [2, 13]. However it is widely assumed that the environment changes gradually, so if we handle each change as a different problem we are probably loosing information. Early studies already showed that the reuse of information lead to faster adaptations to changes, and thus, to better solutions. This fact is

---

* This work is supported in part by Projects TIN2011-27696-C02-01, Spanish Ministry of Economy and Competitiveness and P11-TIC-8001 from the Andalusian Government (including FEDER funds from the European Union).

G. Terrazas et al. (eds.), *Nature Inspired Cooperative Strategies for Optimization (NICSO 2013)*, Studies in Computational Intelligence 512,
DOI: 10.1007/978-3-319-01692-4_7, © Springer International Publishing Switzerland 2014

specially important in many practical cases, where the time constraints makes the exhaustive exploration of the search space not feasible.

Among the methods proposed to solve DOPs, we found Evolutionary Algorithms [19], Multi-Swarm methods [3], Ant Colony Optimization [9, 15] or cooperative strategies [7, 8]. A recent trend in DOPs resolution is the use of learning schemes to adapt the configuration of the solver (parameters, operators, etc.) during the search [7, 8, 14, 15, 17, 18]. The main reason behind this trend is the difficulty of finding an appropriate configuration for the solver -being even harder than in static problems- since environment changes over time. Despite this difficulty, works as those mentioned show that the learning increases the robustness and performance of the methods.

An important question that arises in the design of adaptive mechanisms for DOPs is what to do with the knowledge acquired once a change in the environment occurs: should we reuse it for the next changes or "forget it"? Although we found methods that works in one sense (reuse it) [14, 17] and the other (forget it)[8, 18], as far as we know, this is still an open question.

In this work, the knowledge is associated with the selection probability of two local search operators in the Adaptive Hill Climbing Memetic Algorithm (AHMA)[17]. When a problem change is detected, those probability values can be restarted or maintained.

In this work, we aim to shed light on this issue using as example the Adaptive Hill Climbing Memetic Algorithm (AHMA)[17]. The knowledge acquired is associated with the selection probability of two local search operators available in (AHMA). When a problem change is detected, those probability values can be restarted (i.e. forgetting) or maintained (keeping). The benchmark is composed by the dynamic versions of the knapsack (three instances), OneMax, Plateau, RoyalRoad and Deceptive problems.

The rest of this paper is organized as follows. Firstly, Section 2 presents AHMA and the variants we study in this work. Then, in Section 3, we show the experimental framework, describing the benchmark, performance measures, comparison methods and the experimental set-up. Section 3.2 analyses the results obtained and finally, Section 5 gives the main conclusions and further work.

2   Adaptive Hill Climbing Memetic Algorithm

The Adaptive Hill Climbing Memetic Algorithm (AHMA) proposed by Yang[17] was originally designed for combinatorial DOPs. AHMA combines a genetic algorithm with a local search that uses two operator according to a probability distribution that is adjusted during the search. This idea of adapting the selection of the local search operator in memetic algorithms was previously tested with success in static problems [11, 12].

Initially, AHMA generates a random population of pop_{size} individuals. Then, in each generation, the method selects individuals using the roulette wheel method; applies uniform crossover operator to them with probability p_c; performs a uniform mutation to the resultant individuals with probability p_m;

accomplishes a steady-state replacement that allow survive the best pop_{size} individuals among parents and offspring; and applies a local search to the best individual of the population (elite).

AHMA also includes two methods to generate diversity in the current population: Adaptive Dual Mapping (ADM) and Triggered Random Immigrants (TRI). The ADM, performed before local search, generates a new solution from the elite individual by applying to each variable the operator $1 - x_i$. The new solution replaces the elite if it is better. TRI is used after the local search to replace a certain percentage of the worst elements of the population by randomly generated individuals.

As mentioned above, the local search uses two operators: Greedy Crossover-base Hill Climbing (GCHC) and Steepest Mutation Hill Climbing (SMHC). GCHC applies uniform crossover operator to the elite solution and another individual from the current population selected by the roulette wheel method. Then, the offspring obtained replaces the elite solution if it is better. This process is repeated ls_size times. SMHC selects randomly a number of bits of the solution and flips them. As in the former operator, if the resulting solution improves the elite then it replaces it, and the process is repeated ls_size times. Every time the local search is executed, AHMA selects randomly one of these two operators according to a probability distribution that is adapted along the search with the aim of giving a higher probability to the best performing operator. The learning scheme that adjust the probability distribution is explained below.

2.1   Adaptive Scheme

Let p_{gchc} and p_{smhc} be the probabilities of applying GCHC and SHMC, respectively ($p_{gchc} + p_{smhc} = 1$). At the beginning of the search, both probabilities are initialized to 0.5, to give them the same chances of being selected. The learning scheme adjusts the values of p_{gchc} and p_{smhc} depending on the improvement degree (η) they produce, that is calculated as follows:

$$\eta = \frac{|f_{imp} - f_{ini}|}{f_{ini}} \tag{1}$$

where f_{ini} and f_{imp} are the fitness of the elite solution before and after the application of the local search, respectively. In the generation t, the values of p_{gchc} and p_{smhc} are adjusted as follows:

$$p_{gchc}(t) = p_{gchc}(t - 1) + \triangle \cdot \eta_{gchc}(t) \tag{2}$$

$$p_{smhc}(t) = p_{smhc}(t - 1) + \triangle \cdot \eta_{smhc}(t) \tag{3}$$

$$p_{gchc}(t) = \frac{p_{gchc}(t)}{p_{gchc}(t) + p_{smhc}(t)} \tag{4}$$

$$p_{smhc}(t) = 1 - p_{gchc}(t) \tag{5}$$

where $\triangle$ is the relative influence of the improvement degree on the selection.

With the aim of studying what we should do with the knowledge acquired once a change in the environment occurs, in this paper we will analyse three variants for the adaptive scheme of AHMA:

- AHMA: this is the original approach. Probabilities p_{gchc} and p_{smhc} are adjusted as explained above, so the probabilities learned are kept after the environment changes.
- AHMA Reset (AHMAR): p_{gchc} and p_{smhc} are reset to 0.5 after each change to "forget" the learning gained.
- AHMA Without Learning (AHMAWL): this is the baseline for comparison. In this variant p_{gchc} and p_{smhc} are fixed to 0.5 during the whole search process.

3   Experimental Study

This section describes the test problems, the performance measure used, how the comparisons are made and the details of the computational experiments performed.

3.1   Dynamic Test Problems

In order to make a problem dynamic, we use the XOR-DOP generator [18]. This XOR-DOP can generate dynamic environments from any binary-encoded stationary function $f(x)(x \in \{1, 0\})$ by a bitwise exclusive-or (XOR) operator. The environment is changed every t generations. For each environmental period k, an XOR mask $M(k)$ is incrementally generated as follows:

$$M(k) = M(k-1) \oplus T(k)$$

where $\oplus$ is the XOR operator (i.e., $a \oplus b = 1 \iff a \neq b$) and $T(k)$ is an intermediate binary template randomly created with ρm ones for the kth environmental period. For the first period $k = 1$, $M(1) = \{0 \ldots 0\}$. Then, the population at generation t is evaluated as:

$$f(x, t) = f(x \oplus M(k))$$

where $k = \lceil t/\tau \rceil$ is the environmental change index. One advantage of this XOR generator lies in that the speed and severity of environmental changes can be easily tuned. The parameter τ controls the speed of changes while $\rho \in [0.0, 1.0]$ controls their severity. A bigger ρ means more severe changes while a smaller τ means more frequent changes.

The benchmark used for the experimentation is composed by dynamic optimization problems generated from well-known static binary problems by means of the XOR-DOP generator shown above. The next part of this subsection is devoted to describe these problems.

Dynamic Knapsack Problem. The knapsack problem is a well known NP-Hard combinatorial optimization problem [10]. Given a set of m elements the knapsack problem is described as follows:

$$\max f(x) = \sum_{i=1}^{m} p_i x_i$$

$$\text{subject to } \sum_{i=1}^{m} w_i x_i \le C$$

$$x_i \in \{0, 1\} \ i = 1 \ldots m$$

where $x = (x_1, ..., x_m)$ and $x_i = 1$ if item i is selected or $x_i = 0$, otherwise. Values p_i and w_i represent the profit and the weight of item i, respectively and C is the capacity of the knapsack, which cannot be exceeded.

The instances of the knapsack problem used here have $m = 100$, while the weights, profits and capacity are defined as follows:

$$\text{Capacity } C = 0.6 \times \sum_{i=1}^{m} w_i$$

We will deal with three types of instances, depending on the level of correlation between the profits and weights [16]:

Type	Weight	Profit
No Correlation (NC)	$w_i = U(1, 50)$	$p_i = U(1, 55)$
Weak Correlation (WC)	$w_i = U(1, 50)$	$p_i = 0.5w_i + U(1, 5)$
Strong Correlation (SC)	$w_i = U(1, 50)$	$p_i = w_i + U(1, 5)$

where $U(a, b)$ is a function that returns a random integer between a and b coming from uniform distribution. In this contribution, we generate one instance per type.

It is believed that knapsack is one of the easiest NP-Hard problem. Several exact algorithms are available and for them, the hardness of random instances, increases with the correlation between weights and profits [16].

In order to deal with unfeasible solutions during the optimization process, we use the same penalization scheme as [18].

$$f(x) = \begin{cases} \sum_{i=1}^{100} p_i x_i & if \ \sum_{i=1}^{100} w_i x_i \le C \\ 10^{-10} \times \left(\sum_{i=1}^{100} w_i - \sum_{i=1}^{100} w_i x_i \right) & otherwise \end{cases}$$

Dynamic Test Functions. The other test problems used on this paper are four binary-encoded combinatorial optimization funtions: OneMax, Plateau, Royal Road and Deceptive. All of them consists of finding solutions that match all the bits of a target optimal solution. This target solution is initially considered

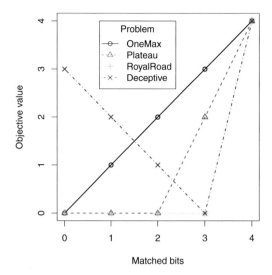

Fig. 1. Contribution of the 4-bits block to the corresponding cost function with respect to the number of correctly matched bits

to be the solution where all its bits are set to 1. To evaluate a solution, we consider blocks of 4 bits where each block contributes a given amount to the final objective value. The contribution of every block of 4 bits for each of the considered functions is computed as follows:

- **OneMax**: Each matched bit adds 1 to the fitness.
- **Plateau**: Three matched bits add 2 to the fitness while four matched bits add 4 and any other amount of bits matched leads to a 0 contribution.
- **RoyalRoad**: Each perfectly matched block adds 4 to the fitness. Partially matched blocks have 0 fitness.
- **Deceptive**: Fitness is 4 if all the 4 bits are matched. If not, the fitness for the block is 3 minus the number of matched bits.

This information is graphically shown in Fig. 1.

The dimension of these problems is defined as l (where l is divisible by 4), and given the aforementioned rules to compute the fitness, the optimum value for any of the problems is also l.

3.2   Performance Measures

To assess the performance of the algorithms, we have used the offline performance [4] that is defined as the best-of-generation fitness averaged across the number of total runs and then averaged over the data gathering period, as formulated below:

$$F_{BG} = \frac{1}{G} \sum_{i=1}^{G} \left(\frac{1}{N} \sum_{j=1}^{N} F_{BG_{ij}} \right) \tag{6}$$

where G is the number of generations (i.e., $G = 10 \times \tau$), N is the total number of runs, and $F_{BG_{ij}}$ is the best-of-generation fitness of generation i of run j.

3.3   Experimental Set Up

We performed 30 independent runs with 100 fitness function changes, for every instance, period, severity and algorithm variant. We considered five different periods of change ($\tau \in \{10, 25, 50, 75, 100\}$ generations) and four severities ($\rho \in \{0.1, 0.2, 0.5, 0.9\}$).

The parameters of AHMA were configured according to the guidelines given in the original work by Wang [17].

3.4   About the Comparison of the Methods

Since the results for every possible algorithm/adaptive scheme and problem configuration would be difficult to interpret if displayed as a numerical table, we used the "Statistical Ranking Color Scheme" (SRCS) [1].

This technique uses non-parametric tests (Kruskal-Wallis and Mann-Whitney Wilcoxon with Holm's correction for multiple comparisons) to assess the statistical significance of the individual differences among each pair of algorithms on every problem configuration. If the test concludes that there is enough evidence for statistical differences, the algorithm with the highest overall offline performance adds 1 to its rank, and the other adds -1. In case of a tie, both receive a 0. The range of the rank values for n algorithms for any specific problem, period and severity will therefore be in the $[-n+1, n-1]$ interval. The higher the rank obtained, the better an algorithm can be considered in relation to the other ones. Those ranks are displayed as colors with the highest rank value $(n-1)$ being displayed as white and the lowest rank value $(-n+1)$ having the darkest color. The remaining rank values will be assigned a progressively darker color as the rank decreases. If we group together the ranks of an algorithm for a given problem with every possible different period and severity we can obtain a colored matrix, where it is easy to observe how the algorithm performs for that specific problem. A white color in a given cell indicates that the algorithm is statistically better than all of the other algorithms for that specific problem configuration. If the color gets darker means that the algorithm starts to be statistically equal or worse than some other algorithms. The worst case for a given algorithm occurs when its cell has the darkest possible color, meaning that the algorithm is statistically worse than all of the others for that problem configuration.

4   Results

4.1   On the Influence of the Adaptive Schemes

The results of the AHMA and its variants appear in the Fig. 2 that was obtained using the SRCS [1] technique.

Results should be analyzed per test problem. Every matrix (one per algorithm) in the corresponding column contains the ranking of such algorithm in all the scenarios tested. In order to compare the results from different algorithms in a given scenario we should look at the same matrix position in the three matrices. For example, consider the SC instance of the knapsack problem (first column), severity = 0.9 and change period = 10. The positions to observe are marked with a square in the figure. AHMA has a white colour in that cell, meaning that its rank is 2: it is significantly better than the other two alternatives. AHMAR and AHMAWL have an orange colour, associated with rank -1: they are significantly worse than one of the other two alternatives. The figure can also be read in a more general way. For example, again in the SC instance, the algorithm AHMA has an almost white matrix, meaning that it is better than the other two algorithms in most of the scenarios tested.

We will start the analysis considering the point of view of adapation/no adaptation, or learning vs. no learning.

The original algorithm AHMA obtains, in general, better results than AHMAWL. This is clear for SC and WC instances and considering high values of severity. AHMAWL is better than AHMA on the NC instance. When considering the other problems (OneMax, Plateau, RoyalRoad and Deceptive), the advantage of AHMA over AHMAWL is clear over all the scenarios tested.

When considering AHMAR vs AHMAWL, we can observe that both methods have a pretty similar behaviour over the three knapsack instances (and over all the configurations). Again, over the NC instance and considering higher values of severity AHMAWL is better than the alternative including learning (AHMAR). The plots are quite similar for the rest of problems. AHMAR achieved better or equal performance than AHMAWL. Similar behaviour is observed for low values of severity while for higher values, AHMAR achieved a higher position in the ranking (lighter colour in the corresponding cells).

We will finally compare the two adaptive schemes. Recall that the difference among them is what to do with the knowledge acquired when a change arrived. In one scheme (AHMAR), everything is forgotten, while in the other, no action is taken (AHMA) so the knowledge is accumulated from one change to the other.

Clearly, forgetting is not a good way to proceed when considering SC and WC instances (AHMA is better than AHMAR). However, on the NC instance, AHMAR is better than AHMA, specially when higher severities are considered. In the other group of problems the situation is clear: AHMA is definitely better than AHMAR.

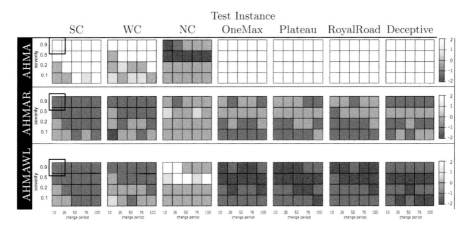

Fig. 2. Performance of the algorithms AHMA, AHMAR, AHMAWL over the test problems. Lighter colors indicates better results.

4.2   Analysis Matched Pairs Non-parametric Tests

The SRCS technique gives statistical information about the ranking of a set of methods in just one configuration of one problem. However, it is also interesting to compare the results of the methods over a *set* of configurations from one or various problems to have a more comprehensive view. In this sense, García et al. proposed in [6] the directives to carry out this global analysis using paired non-parametric statistical tests. These tests allow to rank the performance of two or more methods over a set problem configurations and to determine if the difference in performance are significant or not. The significance of the differences is assessed by comparing estimators (usually the mean/median fitness or error value) of the performance of the algorithms over each problem configuration.

Matched pairs non-parametric tests allow, for example, to determine the best global method over all the configurations of the Knapsack problem for the three instance or over the 60 problem configuration considered in this work. The specific matched pairs non-parametric tests performed here were done with the SPSS statistical software. We used *Friedman's test* to determine if there are significant differences among the performance of two or more methods, and *Wilcoxon paired test* to check if there exist significance differences between the performance of two algorithms.

Firstly, we performed a global analysis of the algorithmic variants over all the test problems and configurations, using the Friedman test. Results indicated a statistical significant difference among the algorithms, and then we performed a pairwise comparison using the Wilcoxon's test. The results are shown below (given $Algorithm_1$ vs $Algorithm_2$, '>' stands for $Algorithm_1$ is significantly better than $Algorithm_2$):

AHMA vs. AHMAR	AHMA vs. AHMAWL	AHMAR vs. AHMAWL
>	>	>

Table 1. Results of the pairwise comparison using the Wilcoxon's test over all the test problems. Two levels of severity and change frequency were considered. A '>' sign indicates that the first algorithm was significantly better than the second one. Symbol '-' denotes no significant differences.

	Severity		
Change	*Low*		
	AHMA vs AHMAR	AHMA vs AHMAWL	AHMAR vs AHMARWL
Low	>	>	–
High	>	>	>
Change	*High*		
	AHMA vs AHMAR	AHMA vs AHMAWL	AHMAR vs AHMARWL
Low	>	>	–
High	>	>	>

We can observe that AHMA is better than AHMAR and AHMAWL, and AHMAR is better than AHMAWL. The conclusion is clear: having some sort of adaptation is better than not having it.

Then, we performed a similar analysis but grouping the scenarios in terms of low (0.1,0.2) /high (0.5,0.9) severity and low (10,25) /high (50,75,100) change frequency.

After checking the existence of significant difference among the three methods by Friedman's test, we apply the Wilcoxon's test pairwise and the results are shown in Table 1. A '>' sign indicates that the first algorithm was significantly better than the second one, whereas '-' denotes no significant differences.

The original AHMA is better than AHMAR and AHMAWL over all the combinations tested: low/high change frequency and low/high severity. When considering AHMAR against AHMAWL, no differences were detected when the change frequency is low (and independently of the severity level). In scenarios with a high frequency of changes (with low or high severities), AHMAR was better than AHMAWL.

The last analysis considers the results on every test problem. The application of the non-parametric tests was done in an analogous way and the results are collected in Table 2. The conclusion is clear: AHMA is better than the other two variants in all the problems, excepting the NC instance of the knapsack problem, where the differences were not significant. In this analysis, AHMAR is better than AHMAWL in almost all of the problems; but no statistical significant differences were found over WC and NC knapsack instances.

Table 2. Pairwise comparison of the algorithms on every test problems (each one consisting on twenty scenarios). A '>' sign indicates that the first algorithm was significantly better than the second one. Symbol '-' denotes no significant differences.

Problem	AHMA vs AHMAR	AHMA vs AHMAWL	AHMAR vs AHMAWL
OneMax	>	>	>
Plateau	>	>	>
RoyalRoad	>	>	>
Deceptive	>	>	>
Knapsack SC	>	>	>
Knapsack WC	>	>	–
Knapsack NC	–	–	–

5   Conclusions and Future Work

In this work we have analysed an important issue that arises in the design of adaptive schemes for DOPs: should we forget the knowledge acquired once a change in the environment occurs or should we use it in the new environment? To this end, we have used AHMA, a state-of-the-art algorithm for binary DOPs that incorporates a simple learning or adaptive scheme to adjust the usage probability of two local search operators. Concretely, we analyzed three variants of AHMA's adaptative scheme: AHMAWL , AHMAR and AHMA. The first variant does not perform any learning, the second one forgets the learning after changes and the third one accumulates the knowledge during the whole search process.

The experimentation was done over three instances of a dynamic version of the Knapsack Problem and four dynamic test functions: OneMax, Plateau, Royal-Road y Deceptive. Twenty configurations (four severities × five change frequencies) were considered for each instance. In short, the algorithms were compared over 140 different scenarios.

Two main conclusions can be derived from this work. Firstly, it is clear that including some sort of adaptation is beneficial: AHMA and AHMAR were better than AHMAWL. In some specific scenarios where the severity was low, the performance of AHMAR and AHMAWL was similar.

Secondly, restarting the selection probabilities of the local search operators is not a good strategy. Maintaining the learnt values after a change is clearly beneficial, at least under the scenarios tested.

Now, several lines of research arise. For example, it is not clear what it would happen if more operators are available. If one of them gets a very low probability of selection and these probabilities are kept, then it may be quite difficult to "recover" it.

Another point to analyze is the "speed" of adaptation or learning. In this sense a possible idea is to define an ideal adaptation profile and observe how different learning techniques match this profile.

Finally, we need to explore how these ideas on what to do with the acquired knowledge at the level of parameters may have some parallelism with the changes that are need to perform in the population when a change occurs.

Some of these topics are now being explored.

References

[1] del Amo, I.G., Pelta, D.A., Gonzalez, J.R., Masegosa, A.D.: An algorithm comparison for dynamic optimization problems. Applied Soft Computing 12(10), 3176–3192 (2012),
http://www.sciencedirect.com/science/article/pii/S1568494612002700,
doi:10.1016/j.asoc.2012.05.021

[2] Blackwell, T.: Particle swarm optimization in dynamic environments. In: Yang, S., Ong, Y.-S., Jin, Y. (eds.) Evolutionary Computation in Dynamic and Uncertain Environments. SCI, vol. 51, pp. 29–49. Springer, Heidelberg (2007),
http://dx.doi.org/10.1007/978-3-540-49774-5_2

[3] Blackwell, T., Branke, J.: Multiswarms, exclusion, and anti-convergence in dynamic environments. IEEE Transactions on Evolutionary Computation 10(4) (2006)

[4] Branke, J., Schmeck, H.: Designing evolutionary algorithms for dynamic optimization problems. In: Advances in Evolutionary Computing. Natural Computing Series, pp. 239–262 (2003), http://dx.doi.org/10.1007/978-3-642-18965-4_9,
doi:10.1007/978-3-642-18965-4_9

[5] Cruz, C., González, J.R., Pelta, D.A.: Optimization in dynamic environments: a survey on problems, methods and measures. Soft Computing 15(7), 1427–1448 (2011)

[6] García, S., Molina, D., Lozano, M., Herrera, F.: A study on the use of nonparametric tests for analyzing the evolutionary algorithms' behaviour: a case study on the CEC 2005 special session on real parameter optimization. Journal of Heuristics 15(6), 617–644 (2009), http://dx.doi.org/10.1007/s10732-008-9080-4

[7] Gonzalez, J.R., Masegosa, A.D., del Amo, I.G.: A cooperative strategy for solving dynamic optimization problems. Memetic Computing 3, 3–14 (2011),
http://dx.doi.org/10.1007/s12293-010-0031-x

[8] González, J.R., Cruz, C., del Amo, I.G., Pelta, D.A.: An adaptive multiagent strategy for solving combinatorial dynamic optimization problems. In: Pelta, D.A., Krasnogor, N., Dumitrescu, D., Chira, C., Lung, R. (eds.) NICSO 2011. SCI, vol. 387, pp. 41–55. Springer, Heidelberg (2011),
http://www.springerlink.com/content/c1006661q2h06577/

[9] Guntsch, M., Middendorf, M., Schmeck, H.: An ant colony optimization approach to dynamic tsp. In: Proceedings of the Genetic and Evolutionary Computation Conference, pp. 860–867 (2001)

[10] Kellerer, H., Pferschy, U., Pisinger, D.: Knapsack problems. Springer (2004)

[11] Krasnogor, N., Smith, J.: A memetic algorithm with self-adaptive local search: TSP as a case study. In: Proceedings of the Genetic and Evolutionary Computation Conference, pp. 987–994 (2000)

[12] Krasnogor, N., Smith, J.: Emergence of profitable search strategies based on a simple inheritance mechanism. In: Proceedings of the Genetic and Evolutionary Computation Conference, pp. 432–439 (2001)

[13] Li, C., Yang, S.: Fast multi-swarm optimization for dynamic optimization problems. In: Natural Computation, ICNC 2008, vol. 7, pp. 624–628 (2008)

[14] Liu, L., Wang, D., Ip, W.: A permutation-based dual genetic algorithm for dynamic optimization problems. Soft Computing 13(7), 725–738 (2009), http://dx.doi.org/10.1007/s00500-008-0345-5

[15] Mavrovouniotis, M., Yang, S.: A memetic ant colony optimization algorithm for the dynamic travelling salesman problem. Soft Computing 15(7), 1405–1425 (2011), http://dx.doi.org/10.1007/s00500-010-0680-1

[16] Pisinger, D.: Where are the hard knapsack problems? Computers and Operations Research 32(9), 2271–2284 (2005), http://www.sciencedirect.com/science/article/pii/S030505480400036X

[17] Wang, H., Wang, D., Yang, S.: A memetic algorithm with adaptive hill climbing strategy for dynamic optimization problems. Soft Computing 13(8-9), 763–780 (2009), http://www.springerlink.com/content/5vu17644466p7200/, doi:10.1007/s00500-008-0347-3

[18] Yang, S., Yao, X.: Experimental study on population-based incremental learning algorithms for dynamic optimization problems. Soft Computing 9(11), 815–834 (2005), http://dx.doi.org/10.1007/s00500-004-0422-3, doi:10.1007/s00500-004-0422-3

[19] Yang, S., Ong, Y., Jin, Y.: Evolutionary Computation in Dynamic and Uncertain Environments. SCI, vol. 51. Springer, Heidelberg (2007)

Using Base Position Errors in an Entropy-Based Evaluation Function for the Study of Genetic Code Adaptability

Lariza Laura de Oliveira and Renato Tinós

Department of Computing and Mathematics
FFCLRP, University of São Paulo (USP)
14040-901, Ribeirão Preto, SP, Brazil
larizalaura@usp.br, rtinos@ffclrp.usp.br

Abstract. The canonical genetic code is almost universal. An intriguing question is why the canonical genetic code is used instead of another genetic code. Some researchers have proposed that the canonical genetic code is a product of natural selection. This hypothesis is supported by its robustness against mutations. In this paper, we propose a new evaluation function based on entropy and robustness against base position errors for the study of genetic code adaptability. In order to find the best hypothetical genetic codes in the search space, we use a genetic algorithm (GA). The experimental results indicate that, when the proposed evaluation function is compared to the standard evaluation function based only on robustness, the difference between the fitness of the best hypothetical codes found by the GA and the fitness of the canonical genetic code is smaller.

Keywords: Genetic Algorithms, Genetic Code Adaptability, Base Position Errors.

1 Introduction

The genetic information is stored in living organisms as DNA molecules. In one of the gene expression steps, the DNA is copied into messenger RNA, this process is called transcription. The messenger RNA dictates the amino acid sequence of a protein, in a process called translation. Although, the alphabet of the acids nucleic RNA and DNA is composed of 4 letters (4 nucleotides), whereas proteins are encoded by 20 amino acids. During the translation process each triple of nucleotide, which is called codon, is mapped in an amino acid, according to the genetic code [3]. Therefore, the genetic code is an interface between genetic information and the proteins, which are the macromolecules essential to most biological processes in living organisms [1].

A genetic code maps each one each one of the 64 (4^3) codons into one of the amino acids used in living organisms. Considering all possible codes mapping the 64 codons into 21 amino acids, more than 1.51×10^{84} possible genetic codes can be generated [4]. However, only one genetic code, named canonical genetic

G. Terrazas et al. (eds.), *Nature Inspired Cooperative Strategies for Optimization
(NICSO 2013)*, Studies in Computational Intelligence 512,
DOI: 10.1007/978-3-319-01692-4_8, © Springer International Publishing Switzerland 2014

code, is used in almost all complex living organisms. Why exactly this code was selected over this large number of possible codes is a question that has intrigued researchers for decades [5], [6], [7], [8], [9].

The canonical code's organization remains under discussion, but many researchers argue that the genetic code is a product of natural selection, instead of random product [6]. This hypothesis is supported by its robustness against mutations when amino acids properties like polar requirement are considered [5]. Haig and Hurst [10], and some other authors after them [8], [11], showed that a very small percentage of random codes are better than the canonical code in minimizing the deleterious effects of errors in the translation process.

According to the authors in [11], two approaches can be used to analyse the genetic code adaptability. In the statistical approach, the number of random codes better than the canonical genetic code is estimated, using a given mathematical function to evaluate a possible genetic code. In the engineering approach, the canonical code is compared with the best code obtained by an optimization algorithm. The engineering approach allows to identify regions of the genetic code space where best codes, according to a given evaluation function, can be found.

Following the engineering approach, Santos and Monteagudo [12] used a Genetic Algorithm (GA) with the standard evaluation function based on robustness considering the polar requirement of the amino acids. They also used two types of encodings: the first one is a non-restrictive encoding, where the allowed genetic codes map the 64 codons into the 21 amino acids; the second is a restrictive encoding, where only genetic codes with codons grouped in the same way as the canonical code are allowed. One can observe that canonical code information is used in the restrictive encoding in order to drastically reduce the number of possible genetic codes found in the non-restrictive encoding. As a consequence, best results are obtained with the restrictive encoding when the standard evaluation function based on the polar requirement is used (the authors in [12] also considered other amino acids properties in the evaluation function, obtaining similar results).

One of the problems with the non-restrictive encoding is that the best codes, according to the standard evaluation function, are those with high frequency of some few amino acids associated to codons. When the frequencies of codons associated to amino acids in the canonical genetic code are plotted, one can observe that most amino acids are codified by two or more codons. This property is not observed in the best codes obtained by an optimization procedure when polar requirement and non-restrictive encoding are considered, as we show in next section. In those best codes, which are better than the canonical genetic code when the standard robustness-based evaluation is considered, few amino acids are associated with more than one codon.

In order to solve this problem, we proposed in a previous work [16] an entropy-based evaluation function, considering that the genetic code was optimized not only according to its robustness, but also according to the frequencies of the codons associated to each amino acid. When the distribution of these frequencies is uniform, the amino acids have shorter distances to all others, i.e. the number of changes necessary to replace one amino acid for another is smaller. As consequence, it is easier for evolution to incorporate an amino acid to a protein as the number of possibilities to codify an amino acid is higher. The entropy-based function is composed of the robustness against errors considering the polar requirement [8], [10], [12], [13] and the entropy. Maximizing the entropy, uniform distributions of the frequencies of the codons are preferable than distributions where the frequency of codons codifying the amino acids are very unequal.

In [16], when the evaluation function is computed, the robustness in all base positions of the codons are considered with the same probability. However, experimental data indicated that errors in the translational process vary according to base position within a codon [12]. Hence, in this paper, we investigate a new functions where the mistranslations and base position errors are considered in the robustness term of the evaluation function, following the methodology presented in [12]. Here, we also use a GA to search the best genetic codes in order to compare them to the canonical code. Using the new entropy-based function, we obtained better results, even when compared to the restrictive encoding presented in [12]. In this sense, we show that a restrictive approach is not essential in the investigation of the genetic code adaptability.

This paper is organized as follows: the standard and the new evaluation functions are presented in next section, as well as the methods employed here; the experimental results are presented in Section 3; finally, the conclusions are presented in Section 4.

2   Methods

The GA used here is implemented using C++ programming language. The non-restrictive encoding, where each individual of the GA's population is composed of 61 positions, each one related to one codon, is employed. Each position corresponds to one of 20 labels, each one representing an amino acid (the stop codons are not considered, i.e., they remain fixed). Thus, each GA's individual encodes a hypothetical genetic code. Figure 1(a) shows a fragment of a hypothetical genetic code, where each codon is associated to an amino acid.

The GA uses two reproduction operators: swap and mutation [12]. The first one interchanges amino acids associated to two codons, i.e., two positions are randomly selected and their amino acids are swapped as shown in Figure 1(b).

In the mutation operator, a position is selected in the code (individual) and its corresponding amino acid is replaced by another one, selected among the 20 possible amino acids (Figure 1(c)). The position and the new amino acid are randomly selected using a uniform distribution.

In order to select the individuals to be reproduced, tournament selection is employed. In this technique, a percentage of individuals is randomly selected and the individual with the best fitness is chosen. Furthermore, elitism is also used to preserve the best individual found on previous generations. In the experiments

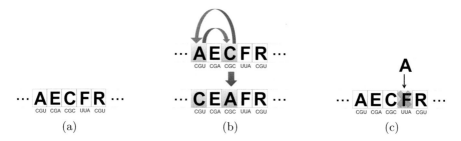

Fig. 1. a) Individual's encoding. Each GA's individual represents a hypothetical genetic code, which is composed by codons associated to amino acids. b) Swap operator. c) Mutation operator.

presented in Section 3, the population size is equal to 500, the mutation rate is 0.01, swap rate is 0.5, and the tournament size is 3%. The GA is executed 10 times during 1000 generations with different random initial populations.

2.1 Robustness-Based Evaluation Function

The standard evaluation function which is commonly employed in literature is the mean square ($M_s(C)$) change in an amino acid property. This measure computes all possible changes to each base of all codons for a given code C [8], [10],[12], [14], [15]. In general, polar requirement is considered the most substantial property when $M_s(C)$ is computed. The measure $M_s(C)$ is defined as:

$$M_s(C) = \frac{\sum_{ij}(X(i,C) - X(j,C))^2}{\sum_{ij} N(i,j,C)} \tag{1}$$

where $X(k,C)$ is the amino acid property value (in this paper, the polar requirement) for the amino acid codified by the k-th codon of the genetic code C, and $N(i,j,C)$ is the number of possible replacements between codons i and j.

A lower value of $M_s(C)$ means that the code C is robust, i.e., a change in a codon base will not cause a drastic change in the amino acid property considered. Ideally, the amino acid of a mutated codon is not replaced or replaced by another with similar properties.

The fitness of the best generated code and the fitness of the canonical genetic code are closer when compared to the average fitness of the random codes, i.e., the *pmd* indicates how close is the fitness of the canonical code to the fitness of the best code found by the GA when compared to the average fitness of the random codes.

Intuitively, if we think which amino acids are most important for minimizing Eq. 1, we will conclude that those with shorter mean distances to all others are the most important, i.e., those with intermediate values of polar requirement. Table 1 shows the polar requirement of each amino acid and the mean distance among the amino acids, considering the polar requirement . One can observe that Alanine, Glycine, and Serine have the shorter mean distances.

Table 1. Polar requirement and mean distance among amino acids

Amino Acid	Cys	Leu	Ile	Phe	Trp	Met	Tyr	Val	Pro	Thr
Polar requirement	4.8	4.9	4.9	5	5.2	5.3	5.4	5.6	6.6	6.6
Mean distance	12.86	12.35	12.35	11.85	10.93	10.49	10.08	9.31	6.67	6.67
Amino Acid	Ala	Ser	Gly	His	Gln	Arg	Asn	Lys	Glu	Asp
Polar requirement	7	7.5	7.9	8.4	8.6	9.1	10	10.1	12.5	13
Mean distance	6.17	6.00	6.23	6.96	7.39	8.82	12.65	13.18	31.80	37.13

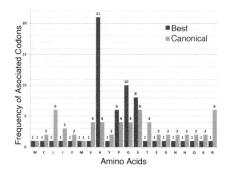

Fig. 2. Frequencies of codons associated to amino acids in best code found in a run of the GA with non-restrictive encoding

In this way, hypothetical genetic codes with higher number of codons associated to Alanine, Glycine, and Serine present best values for the evaluation function given by Eq. 1. Running the GA with non-restrictive encoding and using Eq. 1 reinforces this conclusion, as shown in Figure 2, which shows the distribution of frequencies for the best genetic code obtained by a run of the GA. One can observe that the code, which presents better robustness than the canonical code, showed in Figure 2 has a non-uniform frequency distribution, with most amino acids codified by only one codon, differently from the canonical code.

In [16], we proposed to add an entropy-based term to $M_s(C)$ in the evaluation function of the genetic codes in order to minimize this problem (see next section). Our hypothesis is that, having more codons codifying an amino acid in a genetic code, it becomes easier to incorporate an amino acid to a protein, as there are more ways to change a current codon to obtain codons corresponding to new amino acids. In other words, the mean distance between the amino acids, in terms of number of changes necessary to replace one for another, decreases as the codons distribution becomes more uniform. When entropy is maximized, possible changes in one amino acid to another occur without excessive cost, what is certainly useful in the biological evolution.

In order to show how uniform distributions of the frequencies of codons can be useful in this problem, a simple is presented. In this example, a standard GA is used to optimize the following fitness function:

$$f(\mathbf{y}) = \|\mathbf{d} - \mathbf{y}\| \tag{2}$$

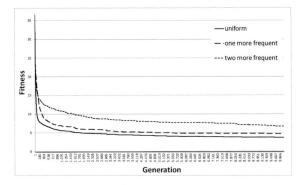

Fig. 3. Average best fitness for the sets of runs with different mapping tables

where **d** is the desired integer vector and **y** is integer vector representing the phenotype of a GA's individual. In this experiment, each individual of GA is composed by a binary vector (individual genotype). Each 5 bits of the genotype codifies one element of the vector **y** (each element of the vector **y** is a integer). The genotype is converted into phenotype **y** by means of a mapping table (similar to a genetic code), which maps each 5 bits to a integer value. In the experiment, the vector **d** is randomly generated in the initial population. The vectors **d** and **y** are composed by values between 1 to 10 and the mapping table matches all 5-bit possibilities, which are 32, to these values. In this way, we can have more than one of these 5-bit elements associated to the same value. Three mapping tables with different distributions are used in this example: the first one is uniform, the second one has one integer value more frequent than the others, i.e. a value between 1 to 10 associated with most of the 5-bit elements, and the third table has two more frequent integer values. The standard GA with crossover, mutation, elitism, and tournament selection is used to optimize the vector **y**. The parameters used in the experiment are: size of the population equal to 100 individuals, crossover rate equal to 0.6, mutation rate equal to 0.01, and number of generations equal to 10000. Figure 3 presents the average best fitness for three sets of 10 runs of the GA, each one using a different mapping table.

One can observe that the fitness decreases faster when the uniform mapping table is used. This result can be explained by the fact that it is easier for the GA to change a 5-bit element of the genotype to reach the desired solution when the distribution used in the mapping table is uniform and, as a consequence, the mean distance between all 5-bit elements is smaller. One way to evaluate the uniformity of the distribution, and as a consequence, the average mean distance between the amino-acids (i.e., the number of changes in the codons of an amino-acid to the codons of other amino-acid), is to compute the entropy of the distribution. In fact, in the results presented in Figure 3, higher entropy means faster convergence for the algorithm (in the runs, the uniform distribution had the highest entropy). However, higher entropy means, in general, a smaller robustness against mutations. In this way, a multi-objective evaluation function is presented in next section.

2.2 Entropy-Based Evaluation Function

The entropy-based evaluation function proposed in [16] is given by:

$$f(C) = aM_s(C) + (1 - a)\frac{1}{S(C)} \tag{3}$$

where the real weight $a \in [0, 1]$ and $S(C)$ is the entropy of genetic code C, given by:

$$S(C) = -\sum_k p(k, C) \log p(k, C) \tag{4}$$

where $p(k, C)$ is the relative frequency of the k-th amino acid in the genetic code C. The term $M_s(C)$ is computed using Eq. 1. Normalized values of $M_s(C)$ and entropy are used here. One can observe that, when a is equal to 1.0, Eq. 3 reproduces Eq. 1. Smaller values of a yield in more uniform genetic codes. When a is equal to 0.0, only entropy is considered and the best genetic codes are those where all amino acids are codified by almost the same number of codons.

2.3 Proposed Evaluation Function

Nucleotides are composed by a nitrogenous base, a pentose, and a phosphate. The nitrogenous bases are classified in purines and pyrimidines according to their structure [3]. The purines Adenine (A) and Guanine (G) have a pair of fused rings, while the bases Cytosine (C), Thymine (T), and Uracil (U) contain a single ring [2]. Transition errors occur when a purine is replaced by another purine or a pyrimidine is replaced by a pyrimidine. On the other hand, transversion errors occur when a purine is replaced by a pyrimidine or vice versa. Experimental data show that errors in the translational process occur in a complex manner [17], but, in general, mistranslation accuracy varies according to base position within a codon. Freeland and Hurst [8] summarized this knowledge in the following rules:

- Mistranslation of the second base is much less frequent than mistranslation in the other two bases, whereas mistranslation of the first base is less frequent than mistranslation of the third base.
- Most mistranslations of the second base are transitional.
- Most mistranslations of first base are transitional.
- The transition bias is very small in the third base mistranslation.

Freeland and Hurst [8] propose that this information should be added to the evaluation function when analyzing the genetic code adaptability. For this purpose, a mistranslation weight matrix is used, as shown in Table 2. The $M_s(C)$ computed with mistranslation weights is called $M_{st}(C)$.

Table 2. Weights used in M_{st} calculation

Weight	Fist base	Second base	Third base
Transitions	1	0.5	1
Transversions	0.5	0.1	1

In this paper, we propose to use $M_{st}(C)$ and entropy to evaluate the hypothetical and canonical genetic codes. In this way, the proposed evaluation function, based on Eq. 3, is given by:

$$f(C) = aM_{st}(C) + (1 - a)\frac{1}{S(C)} \tag{5}$$

2.4   Comparison of the Evaluation Functions

In order to compare the canonical genetic code to the best codes obtained by the GA using Eq. 5 with different values of a as fitness function (engineering approach), we use two measures:

– Percentage of Minimization Distance (pmd), as described in [7];
– Improvement, as mentioned in [12];

The pmd is computed as follows:

$$pmd = 100\frac{\Delta_{mean} - \Delta_{code}}{\Delta_{mean} - \Delta_{low}} \tag{6}$$

where Δ_{mean} is the average fitness of genetic codes randomly generated, Δ_{code} is the fitness of the canonical genetic code, and Δ_{low} is the fitness of the best code found by the GA.

Higher values of pmd means that the fitness of the best generated code and the fitness of the canonical genetic code are closer when compared to the average fitness of the random codes, i.e., the pmd indicates how close is the fitness of the canonical code to the fitness of the best code found by the GA when compared to the average fitness of the random codes.

The other measure, improvement, gives the percentage of the best code improvement in relation to the canonical genetic code fitness, i.e.,

$$Improvement = 100\frac{\Delta_{code} - \Delta_{low}}{\Delta_{code}} \tag{7}$$

Improvement decreases as the pmd increases, providing a measurement of how the best code found improved the fitness compared to the canonical genetic code.

The average fitness obtained for different values of a in Eq. 5 is also presented in the tables of results.

Besides, we also use the statistical approach (see 1) to analyse the proposed evaluation function. In this approach, the GA is not used. Instead, a large number of random codes is generated and they are compared to the canonical code using the proposed evaluation function with different values of a.

Table 3. Fitness of the best individuals (over 10 runs) for different values of a

a	mean	std	median	minimum
0.4	0.635	0.005	0.634	0.630
0.5	0.540	0.004	0.539	0.534
0.6	0.442	0.003	0.442	0.438
0.7	0.349	0.007	0.347	0.342
0.8	0.253	0.008	0.253	0.246
0.9	0.154	0.005	0.154	0.148
1.0	1.138	0.083	1.130	1.021

Table 4. Mean and minimum values of pmd and improvement (over 10 runs)

a	pmd (mean)	pmd (best)	Impr.(mean)	Impr. (best)
0.4	80.4	78.8	6.02	6.62
0.5	83.1	81.5	7.4	8.2
0.6	84.3	83.1	9.7	10.6
0.7	86.0	84.4	12.2	13.9
0.8	87.0	85.4	16.9	19.2
0.9	87.0	85.8	27.3	30.2
1.0	86.4	85.5	56.7	61.2

3   Results

The results of the GA with fitness given by the proposed evaluation function (Eq. 5) are presented in Section 3.1 (engineering approach). The results of the statistical approach with the proposed evaluation function are presented in Section 3.2.

3.1   Engineering Approach

Tables 3 and 4 show the results for the proposed evaluation function (Eq. 5) for different values of a.

One can observe in the tables that the pmd and the improvement of the experiments with the proposed evaluation function for $a = 0.9$ are better than the values for the standard evaluation function (i.e, when $a = 1.0$). When a is equal to 0.9, the pmd is 85.8%, while, when entropy is not considered ($a = 1.0$), the pmd is 85.5%. The fitness of the best code, when a is equal to 1, is 1.021, while the fitness of the canonical code is 2.63, which is very close to the value obtained in [12]. The pmd obtained when $a = 1.0$ is 85% in [12], also similar to the value obtained here.

The frequencies of codons associated with each amino acid for the best code found for the runs with $a = 0.9$ is presented in Figure 4. One can observe that the frequency distribution of codons associated with the amino acids is more uniform than the distribution for the best code obtained for the standard evaluation function ($a = 1.0$) shown in Figure 2. One can observe that the distribution for $a = 0.9$ is similar to the distribution of the canonical code.

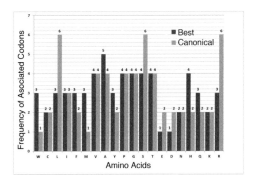

Fig. 4. The best code found by GA optimization when $a = 0.9$ and using $M_{st}(C)$, in comparison with the canonical genetic code

3.2 Statistical Approach and Frequency Code Analysis

Here, the results of the comparison of the canonical code with random codes using the proposed evaluation function with different values of a are presented. This experiment intends to find how many random codes are better than the canonical code when the evaluation function proposed in [16] (Eq. 3) and the proposed evaluation function (Eq. 5) are employed for different values of a. In the experiments presented here, 1 billion random codes are generated. The results are presented in Table 5. One can remember that, when a is equal to 0, the evaluation fitness depends only on the entropy, and as consequence, a large number of random codes have almost uniform distributions, generating a large number of random codes better than the canonical code. On the other hand, when a is equal to 1, the standard evaluation function, where only robustness is considered, is reproduced. In this case, similar results to those found in the literature were found in the experiments. One can observe that very few random codes are better than the canonical codes when a is in the interval $[0.4, 0.7]$ for the evaluation function proposed in [16], indicating the good quality of the evaluation function when compared to the standard evaluation function ($a = 1$).

One can still observe that, when mistranslation is used (proposed evaluation function), there is no random codes better than canonical code in the interval $[0.2, 1.0]$, indicating that the use of the error as a function of base position increases codes robustness. These statistical approach results reinforces the better results of the proposed evaluation function, when compared to the function presented in [16], also presented when the GA was employed (engineering approach).

In addition, the histograms for the evaluation value of 1 billion genetic codes randomly generated are presented in figures 5(a)-6(b) . The horizontal axis represents small ranges of evaluation values, while the vertical axis gives the number of random codes in each range. The evaluation value for the canonical genetic code is also presented. Figures 5(a) and 5(b) show the histograms for the evaluation function proposed in [16] (Eq. 3) for $a = 1.0$ and $a = 0.7$ (best value found), while figures 6(a) and 6(b) show the histograms for the proposed

Table 5. Number of random codes (over 1 billion codes) better than the canonical code using evaluation functions given by equations 3 and 5 with different values of a

(a)	Eq. 3	Eq. 5
0	778478253	778478253
0.1	11491417	58711
0.2	5583	0
0.3	75	0
0.4	7	0
0.5	6	0
0.6	4	0
0.7	5	0
0.8	9	0
0.9	13	0
1.0	18	0

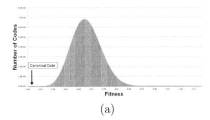
(a)

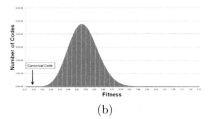
(b)

Fig. 5. a) Empirical distribution for the evaluation function proposed in [16] with $a = 0.7$. b) Empirical distribution for the evaluation function proposed in [16] with $a = 1.0$.

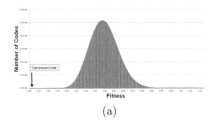
(a)

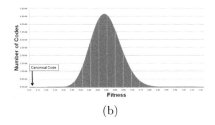
(b)

Fig. 6. a) Empirical distribution for the proposed evaluation function with $a = 0.9$. b) Empirical distribution for the proposed evaluation function with $a = 1.0$.

evaluation function (Eq. 5) for $a = 1.0$ and $a = 0.9$. Those results reinforce the importance of the engineering approach, as it allows identifying regions of the optimization space where best codes can be found, i.e., it allows to find the best codes without generating a huge number of random genetic codes.

4   Conclusion

In this paper, we to use base position errors in an entropy-based evaluation function for the study of genetic code adaptability. In the proposed evaluation function, robustness should be minimized while entropy should be maximized. Using both the engineering and statistical approach, better results were obtained with the proposed evaluation function when compared to the standard evaluation function based only on robustness and to the evaluation function proposed in [16]. It is important to observe that, despite closer results of the fitness were obtained between the best codes found by the GA and the canonical codes, the codes are different. In this way, a future work must investigate if the canonical genetic code represents a local optimum for the proposed evaluation function. Hence, another important future work is to investigate new evaluation functions. Additionally, a term to represents the noise of evolutionary process can be considered in the fitness function. Polar requirement seems to be a important property to be taken into account, but cannot not be the only. Other amino acids properties can be investigated using a multiobjective approach.

Acknowledgments. The authors would like to thank Fapesp and CNPq for the financial support to this work. Also, the authors are grateful to Prof. José Santos Reyes for his suggestions about the proposed work.

References

[1] Freeland, S.J.: The Darwinian genetic code: an adaptation for adapting? Genetic Programming and Evolvable Machines 2, 113–127 (2002)
[2] Lodish, H., Berk, A., Zipursky, S., Lawrence, S., Kaiser, C.A., Krieger, M., Scott, M.P., Bretscher, A., Ploegh, H., Matsudaira, P.: Molecular Cell Biology. W.H. Freeman (2007)
[3] Lehninger, A.L., Nelson, D.L., Cox, M.M.: Lehninger Principles of Biochemistry. W.H. Freeman (2005)
[4] Schoenauer, S., Clote, P.: How optimal is the genetic code. In: Proceedings of the German Conference on Bioinformatics (GCB 1997), pp. 65–67. IEEE Press, New York (1997)
[5] Woese, C.R.: On the evolution of the genetic code. Proceedings of the National Academy of Sciences of the United States of America 54, 1546–1552 (1965)
[6] Crick, F.H.: The origin of the genetic code. J. Mol. Biol. 38, 367–379 (1968)
[7] Di Giulio, M.: The extension reached by the minimization of the polarity distances during the evolution of the genetic code. Journal of Molecular Evolution 29, 288–293 (1989)
[8] Freeland, S.J., Hurst, L.D.: The genetic code is one in a million. Journal of Molecular Evolution 47, 238–248 (1998)
[9] Di Giulio, M.: The origin of the genetic code: theories and their relationships, a review. Biosystems 2, 175–184 (2005)
[10] Haig, D., Hurst, L.D.: A quantitative measure of error minimization in the genetic code. J. Mol. Biol. 33, 412–417 (1991)

[11] Knight, R.D., Freeland, S.J., Landweber, L.F.: Selection, history and chemistry: the three faces of the genetic code. Trends in Biochemical Sciences 24, 241–247 (1999)

[12] Santos, J., Monteagudo, Á.: Study of the genetic code adaptability by means of a genetic algorithm. Journal of Theoretical Biology 264, 854–865 (2010)

[13] Santos, J., Monteagudo, Á.: Simulated evolution applied to study the genetic code optimality using a model of codon reassignments. BMC Bioinformatics 12, 56 (2011)

[14] Di Giulio, M., Capobianco, M.R., Medugno, M.: On the optimization of the physicochemical distances between amino acids in the evolution of the genetic code. Journal of Theoretical Biology 168, 43–51 (1994)

[15] Goldman, N.: Further results on error minimization in the genetic code. Journal of Molecular Evolution 37, 662–664 (1993)

[16] Oliveira, L.L., Tinós, R.: Entropy-based evaluation function for the investigation of genetic code adaptability. In: BCB 2012: Proceedings of the ACM Conference on Bioinformatics, Computational Biology and Biomedicine, Orlando, pp. 558–560 (2012)

[17] Parker, J.: Errors and alternatives in reading the universal genetic code. Microbiology and Molecular Biology Reviews 53, 273 (1989)

An Adaptive Multi-Crossover Population Algorithm for Solving Routing Problems

E. Osaba, E. Onieva, R. Carballedo,
F. Diaz, and A. Perallos

Deusto Institute of Technology (DeustoTech), University of Deusto,
Av. Universidades 24, Bilbao 48007, Spain
{e.osaba,enrique.onieva,roberto.carballedo,
fernando.diaz,perallos}@deusto.es

Abstract. Throughout the history, Genetic Algorithms (GA) have been widely applied to a broad range of combinatorial optimization problems. Its easy applicability to areas such as transport or industry has been one of the reasons for its great success. In this paper, we propose a new Adaptive Multi-Crossover Population Algorithm (AMCPA). This new technique changes the philosophy of the basic genetic algorithms, giving priority to the mutation phase and providing dynamism to the crossover probability. To prevent the premature convergence, in the proposed AMCPA, the crossover probability begins with a low value, and varies depending on two factors: the algorithm performance on recent generations and the current generation number. Apart from this, as another mechanism to avoid premature convergence, our AMCPA has different crossover functions, which are used alternatively. We test the quality of our new technique applying it to three routing problems: the Traveling Salesman Problem (TSP), the Capacitated Vehicle Routing Problem (CVRP) and the Vehicle Routing Problem with Backhauls (VRPB). We compare the results with the ones obtained by a basic GA to conclude that our new proposal outperforms it.

Keywords: Adaptive Population Algorithm, Genetic Algorithm, Routing Problems, Combinatorial Optimization, Intelligent Transport Systems.

1 Introduction

Since its proposal in the '70s, genetic algorithm (GA) has become one of the most successful meta-heuristic techniques for solving combinatorial optimization problems. GAs are based on the genetic process of living organisms and in the law of the species evolution, proposed by Darwin. The basic principles of this technique were proposed by Holland [16], trying to imitate the natural selection process and the strongest specimens survival. Even though, its practical use for solving complex problems was shown later by De Jong [8] and Goldberg [12]. From that moment, GAs has been the focus of a large number of papers and

G. Terrazas et al. (eds.), *Nature Inspired Cooperative Strategies for Optimization*   113
(NICSO 2013), Studies in Computational Intelligence 512,
DOI: 10.1007/978-3-319-01692-4_9, © Springer International Publishing Switzerland 2014

books [1, 15], and they have been applied in a wide range of fields, like transport [24], software engineering [22] or industry [11].

In this paper, we present an Adaptive Multi-Crossover Population Algorithm (AMCPA) for solving routing problems. This new meta-heuristic is a variant of the basic GA. It prioritizes the local optimization (mutation), applying crossover operators only when they would be beneficial to the search process. In our AMCPA the crossover probability varies, depending on the search performance on recent generations and the current generation number. This dynamism helps our technique to prevent premature convergence. Apart from this, the proposed AMCPA has multiple crossover functions, which are applied alternatively.

Adjusting the control parameters of the GAs has always been one of the most controversial questions in the field of genetic algorithms. Related works have been done since the 80's [14] until today [9]. Concretely, the idea of adapting crossover and mutation probabilities (p_c and p_m) to improve the performance of GAs has been studied since long time ago, for example in [34] and [7], but it is also subject of many studies nowadays. Below were mentioned several examples of works on this topic, being the whole literature for this field much larger. In [37], for example, a genetic algorithm that adapts its p_c and p_m in function of the population fitness difference and the maximum fitness value is presented. In [45] and [46], a GA that uses fuzzy logic to adaptively tune p_c and p_m is introduced. In these proposals, a clustering technique is used to split the population in clusters. Then, a fuzzy system determines the p_c and p_m depending on the best and worst chromosome of each cluster. In [40] is proposed a GA that, besides adapting the p_m, determines the types of replacing genes in the mutation procedure. In [44] some improvements on adaptive GAs for reliability-related applications are introduced. In that work, the authors present a simple parameter-adjusting method, using the fitness average and variance of the population. In [43] an adaptive algorithm for optimizing the design of high pressure hydrogen storage vessel is presented. That algorithm adapts p_c and p_m depending on the fitness value of each individual. Finally, another example of adapting p_c and p_m is the one presented in [42]. In that work an improved adaptive genetic algorithm based on hormone modulation mechanism to solve the job-shop scheduling problem is proposed.

Regarding the multi-crossover, this mechanism has been used less than the previously explained one. Anyway, it has also been studied before, long time ago and nowadays. In [36], for example, an adapting crossover technique for an population algorithm is presented, which varies the crossover operator and its utilization frequency. The algorithm proposed in that work uses two crossovers functions depending on the population situation in the solution space. Another example is [25]. In this work a strategy adaptive genetic algorithm is proposed for solving the well-known Traveling Salesman Problem (TSP) [18]. This algorithm works with three different crossover functions. The choice of the function is decided partly by the quality of each of them and partly at random.

After a brief analysis of the state of the art, we detail the most innovative aspects of the new technique we propose:

- Our AMCPA reverses the philosophy of conventional GAs. It starts with a high values of p_m and a very low or null value for p_c. This fact is based on our previous work [28].
- Our proposal adapts its p_c depending on the current generation number and the search performance in recent iterations, instead of relying on the population fitness, as most previous studies.
- The proposed algorithm combines the p_c adaptation and the multi-crossover mechanism, something that has not been done frequently before.
- The introduced AMCPA is tested with routing problems. Traditionally, adaptive algorithm has not been applied to this type of problem.

This paper is structured as follows. In the following section we introduce our proposed technique. Then, we will show the results of our AMCPA applied to three different well known routing problems: TSP, Capacitated Vehicle Routing Problem (CVRP) [17] and Vehicle Routing Problem with Backhauls (VRPB) [13]. In the same section we compare the results obtained by our algorithm with the results of a basic GA. We finish this work with the conclusions and future work.

2   Our Adaptive Multi-Crossover Population Algorithm

As already mentioned, our AMCPA is a variant of a conventional GA. Its flowchart can be seen in figure 1. The proposed technique reverses the philosophy of conventional GAs, giving higher priority to the individual optimization,

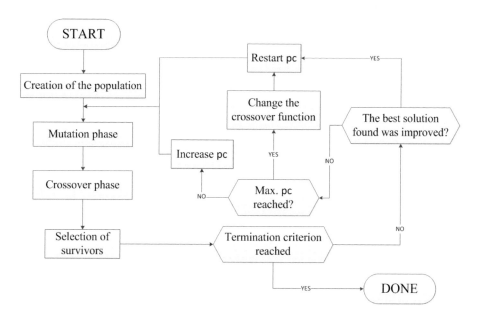

Fig. 1. Flowchart of the algorithm

provided by the mutation phase, and giving less importance to the crossovers phase. These fundamentals are based on our recently published study [28], in which we analyze the blind crossover suitability in GAs solving routing problems. In that work we check our theory, which stands that the crossover phase is not efficient for the optimization capacity of the technique when it is applied to routing problems using path encoding. For this reason, the proposed AMCPA offers a greater role to the mutation phase. Despite this, we consider that the crossovers between different individuals can be beneficial to maintain the population diversity. Therefore, in the proposed AMCPA we try to fit the p_c to the search process needs. Apart from that, as an additional tool to avoid the premature convergence, our AMCPA has a multi-crossover mechanism, which changes the crossover operator of the technique for all the population. These changes are made based on various concepts which will be explained later. Below, we will describe these mechanisms.

2.1   Adaptive Mechanism

Regarding the p_m, in our AMCPA, all individuals in the population go through the mutation process every generation. This would be equivalent to having a p_m equals to 1.0. On the other hand, in the proposed method the p_c starts with a very low value, close to 0.0. The latter parameter is modified as search progresses, increasing or restarting its value to 0. The modification is performed based on the improvement in the best solution found in the last generation. This modification is based on the following criteria:

- *The best solution found by the technique has been improved in the last generation*: This means that the search process evolves correctly and that it is not necessary to diversify the population. In this case, the value of p_c is restarted.
- *The best solution found by the technique has not been improved in the last generation*: In this case, it could be considered that the search is in a bump. This means that the search process could be trapped in a local optimum, or that the population could be concentrated in the same region of the solution space. At this time, increasing the population diversification using crossover operators would be beneficial. With this intention p_c is increased.

Whenever the best solution found has not been improved over the previous generation, p_c increases based on the following function, where N represents the number of generations executed without improvements, NG the total number of generations executed and NMF represents the size of the mutation operator neighborhood:

$$p_c = p_c + \frac{N^2}{NMF^2} + \frac{NG}{NMF^2}$$

As seen in the formula above, p_c increases proportionally to the total number of generations (NG) and the number of generations without any improvement in the best solution (N).

2.2   Multi-Crossover Mechanism

In relation to the multi-crossover feature, as we have already said, our AMCPA has more than one crossover operator which are alternated during the execution. At the beginning, one operator is assigned at random. Along the execution, this function will be randomly replaced by another available, allowing repetitions. For this purpose, a maximum p_c value is defined. If over the generations the p_c value exceeds that maximum, the crossover function will be replaced at random by another one, and p_c will be restarted with the initial value.

The maximum p_c value is an adjustable parameter, which has to be high enough to prevent a premature function change. Furthermore, to avoid an excessive runtime waste, the value cannot be too high.

This mechanism allows a diversification of the population much more efficient than other similar techniques. That is, prevent the algorithm from being trapped in a local optimum.

3   Experimentation

In this section we show in detail the results of applying our AMCPA three well-known combinatorial optimization problems. As we have mentioned, the technique proposed in this paper is a basic GA variation. For that reason, we compare the results obtained by a traditional GA, and our new AMCPA. For both algorithms we have used similar functions and parameters, so that the only difference between them is their working way. This method of comparing meta-heuristics is the most reliable way to determine which technique gets better results. The tests were performed with the three different problems that have been mentioned in the introduction: TSP, CVRP and VRPB. All these problems are well-known in combinatorial optimization and they are used in many studies annually [3, 4, 10, 19, 20, 23, 26, 33].

3.1   Parameters of the Algorithms

For both algorithms, the population is composed by 50 individuals, which are created randomly. The aim of this study is to make a comparison between our AMCPA and a GA, for that reason the population size is not very important, as long as the two meta-heuristics have the same. Regarding the selection and survivor phases, same function is used for both in all instances, which is the 0.5 elitist - 0.5 random. About the ending criteria, the execution of both algorithms finishes when there are a generation number proportional to the size of the neighborhood (obtained by the mutation operator) without improvements in the best solution found. The individuals encoding mode is the *Path Encoding*.

For the GA, the p_m is 0.05 while the p_c is 0.95. In the case of the proposed AMCPA, the p_c starts at 0.0. When the best solution found is not improved, the p_c increases following the formula shown in Section 2.1, otherwise, it returns to 0.0.

For the TSP, the crossover functions used for our AMCPA are Order Crossover (OX) [6], Modified Order Crossover (MOX) [30] and Order Based Crossover (OBX) [38]. These functions have been widely used since their creation [2, 29, 32, 35, 41]. On the other hand, OX is used as crossover function for the GA. The mutation function for both techniques is the 2-opt [21], which has been very used since its formulation [5, 39].

For the CVRP and VRPB, the crossover functions used for the proposed AM-CPA are the *Half Crossover* (HX) and *Half Random Crossover* (HRX). These functions are a particular case of the traditional crossover, in which the cut point is made always in the middle of the path. With HX, first, the 50% of the best routes in one randomly chosen parent are selected and inserted in the child. Then, the nodes already inserted are removed from the other parent. Finally, the remaining nodes are inserted in the same order in the final solution, creating new routes. The HRX working way is similar to HX. In this case, in the first step, the routes selected from one of the parents are chosen randomly, instead of selecting the best ones. For the GA the crossover function used is the HX.

Continuing with the CVRP and VRPB, regarding the mutation function, we have used for both techniques the called *Vertex Insertion Routes*. This function selects and extracts one random node from a random route. Then, the node is re-inserted in a random position in another randomly selected route. New routes creation is possible with this function.

3.2   Results

All the tests were performed on an Intel Core i5 2410 laptop, with 2.30 GHz and a 4 GB of RAM. For each run we display the total average, the best result obtained and the standard deviation. The objetive function used in the three problems is the total traveled distance. We also show the average runtime, in seconds. In order to determine if our AMCPA average is significantly different than the averages obtained by GA, we perform Students t-test. The t statistic has the following form [27]:

$$ t = \frac{\overline{X_1} - \overline{X_2}}{\sqrt{\frac{(n_1-1)SD_1^2+(n_2-1)SD_2^2}{n_1 n_2 - 2}} \frac{n_1+n_2}{n_1 n_2}} $$

where:

$\overline{X_1}$: Average of our AMCPA,
SD_1: Standar deviation of our AMCPA,
$\overline{X_2}$: Average of GA,
SD_2: Standar deviation of GA,
n_1: Our AMCPA size,
n_2: GA size,

The t values shown can be positive, neutral, or negative. The positive value of t indicates that our proposal is significantly better than GA. In the opposite case, GA obtains better solutions. If t is neutral, the difference between the two algorithms is not significant. We stated confidence interval at the 95% confidence level ($t_{0.05} = 2.021$).

Each experiment is repeated 20 times. Instances for the TSP were obtained from the TSPLIB Benchmark [31]. For the CVRP, the instances were picked from the CVRP set of Christofides and Eilon (*http://neo.lcc.uma.es/vrp*[1]). The name of each TSP and CVRP instances has a number that displays the number of nodes it has. Tables 1 and 2 show the results for these problems.

Table 1. Results of our AMCPA and GA for the TSP

Instance		Proposed AMCPA				Genetic Algorithm				t test
Name	Optima	Avg.	S. dev.	Best	Time	Avg.	S. dev.	Best	Time	t
Oliver30	420	**427.5**	4.4	420	0.08	435.3	15.3	420	0.19	+
Eilon50	425	**440.5**	6.6	430	0.42	469.9	17.5	435	1.59	+
Eil51	426	**445.0**	5.9	441	0.36	465.7	10.5	441	1.33	+
Berlin52	7542	**7805.2**	284.7	7542	0.29	8040.1	188.4	7745	1.36	+
St70	675	**706.5**	16.1	692	0.81	750.2	30.1	707	4.18	+
Eilon75	535	**575.0**	9.5	547	1.27	615.4	14.7	585	6.22	+
Eil76	538	**578.1**	12.2	566	1.28	610.6	12.2	558	6.96	+
KroA100	21282	**22125.3**	460.3	21608	2.25	22270.4	711.0	21566	14.84	+
KroB100	22140	**23043.7**	355.6	22536	2.28	23565.4	489.3	23253	13.54	+
KroC100	20749	**21550.8**	355.6	20785	2.40	22572.2	713.6	22271	16.52	+
KroD100	21294	**22125.5**	457.0	21725	2.25	23246.8	424.7	22162	12.92	+
KroE100	22068	**23196.7**	484.9	22611	2.54	23329.6	712.4	22412	12.72	+
Eil101	629	**678.1**	13.6	657	4.11	725.8	22.2	696	18.38	+
Pr107	44303	**45361.2**	953.1	44438	4.50	46742.2	1404.7	45833	18.21	+
Pr124	59030	**60578.6**	752.4	59030	6.86	62203.0	1223.3	60127	22.34	+
Pr136	96772	**101712.4**	1548.1	98125	7.42	104308.7	2425.1	99835	46.79	+
Pr144	58537	**60259.8**	1128.6	59061	9.57	62892.2	2552.9	60275	50.12	+
Pr152	73682	**76225.4**	1138.0	74518	10.32	77925.1	2862.3	74250	57.25	+

For the VRPB we have used 10 instances. The first 6 were obtained from the VRPTW Benchmark of Solomon (*http://neo.lcc.uma.es/vrp*). In this case, the time constraints have been removed, but vehicle capacities and the amount of customer demands are retained. Apart from this, we also have been modified the demands nature with the aim of creating pickup and deliveries. The remaining 4 instances were obtained from the CVRP set of Christofides and Eilon. In these instances, the vehicle capacities and the number of nodes have been maintained, but the demand types have been also changed to have pickups and deliveries. For these cases the optimums are not shown, since they are not typical VRPB instances, therefore, these values are unknown. The last table (table 3) shows the results for this problem.

---

[1] Last update: January 2013.

Table 2. Results of our AMCPA and GA for the CVRP

Instance		Proposed AMCPA				Genetic Algorithm				t test
Name	Optima	Avg.	S. dev.	Best	Time	Avg.	S. dev.	Best	Time	t
En22k4	375	395.6	7.1	375	2.08	**388.4**	15.1	375	3.78	-
En23k3	569	**611.8**	43.3	569	2.24	646.5	38.6	592	3.78	+
En30k3	534	**560.7**	25.8	534	3.02	570.8	25.1	535	6.74	+
En33k4	835	**903.5**	20.2	869	3.15	921.1	27.2	882	7.40	+
En51k5	521	**617.6**	27.2	587	4.56	680.8	47.1	604	18.17	+
En76k7	682	**813.0**	62.8	762	10.17	878.9	44.8	793	57.17	+
En76k8	735	**876.5**	32.1	819	10.67	953.4	46.5	920	53.86	+
En76k10	830	**965.8**	17.6	921	11.04	1029.6	34.4	956	55.27	+
En76k14	1021	**1170.7**	46.3	1135	7.39	1191.6	35.0	1125	80.54	+
En101k8	815	**1012.0**	59.6	916	17.95	1081.0	43.3	1011	100.25	+
En101k14	1071	**1272.6**	47.1	1201	18.60	1369.7	49.8	1308	120.84	+

Table 3. Results of our AMCPA and GA for the VRPB

Instance	Proposed AMCPA				Genetic Algorithm				t test
Name	Avg.	S. dev.	Best	Time	Avg.	S. dev.	Best	Time	t
C101	724.4	41.9	627	8.25	**723.0**	71.9	624	34.93	*
C201	**652.4**	12.4	617	4.04	849.9	98.8	744	25.43	+
R101	**962.2**	38.7	875	5.86	1081.2	70.7	957	29.51	+
R201	**1105.2**	41.7	1021	12.54	1335.7	112.4	1224	56.94	+
RC101	**595.2**	47.1	529	2.13	659.4	64.2	563	5.45	+
RC201	**1221.6**	90.8	1167	18.35	1505.3	92.2	1367	57.59	+
En30k4	**534.5**	27.8	500	1.57	583.4	67.0	520	3.53	+
En33k4	**812.9**	45.8	787	1.85	848.9	44.4	751	4.64	+
En51k5	**688.1**	34.8	636	4.01	727.6	30.6	680	9.41	+
En76k8	**912.8**	43.5	798	7.55	1008.8	40.5	927	28.61	+

3.3 Analysis of Results

Viewing the results obtained the conclusion that can be drawn is clear. The proposed technique outperforms the GA in terms of solution quality and runtimes. The reason why our algorithm needs lower runtime is logical. If mutation and crossover functions are compared, the last ones needs more time to execute, since they operate with two different solutions, and their working way is more complex that the mutation. On the other hand, the mutation operates with one solution and it is a simple modification in a chromosome which can be made in a minimum time. Our AMCPA makes fewer crossovers than the GA. This fact is perfectly reflected in the runtimes, giving a great advantage to our technique.

The reason why the proposed AMCPA gets better results can also be explained, and it is based on the conclusions obtained in our recent study [28]. Crossovers between different individuals are very useful resources if we want to make jumps in the solution space. Using crossovers helps a broad exploration of the solution space, but does not help to make an exhaustive search. To get a deeper search, the existence of a function that takes care of optimizing the solutions independently becomes necessary. The mutation function can handle this goal easily.

With all this, our AMCPA is a technique that is able to perform a thorough and intense search in promising regions of the solution space using the mutation function. While it do this, it uses the crossover function in case the search is in a bump, in order to avoid local optimums. Using the crossovers, the current population is expanded through the entire solution space, and will be easier to find regions that allow the search to reach better results. This diversification is enhanced thanks to the multi-crossover, allowing a broader exploration.

By contrast, with the GA basic structure, the search performed by the algorithm comprises a large percentage of the solution space, but has a smaller capacity to deepen in those areas which are most promising. This means that, finally, the GA obtains worse results than the AMCPA.

4   Conclusions and Further Work

In this paper we have presented an Adaptive Multi-Crossover Population Algorithm for solving routing problems, which is a variation of the conventional genetic algorithm. Our AMCPA reverses GAs conventional philosophy, giving priority to the individual autonomous improvement, making crossovers only when they are beneficial for the search process. The proposed technique has two mechanisms to avoid the premature convergence, helping to the population diversity. These mechanisms are the crossover probability adaption and the use of multiple crossover operators.

Initially we have introduced our new meta-heuristic, explaining how it works. Then, we have shown the results obtained by applying it to three different routing problems. We have compared these outcomes with obtained by a basic GA, to conclude that our method gets better results. Finally, we have reasoned why our new technique is better than the GA.

As future work, we will compare the performance of our technique with other approaches of similar philosophy that we can find in the literature. In addition, we are planning to apply our new proposal to real life routing problems. At this time, we are planning its application to a dynamic distribution system of car windscreen repairs. In this case the problem is designed as a dynamic CVRP, wherein the routes may be re-planned according to the needs of the customers. Apart from this, we are planning to extend our technique, turning it into an island-based meta-heuristic. This new technique will use different crossover functions for the different populations evolving in each island, and will make transfers of individuals between them.

References

[1] Affenzeller, M., Wagner, S., Winkler, S.: Genetic algorithms and genetic programming: modern concepts and practical applications, vol. 6. Chapman & Hall/CRC (2009)

[2] Albayrak, M., Allahverdi, N.: Development a new mutation operator to solve the traveling salesman problem by aid of genetic algorithms. Expert Systems with Applications 38(3), 1313–1320 (2011)

[3] Anbuudayasankar, S., Ganesh, K., Lenny Koh, S., Ducq, Y.: Modified savings heuristics and genetic algorithm for bi-objective vehicle routing problem with forced backhauls. Expert Systems with Applications 39(3), 2296–2305 (2012)

[4] Bae, J., Rathinam, S.: Approximation algorithms for multiple terminal, hamiltonian path problems. Optimization Letters 6(1), 69–85 (2012)

[5] Bianchessi, N., Righini, G.: Heuristic algorithms for the vehicle routing problem with simultaneous pick-up and delivery. Computers & Operations Research 34(2), 578–594 (2007)

[6] Davis, L.: Applying adaptive algorithms to epistatic domains. In: Proceedings of the International Joint Conference on Artificial Intelligence, vol. 1, pp. 161–163 (1985)

[7] Davis, L.: Adapting operator probabilities in genetic algorithms. In: Proceeding of the Third International Conference on Genetic Algorithms, pp. 61–69 (1989)

[8] De Jong, K.: Analysis of the behavior of a class of genetic adaptive systems. PhD thesis, University of Michigan, Michigan, USA (1975)

[9] Fernandez-Prieto, J., Gadeo-Martos, M., Velasco, J.R., et al.: Optimisation of control parameters for genetic algorithms to test computer networks under realistic traffic loads. Applied Soft Computing 11(4), 3744–3752 (2011)

[10] Gajpal, Y., Abad, P.: Multi-ant colony system (macs) for a vehicle routing problem with backhauls. European Journal of Operational Research 196(1), 102–117 (2009)

[11] Gao, J., Gen, M., Sun, L., Zhao, X.: A hybrid of genetic algorithm and bottleneck shifting for multiobjective flexible job shop scheduling problems. Computers & Industrial Engineering 53(1), 149–162 (2007)

[12] Goldberg, D.: Genetic algorithms in search, optimization, and machine learning. Addison-Wesley Professional (1989)

[13] Golden, B., Baker, E., Alfaro, J., Schaffer, J.: The vehicle routing problem with backhauling: two approaches. In: Proceedings of the Twenty-first Annual Meeting of SE TIMS, South Carolina, USA, pp. 90–92 (1985)

[14] Grefenstette, J.J.: Optimization of control parameters for genetic algorithms. IEEE Transactions on Systems, Man and Cybernetics 16(1), 122–128 (1986)

[15] Harvey, I.: The microbial genetic algorithm. In: Kampis, G., Karsai, I., Szathmáry, E. (eds.) ECAL 2009, Part II. LNCS, vol. 5778, pp. 126–133. Springer, Heidelberg (2011)

[16] Holland, J.H.: Adaptation in natural and artificial systems: an introductory analysis with applications to biology, control, and artificial intelligence. MIT Press (1975)

[17] Laporte, G.: The vehicle routing problem: An overview of exact and approximate algorithms. European Journal of Operational Research 59(3), 345–358 (1992)

[18] Lawler, E., Lenstra, J., Kan, A., Shmoys, D.: The traveling salesman problem: a guided tour of combinatorial optimization, vol. 3. Wiley, New York (1985)

[19] Li, W., Shi, Y.: On the maximum tsp with γ-parameterized triangle inequality. Optimization Letters 6(3), 415–420 (2012)

[20] Liefooghe, A., Humeau, J., Mesmoudi, S., Jourdan, L., Talbi, E.: On dominance-based multiobjective local search: design, implementation and experimental analysis on scheduling and traveling salesman problems. Journal of Heuristics 18(2), 317–352 (2012)

[21] Lin, S.: Computer solutions of the traveling salesman problem. Bell System Technical Journal 44(10), 2245–2269 (1965)

[22] Martínez-Torres, M.: A genetic search of patterns of behaviour in oss communities. Expert Systems with Applications 39(18), 13,182–13,192 (2012)

[23] Mattos Ribeiro, G., Laporte, G.: An adaptive large neighborhood search heuristic for the cumulative capacitated vehicle routing problem. Computers & Operations Research 39(3), 728–735 (2012)

[24] Moon, I., Lee, J.H., Seong, J.: Vehicle routing problem with time windows considering overtime and outsourcing vehicles. Expert Systems with Applications 39(18), 13,202–13,213 (2012)

[25] Mukherjee, S., Ganguly, S., Das, S.: A strategy adaptive genetic algorithm for solving the travelling salesman problem. In: Panigrahi, B.K., Das, S., Suganthan, P.N., Nanda, P.K. (eds.) SEMCCO 2012. LNCS, vol. 7677, pp. 778–784. Springer, Heidelberg (2012)

[26] Ngueveu, S., Prins, C., Wolfler Calvo, R.: An effective memetic algorithm for the cumulative capacitated vehicle routing problem. Computers & Operations Research 37(11), 1877–1885 (2010)

[27] Nikolić, M., Teodorović, D.: Empirical study of the bee colony optimization (bco) algorithm. Expert Systems with Applications 40(1), 4609–4620 (2013)

[28] Osaba, E., Carballedo, R., Diaz, F., Perallos, A.: Analysis of the suitability of using blind crossover operators in genetic algorithms for solving routing problems. In: Proceedings of the 8th International Symposium on Applied Computational Intelligence and Informatics, pp. 17–23. IEEE (2013)

[29] Prins, C.: A simple and effective evolutionary algorithm for the vehicle routing problem. Computers & Operations Research 31(12), 1985–2002 (2004)

[30] Ray, S., Bandyopadhyay, S., Pal, S.: New operators of genetic algorithms for traveling salesman problem. In: Proceedings of the 17th International Conference on Pattern Recognition, vol. 2, pp. 497–500. IEEE (2004)

[31] Reinelt, G.: Tsplib: A traveling salesman problem library. ORSA Journal on Computing 3(4), 376–384 (1991)

[32] Rocha, M., Sousa, P., Cortez, P., Rio, M.: Quality of service constrained routing optimization using evolutionary computation. Applied Soft Computing 11(1), 356–364 (2011)

[33] Sarin, S.C., Sherali, H.D., Yao, L.: New formulation for the high multiplicity asymmetric traveling salesman problem with application to the chesapeake problem. Optimization Letters 5(2), 259–272 (2011)

[34] Schaffer, J.D., Morishima, A.: An adaptive crossover distribution mechanism for genetic algorithms. In: Proceedings of the Second International Conference on Genetic Algorithms on Genetic algorithms and Their Application, pp. 36–40. L. Erlbaum Associates Inc. (1987)

[35] Sharma, S., Gupta, K.: Solving the traveling salesmen problem through genetic algorithm with new variation order crossover. In: International Conference on Emerging Trends in Networks and Computer Communications, pp. 274–276. IEEE (2011)

[36] Spears, W.M.: Adapting crossover in evolutionary algorithms. In: Proceedings of the Conference on Evolutionary Programming, pp. 367–384 (1995)

[37] Srinivas, M., Patnaik, L.M.: Adaptive probabilities of crossover and mutation in genetic algorithms. IEEE Transactions on Systems, Man and Cybernetics 24(4), 656–667 (1994)

[38] Syswerda, G.: Schedule optimization using genetic algorithms. In: Handbook of Genetic Algorithms, pp. 332–349 (1991)

[39] Tarantilis, C., Kiranoudis, C.: A flexible adaptive memory-based algorithm for real-life transportation operations: Two case studies from dairy and construction sector. European Journal of Operational Research 179(3), 806–822 (2007)

[40] Vafaee, F., Nelson, P.C.: A genetic algorithm that incorporates an adaptive mutation based on an evolutionary model. In: Proceedings of the International Conference on Machine Learning and Applications, pp. 101–107. IEEE (2009)

[41] Wang, C., Zhang, J., Yang, J., Hu, C., Liu, J.: A modified particle swarm optimization algorithm and its application for solving traveling salesman problem. In: Proceedings of the International Conference on Neural Networks and Brain, vol. 2, pp. 689–694. IEEE (2005)

[42] Wang, L., Tang, D.: An improved adaptive genetic algorithm based on hormone modulation mechanism for job-shop scheduling problem. Expert Systems with Applications 38(6), 7243–7250 (2011)

[43] Xu, P., Zheng, J., Chen, H., Liu, P.: Optimal design of high pressure hydrogen storage vessel using an adaptive genetic algorithm. International Journal of Hydrogen Energy 35(7), 2840–2846 (2010)

[44] Ye, Z., Li, Z., Xie, M.: Some improvements on adaptive genetic algorithms for reliability-related applications. Reliability Engineering & System Safety 95(2), 120–126 (2010)

[45] Zhang, J., Chung, H.S., Zhong, J.: Adaptive crossover and mutation in genetic algorithms based on clustering technique. In: Proceedings of the Conference on Genetic and Evolutionary Computation, pp. 1577–1578. ACM (2005)

[46] Zhang, J., Chung, H.S., Lo, W.L.: Clustering-based adaptive crossover and mutation probabilities for genetic algorithms. IEEE Transactions on Evolutionary Computation 11(3), 326–335 (2007)

Corner Based Many-Objective Optimization

Hélio Freire, P.B. de Moura Oliveira, E.J. Solteiro Pires, and Maximino Bessa

INESC TEC - INESC Technology and Science (formerly INESC Porto, UTAD pole)
Department of Engineering, School of Sciences and Technology,
5001-801 Vila Real, Portugal
freireh@gmail.com, {oliveira,epires,maxbessa}@utad.pt

Abstract. The performance of multi-objective evolutionary algorithms (MOEA) is severely deteriorated when applied to many-objective problems. For Pareto dominance based techniques, available information about optimal solutions can be used to improve their performance. This is the case of corner solutions. This work considers the behaviour of three multi-objective algorithms (NSGA-II, SMPSO and GDE3) when corner solutions are inserted into the population at different evolutionary stages. Corner solutions are found using specific algorithms. Preliminary results are presented concerning the behaviour of the aforementioned algorithms in five benchmark problems (DTLZ1-5).

1 Introduction

In last three decades MOEA have been proposed to solve real-world problems [1–3], involving the simultaneous optimization of several objectives. Most of the objectives are in conflict with each other, meaning that the improvement of one objective value will deteriorate some other objective values. The overall goal is to obtain a set of non-dominated optimal solutions with both good convergence to the true Pareto front and good solution diversity, representing different trade-off among objectives. Some well-known algorithms used in MOEA are the Non-Dominated Sorting Genetic Algorithm (NSGA-II) [4], the Speed-Constrained Multi-objective Particle Swarm Optimization (SMPSO) [5] and the Generalized Differential Evolution (GDE3) [6]. MOEA have been successfully applied in solving problems with two or three objectives [1, 2]. However, when these algorithms are applied to many-objective problems (more than three objectives) [7] some difficulties may arise due to the significant increase of the search complexity [3, 8]. In the next section a perspective of main issues inherent to the evolutionary based many-objective optimization is presented.

An approach based on the introduction of informed solutions in the beginning of the search procedure has been proposed by Dasgupta *et al.* [9], in order to improve many-objective optimization algorithms performance. Examples of informed solutions are the extreme and internal points approximated to the Pareto front. This informed solutions can be found by specific algorithms [9–12]. While the results presented by [9] indicate improvements in the many-objective optimization performance, there are still some open research issues. One of this

G. Terrazas et al. (eds.), *Nature Inspired Cooperative Strategies for Optimization*   125
(NICSO 2013), Studies in Computational Intelligence 512,
DOI: 10.1007/978-3-319-01692-4_10, © Springer International Publishing Switzerland 2014

issues concerns the study of the most appropriate evolution stage to introduce the informed solutions into the search procedure. Indeed, Dasgupta *et al.* [9] solely considered the introduction of informed solutions in the beginning of the search. The introduction of these solutions in a early search stage could lead to an over-dominance effect, preventing a proper initial exploration of the search space. Recently a corner based approach has been proposed to reduce the dimensionality of the many-objective problems [10]. Several techniques based on the Nadir point determination have been proposed, which can be used to determine the corner (or approximated) solutions [8, 13, 14].

The influence of inserting corner solutions in the population at different evolutionary times is the overall objective of this study. Preliminary results regarding the use of corner solutions in different evolutionary stages of the many-objective optimization process are presented. Three different paradigms of evolutionary computation are considered: genetic algorithms, particle swarm optimization and differential evolution through three well known multi-objective algorithms, namely: NSGA-II, SMPSO and GDE3.

The remaining of this article is organized as follows: Section 2 presents introductory notions regarding many-objective optimization; Section 3 presents the problem statement and the research hypothesis under study; Section 4 describes the algorithms, test functions and metrics used. Section 5 presents preliminary results and analysis. Finally, Section 6 concludes the paper and outlines future work.

2   Many-Objective: Introductory Concepts

Many real world engineering applications require solving search and optimization problems with more than three objectives [2]. The majority of MOEA use the domination principle, which together with the concept of diversity maintenance are many-objective MOEA low performance major causes [8]. These challenges motivated significant research efforts in addressing many-objective optimization in the last decade [2, 10, 15]. However despite these efforts, as it will be overviewed for evolutionary based techniques, there are still many open research questions. After the motivation for addressing many-objective optimization problems has been presented, a relevant question is: what makes a many-optimization problem difficult? Some of the key points providing answers to this question are:

- Many-objective problems have an intrinsic tendency to achieve populations in which all (or nearly all) elements represent non-dominated solutions [16] at early search stage. Thus, Pareto based selection pressure is seriously compromised and consequently their convergence properties. This has been typified as deterioration of search ability problem [3, 8].
- A huge number of solutions is necessary to approximate the Pareto front to represent a higher dimensionality. This problem can be attenuated by increasing the population and archive size. However, it has been shown [3]

that there is not an archiving technique which can store all the necessary solutions to characterize the Pareto front, due to the exponential growing factor and considerable increase of the computational effort.

- The visualization process allowing the decision maker to select the non-dominated solution which best suits its needs, is much more difficult for many-optimization techniques [17].

2.1 Approaches to Deal with Many-Objective Problems

Several approaches have been proposed to deal with many-objective problems, which can be classified according to the following categories:

- **Preference-ordering based techniques.** The basic idea is to try to distinguish non-dominated solutions using fitness assignment techniques which improve the many-objective search efficiency, in a way their optimal solutions set is a subset of the Pareto optimal set. Examples of modified non-dominated ranking techniques are [16, 18–20].
- **Objective or dimensionality reduction.** Some high-dimensional problems can be reduced to fewer objectives. In cases where one or more objectives are redundant these can be discarded [10, 21–24].
- **Preference information.** Considering the difficulties inherent to many-objective optimization problems, particularly in representing adequately all Pareto front, some approaches have been proposed which are directed to detect particular regions or points of the Pareto front. The preference information can be incorporated in the many-objective problems by the decision maker, so the algorithms will be focused in a region of interest [25, 26]. Algorithms can also be guided by Pareto front to specific regions, like knee points [11, 12].

The first category gives more importance to some non-dominated solutions when the Pareto dominance considers all non-dominated solutions equally important. For the second category, it is observed that many problems with more than tree objectives have objectives that are not in conflict between them or some objectives are not so important. Therefore, it is possible to exclude those objectives from the problem and if it has less than four objectives, can either be solved using MOEA, or can be a first step to the problem solving with other technique such as preference information [26]. In the preference information category, information can be provided interactively during the search, and often objectives lose relevance throughout the search. The information can also be provided prior to the algorithm execution.

3 Problem Statement

Several studies have explored the use of population initialization techniques to improve the search in many-objective optimization. This is the case of Gutiérrez *et al.*[27]. An approach that appears obvious in improving the optimization procedure is based on the use of optimal solutions or near-optimal solutions in the initial population. This approach was explored by Dasgupta *et al.* [9] and termed informed initialization. The overall goal is to use extreme and internal Pareto front points to improve the search procedure obtained previously using a fast convergence aggregated single objective technique. Preliminary results presented in Dasgupta *et al.* [9] indicate that classical algorithms such as NSGA-II can benefit from the inclusion of informed solutions in the initial population.

Following this line of thought the hypotheses under study here concerns the introduction of corner solutions in different evolutionary stages to evaluate their influence in the MOEA performance.

4   Algorithms, Test Functions and Performance Metrics

4.1   Algorithms

Three well known MOEA were selected to verify if the introduction of corner solutions in the population improves the final result, using the implementation provided by the jMetal framework [28]:

- NSGA-II [4] algorithm is one of the oldest MOEA, with great popularity. It is a genetic algorithm based on Pareto ranking scheme and crowding distance operator.
- SMPSO [5] is a multi-objective particle swarm optimization (PSO) algorithm, based on the OMOPSO [29] algorithm, and like NSGA-II use the crowding distance to choose the archive solutions, however it only considers first rank solutions. It also uses a mutation operator and includes ϵ-dominance.
- GDE3 [6] is a multi-objective variant of differential evolution (DE). GDE3, notable for rotationally invariant operators[30], which produce offspring independent of the orientation of the fitness landscape.

4.2   Test Functions

The test functions, used are DTLZ1-5 [31], which have some important features: i)relatively small implementation effort; ii) can be scaled to any number of decision variables and objectives; iii) the optimal Pareto front is known analytically;

This study considers four to ten objectives. The number of variables and other parameters are used as defined by [31].

4.3   Performance Metrics

The following metrics are deployed to evaluate the results convergence and diversity.

Generational Distance (GD): The concept of generational distance [32] is used to estimate how far elements in the non-dominated solutions set are from those in the optimal Pareto set. The DTLZ1-5 functions Pareto surfaces are easily expressed analytically, and therefore GD is computed as the expected value of the distance between each individual in a population and the Pareto surface. For DTLZ1 problem the equation 1 is used and for DTLZ2-5 problems the following equation 2 is used:

$$GD = \sum_{i=1}^{m} f_i - 0.5, \qquad (1)$$

$$GD = \sum_{i=1}^{m} f_i^2 - 1 \qquad (2)$$

Spacing (SP): Spacing [33] is used to measure the neighboring solutions range variance in the known Pareto front. It is defined as:

$$SP = \sqrt{\frac{1}{n-1} \sum_{i=1}^{n} (\bar{d} - d_i)^2} \qquad (3)$$

where $d_i = \min_j (\sum_{k=1}^{m} |f_m^i - f_m^j|, i, j = 1, ..., n$ and n is the number of non-dominated solutions generated by the algorithm; m is the number of objectives; $\bar{d}$ is the mean of all d_i. A value of zero for this measure indicates that all the non-dominated solutions found are equidistantly spaced.

5   Results and Discussion

The overall goal is to understand how corner solutions inclusion into the population affects algorithms performance. For each combination problem/number of objectives algorithms are run 21 times. Each algorithm gives origin to five test variants. The first is the standard algorithm itself (represented by letter 'N' for NSGAII, 'G' for GDE3 and 'S' for SMPSO). The second is the algorithm when corners are inserted in the initial population, represented by 'N0', 'G0' and 'S0'. In the next three cases corners are inserted at 25%, 50% and 75% of the predefined total number of iterations, respectively referred as: 'N25', 'N50' and 'N75' for NSGA-II algorithm, 'G25', 'G50' and 'G75' for GDE3 algorithm and finally 'S25', 'S50' and 'S75' for SMPSO algorithm.

The population size, the external archive size and the maximum number of iterations are set using the following heuristics:

- Population size = 80 + number of objectives × 10;
- Archive size = Population size;
- Number of iterations = number of objectives × 100.

Table 1. SMPSO Median GD

Problems	Algorithms	Objective Number						
		4	5	6	7	8	9	10
DTLZ1	S	1.64e-01	4.63e-01	1.19e+01	3.03e+01	5.27e+01	5.42e+01	6.89e+01
	S0	**1.20e-01**	2.69e-01	4.21e-01	5.76e-01	7.21e-01	8.77e-01	1.05e+00
	S25	1.23e-01	2.54e-01	4.16e-01	5.70e-01	7.17e-01	**8.24e-01**	1.03e+00
	S50	1.26e-01	2.55e-01	**4.05e-01**	5.60e-01	7.22e-01	9.07e-01	1.05e+00
	S75	1.24e-01	**2.54e-01**	4.12e-01	**5.39e-01**	**7.16e-01**	8.61e-01	**9.71e-01**
DTLZ2	S	1.73e-01	4.90e-01	8.21e-01	1.16e+00	1.41e+00	1.95e+00	2.15e+00
	S0	1.37e-01	**2.66e-01**	**4.42e-01**	6.29e-01	**7.91e-01**	**9.57e-01**	1.17e+00
	S25	1.39e-01	2.75e-01	4.43e-01	6.47e-01	8.24e-01	9.88e-01	**1.17e+00**
	S50	1.32e-01	2.86e-01	4.76e-01	6.20e-01	8.00e-01	9.89e-01	1.21e+00
	S75	**1.30e-01**	2.81e-01	4.58e-01	6.20e-01	8.25e-01	9.82e-01	1.19e+00
DTLZ3	S	2.67e-01	2.76e+00	1.35e+04	4.01e+04	3.86e+04	4.67e+04	5.14e+04
	S0	1.32e-01	3.12e-01	5.43e-01	7.02e-01	9.01e-01	1.11e+00	**1.27e+00**
	S25	1.29e-01	3.20e-01	5.04e-01	7.21e-01	9.00e-01	1.15e+00	1.30e+00
	S50	**1.20e-01**	3.08e-01	4.92e-01	**6.93e-01**	9.21e-01	1.09e+00	1.28e+00
	S75	1.25e-01	**2.70e-01**	**4.76e-01**	6.99e-01	**8.73e-01**	**1.09e+00**	1.30e+00
DTLZ4	S	3.94e-02	**5.43e-02**	**7.58e-02**	1.09e-01	1.51e-01	**1.73e-01**	1.81e-01
	S0	1.07e-01	2.12e-01	3.46e-01	4.91e-01	6.08e-01	8.00e-01	8.75e-01
	S25	1.05e-01	1.94e-01	2.85e-01	4.26e-01	5.56e-01	6.62e-01	5.43e-01
	S50	6.12e-02	1.05e-01	1.58e-01	2.47e-01	2.27e-01	2.82e-01	2.72e-01
	S75	**3.47e-02**	6.79e-02	8.19e-02	1.11e-01	**1.36e-01**	1.76e-01	**1.76e-01**
DTLZ5	S	2.62e+00	4.86e+00	6.28e+00	7.82e+00	8.44e+00	8.51e+00	9.09e+00
	S0	**2.57e+00**	4.84e+00	**6.03e+00**	7.92e+00	**8.19e+00**	8.79e+00	8.99e+00
	S25	2.63e+00	4.87e+00	6.19e+00	**7.54e+00**	8.44e+00	8.71e+00	**8.90e+00**
	S50	2.65e+00	4.86e+00	6.35e+00	7.74e+00	8.33e+00	8.63e+00	9.21e+00
	S75	2.59e+00	**4.75e+00**	6.21e+00	7.65e+00	8.25e+00	**8.49e+00**	9.09e+00

Table 2. SMPSO Median Spacing

Problems	Algorithms	Objective Number						
		4	5	6	7	8	9	10
DTLZ1	S	5.35e-02	1.02e-01	3.71e+00	4.02e+00	7.20e+00	6.87e+00	1.16e+01
	S0	4.76e-02	7.82e-02	1.08e-01	1.29e-01	**1.48e-01**	**1.76e-01**	1.96e-01
	S25	**4.56e-02**	**7.28e-02**	1.04e-01	**1.29e-01**	1.61e-01	1.81e-01	2.05e-01
	S50	4.81e-02	7.29e-02	**1.01e-01**	1.33e-01	1.56e-01	1.87e-01	1.99e-01
	S75	4.86e-02	7.42e-02	1.04e-01	1.30e-01	1.53e-01	1.80e-01	**1.96e-01**
DTLZ2	S	**1.04e-01**	1.80e-01	2.59e-01	3.22e-01	3.77e-01	4.53e-01	4.96e-01
	S0	1.08e-01	**1.56e-01**	2.13e-01	2.65e-01	3.30e-01	3.86e-01	4.23e-01
	S25	1.08e-01	1.62e-01	**2.11e-01**	2.66e-01	**3.24e-01**	**3.75e-01**	4.12e-01
	S50	1.07e-01	1.58e-01	2.20e-01	**2.63e-01**	3.30e-01	3.82e-01	4.26e-01
	S75	1.05e-01	1.60e-01	2.24e-01	2.74e-01	3.26e-01	3.82e-01	4.16e-01
DTLZ3	S	1.16e-01	2.96e-01	2.98e+01	4.09e+01	4.61e+01	5.11e+01	5.57e+01
	S0	1.03e-01	1.63e-01	2.36e-01	**2.66e-01**	3.27e-01	**3.67e-01**	**4.11e-01**
	S25	1.07e-01	**1.58e-01**	2.22e-01	2.81e-01	3.22e-01	3.84e-01	4.19e-01
	S50	**1.01e-01**	1.65e-01	2.23e-01	2.72e-01	**3.24e-01**	3.72e-01	4.20e-01
	S75	1.06e-01	1.65e-01	**2.07e-01**	2.83e-01	**3.20e-01**	3.84e-01	4.22e-01
DTLZ4	S	**8.53e-02**	**1.26e-01**	**1.56e-01**	**1.74e-01**	**1.95e-01**	**2.28e-01**	**2.20e-01**
	S0	1.01e-01	1.52e-01	2.08e-01	2.57e-01	3.06e-01	3.75e-01	4.06e-01
	S25	1.09e-01	1.47e-01	1.95e-01	2.51e-01	3.02e-01	3.43e-01	3.45e-01
	S50	9.48e-02	1.46e-01	1.82e-01	1.99e-01	2.39e-01	2.80e-01	2.72e-01
	S75	9.32e-02	1.29e-01	1.59e-01	1.85e-01	2.03e-01	2.46e-01	2.30e-01
DTLZ5	S	**1.13e-01**	2.35e-01	3.49e-01	4.05e-01	4.71e-01	5.18e-01	5.76e-01
	S0	1.16e-01	**2.27e-01**	3.34e-01	4.22e-01	**4.51e-01**	5.14e-01	**5.42e-01**
	S25	1.13e-01	2.34e-01	3.23e-01	4.10e-01	4.86e-01	5.30e-01	5.70e-01
	S50	1.15e-01	2.29e-01	**3.17e-01**	4.02e-01	4.79e-01	5.07e-01	5.76e-01
	S75	1.13e-01	2.32e-01	3.21e-01	**3.99e-01**	4.59e-01	**5.02e-01**	5.65e-01

Table 3. NSGAII Median GD

Problems	Algorithms	Objective Number						
		4	5	6	7	8	9	10
DTLZ1	N	9.14e-02	2.21e+02	3.66e+02	4.23e+02	4.59e+02	4.71e+02	4.83e+02
	N0	3.08e-03	6.66e-02	2.55e-01	4.17e-01	5.99e-01	7.56e-01	8.97e-01
	N25	**1.87e-03**	5.84e-02	2.62e-01	4.38e-01	5.99e-01	7.78e-01	8.87e-01
	N50	5.74e-03	6.14e-02	2.35e-01	4.13e-01	6.12e-01	7.52e-01	9.10e-01
	N75	9.93e-03	**2.56e-02**	**1.84e-01**	**3.91e-01**	**5.72e-01**	**7.35e-01**	**8.85e-01**
DTLZ2	N	6.44e-02	1.33e+00	6.32e+00	8.47e+00	9.30e+00	9.64e+00	9.91e+00
	N0	4.01e-02	9.61e-02	**2.12e-01**	4.74e-01	7.26e-01	**9.61e-01**	1.29e+00
	N25	**3.44e-02**	9.14e-02	2.21e-01	4.61e-01	7.43e-01	9.99e-01	1.30e+00
	N50	4.01e-02	**8.14e-02**	2.24e-01	4.64e-01	**7.07e-01**	1.01e+00	**1.26e+00**
	N75	3.50e-02	8.97e-02	2.15e-01	**4.52e-01**	7.32e-01	9.97e-01	1.27e+00
DTLZ3	N	1.28e+02	6.99e+05	1.81e+06	2.52e+06	3.15e+06	3.59e+06	3.71e+06
	N0	3.23e-03	3.77e-02	1.77e-01	4.27e-01	6.78e-01	9.03e-01	1.22e+00
	N25	1.64e-03	4.07e-02	1.71e-01	4.51e-01	6.99e-01	9.59e-01	1.17e+00
	N50	**1.35e-03**	2.98e-02	1.77e-01	3.90e-01	7.21e-01	9.44e-01	1.18e+00
	N75	3.43e-03	**1.40e-02**	**8.92e-02**	**3.51e-01**	**6.20e-01**	**8.84e-01**	**1.14e+00**
DTLZ4	N	4.73e-02	2.00e-01	2.09e+00	6.44e+00	8.21e+00	9.13e+00	9.40e+00
	N0	**3.68e-02**	9.09e-02	2.22e-01	4.51e-01	6.97e-01	9.52e-01	1.14e+00
	N25	3.90e-02	9.30e-02	2.10e-01	4.55e-01	6.85e-01	9.84e-01	1.19e+00
	N50	3.71e-02	9.53e-02	2.18e-01	4.36e-01	**6.74e-01**	9.57e-01	1.17e+00
	N75	3.94e-02	**9.08e-02**	**2.09e-01**	**4.34e-01**	6.77e-01	**9.27e-01**	**1.14e+00**
DTLZ5	N	2.69e+00	4.69e+00	6.26e+00	7.86e+00	8.19e+00	8.63e+00	9.00e+00
	N0	2.64e+00	4.66e+00	6.08e+00	7.04e+00	7.63e+00	7.95e+00	8.37e+00
	N25	**2.63e+00**	4.63e+00	**5.84e+00**	7.18e+00	**7.54e+00**	7.96e+00	8.41e+00
	N50	2.70e+00	4.59e+00	5.97e+00	7.04e+00	7.60e+00	8.04e+00	**8.37e+00**
	N75	2.73e+00	**4.50e+00**	6.07e+00	**6.99e+00**	7.58e+00	**7.82e+00**	8.46e+00

Corners solutions were analytically evaluated from the optimal Pareto front. The simulation test results are presented in table 1-6, with the bold numbers representing the best results obtained. Concerning the most appropriate evolutionary stage to introduce corner solutions into the population, to improve diversity and convergence, the results presented in tables 1-6 indicate:

- The majority of tests results clearly show that the introduction of corner solutions improves standard algorithms performance, independently of the stage in which they are introduced. In DTLZ1-3 cases the standard tested algorithms present difficulties regarding many-objective optimization. In these functions the corners insertion improves the search performance, despite the small number of iterations considered.
- The main advantage concerning corners insertion is the improvement of convergence time, as can be observed in the figures 1-6. From these graphics it is also possible to verify that the optimal value obtained is the same for all the insertion evolutionary stages considered.
- In the case of GDE3 algorithm for the DTLZ3, it is important to note that the number of iterations used proved to be insufficient. As DTLZ3 is a highly multi-modal function with multiple local fronts the insertion of the corner set totally dominates all the other solutions, providing results which appears to be excellent regarding the both metrics. However, this is not the case, due to the small number of non-dominated solutions obtained. This small number of solutions can not properly represent the non-dominated front (*e.g.* see all DTLZ3 G75 cases presented in Table 5). In all the other tests results presented the final archive is completely filled with non-dominated solutions.

Table 4. NSGAII Median Spacing

Problems	Algorithms	Objective Number						
		4	5	6	7	8	9	10
DTLZ1	N	5.40e-02	1.82e+01	3.80e+01	5.28e+01	6.06e+01	7.37e+01	8.28e+01
	N0	3.87e-02	5.94e-02	9.89e-02	1.32e-01	1.69e-01	2.04e-01	2.35e-01
	N25	3.90e-02	6.27e-02	1.02e-01	1.35e-01	**1.65e-01**	**1.95e-01**	**2.17e-01**
	N50	3.94e-02	6.42e-02	9.66e-02	**1.31e-01**	1.68e-01	1.97e-01	2.28e-01
	N75	**3.84e-02**	**5.38e-02**	**9.10e-02**	1.31e-01	1.65e-01	2.02e-01	2.36e-01
DTLZ2	N	1.09e-01	2.43e-01	5.06e-01	6.91e-01	8.78e-01	9.61e-01	1.12e+00
	N0	1.13e-01	1.57e-01	**2.16e-01**	**2.87e-01**	**3.57e-01**	4.39e-01	4.86e-01
	N25	**1.08e-01**	1.60e-01	2.26e-01	2.97e-01	3.84e-01	4.39e-01	**4.70e-01**
	N50	1.09e-01	**1.55e-01**	2.20e-01	2.97e-01	3.76e-01	**4.21e-01**	4.82e-01
	N75	1.08e-01	1.60e-01	2.27e-01	2.97e-01	3.72e-01	4.25e-01	4.95e-01
DTLZ3	N	1.27e+00	8.67e+01	1.95e+02	3.12e+02	4.21e+02	5.05e+02	5.86e+02
	N0	1.05e-01	1.57e-01	2.19e-01	2.90e-01	3.61e-01	**4.23e-01**	4.72e-01
	N25	1.09e-01	1.53e-01	2.16e-01	2.95e-01	3.62e-01	4.36e-01	4.73e-01
	N50	1.06e-01	1.53e-01	2.16e-01	2.89e-01	3.67e-01	4.31e-01	**4.72e-01**
	N75	**1.03e-01**	**1.48e-01**	**1.98e-01**	**2.77e-01**	**3.53e-01**	4.32e-01	4.86e-01
DTLZ4	N	**1.03e-01**	1.76e-01	3.52e-01	5.65e-01	7.46e-01	9.15e-01	1.02e+00
	N0	1.10e-01	1.59e-01	2.16e-01	**2.81e-01**	3.61e-01	4.27e-01	4.79e-01
	N25	1.07e-01	1.68e-01	**2.07e-01**	2.87e-01	**3.46e-01**	4.31e-01	**4.72e-01**
	N50	1.09e-01	1.63e-01	2.15e-01	2.82e-01	3.55e-01	**4.08e-01**	4.97e-01
	N75	1.07e-01	**1.55e-01**	2.12e-01	2.84e-01	3.56e-01	4.24e-01	4.78e-01
DTLZ5	N	1.06e-01	2.10e-01	3.31e-01	4.38e-01	5.47e-01	5.73e-01	6.29e-01
	N0	1.16e-01	2.01e-01	3.13e-01	**3.81e-01**	4.86e-01	5.38e-01	5.65e-01
	N25	1.10e-01	2.16e-01	3.12e-01	4.06e-01	**4.45e-01**	**5.13e-01**	5.66e-01
	N50	1.01e-01	1.99e-01	**3.06e-01**	4.06e-01	4.63e-01	5.28e-01	**5.51e-01**
	N75	**9.82e-02**	**1.95e-01**	3.22e-01	4.14e-01	4.73e-01	5.16e-01	5.80e-01

Table 5. GDE3 Median GD

Problems	Algorithms	Objective Number						
		4	5	6	7	8	9	10
DTLZ1	G	7.63e-02	2.29e-01	4.71e-01	6.09e+00	2.97e+01	5.49e+01	1.03e+02
	G0	7.06e-02	2.15e-01	3.60e-01	**5.15e-01**	7.13e-01	8.57e-01	**9.09e-01**
	G25	7.36e-02	2.12e-01	**3.57e-01**	5.36e-01	**6.52e-01**	**8.04e-01**	9.77e-01
	G50	**6.97e-02**	2.12e-01	3.68e-01	5.50e-01	6.77e-01	8.57e-01	9.49e-01
	G75	7.34e-02	**2.09e-01**	3.84e-01	5.44e-01	7.00e-01	8.47e-01	9.72e-01
DTLZ2	G	1.53e-01	3.97e-01	8.29e-01	1.42e+00	2.47e+00	3.96e+00	5.57e+00
	G0	1.39e-01	**3.29e-01**	6.09e-01	**8.38e-01**	1.12e+00	1.36e+00	1.60e+00
	G25	1.41e-01	3.39e-01	5.94e-01	8.69e-01	1.10e+00	1.36e+00	**1.58e+00**
	G50	**1.37e-01**	3.39e-01	5.92e-01	8.57e-01	1.11e+00	**1.33e+00**	1.64e+00
	G75	1.50e-01	3.31e-01	**5.86e-01**	8.50e-01	**1.10e+00**	1.38e+00	1.61e+00
DTLZ3	G	1.10e+02	9.27e+01	1.25e+03	1.54e+04	2.76e+04	8.57e+04	2.05e+05
	G0	1.50e-01	3.44e-01	5.72e-01	8.26e-01	1.06e+00	1.31e+00	1.56e+00
	G25	1.57e-01	3.56e-01	5.90e-01	8.41e-01	1.08e+00	1.30e+00	1.53e+00
	G50	**0.00e+00**	3.46e-01	5.97e-01	8.06e-01	1.09e+00	1.28e+00	1.50e+00
	G75	**0.00e+00**	**1.98e-01**	**0.00e+00**	**0.00e+00**	**2.74e-02**	**7.53e-03**	**9.54e-03**
DTLZ4	G	1.48e-01	3.54e-01	6.04e-01	8.49e-01	1.09e+00	1.34e+00	1.64e+00
	G0	1.49e-01	**3.29e-01**	5.73e-01	8.14e-01	1.04e+00	1.28e+00	1.50e+00
	G25	1.36e-01	3.41e-01	**5.59e-01**	8.15e-01	1.05e+00	1.28e+00	1.51e+00
	G50	1.51e-01	3.48e-01	5.66e-01	8.10e-01	1.05e+00	1.31e+00	**1.50e+00**
	G75	**1.35e-01**	3.38e-01	5.82e-01	**8.05e-01**	**1.02e+00**	**1.26e+00**	1.52e+00
DTLZ5	G	**3.08e+00**	**5.02e+00**	6.29e+00	7.58e+00	8.56e+00	8.94e+00	8.99e+00
	G0	3.23e+00	5.23e+00	6.41e+00	7.48e+00	**7.72e+00**	8.33e+00	8.68e+00
	G25	3.25e+00	5.23e+00	**6.23e+00**	7.46e+00	7.84e+00	**8.31e+00**	8.90e+00
	G50	3.19e+00	5.27e+00	6.39e+00	**7.39e+00**	7.98e+00	8.41e+00	8.68e+00
	G75	3.21e+00	5.21e+00	6.30e+00	7.56e+00	7.93e+00	8.43e+00	**8.67e+00**

Table 6. GDE3 Median Spacing

Problems	Algorithms	Objective Number						
		4	5	6	7	8	9	10
DTLZ1	G	3.59e-02	6.07e-02	9.64e-02	8.26e-01	3.28e+00	5.89e+00	1.38e+01
	G0	3.76e-02	6.04e-02	8.47e-02	1.07e-01	1.24e-01	1.52e-01	1.72e-01
	G25	3.64e-02	**5.76e-02**	**8.15e-02**	1.13e-01	1.35e-01	**1.48e-01**	1.67e-01
	G50	**3.48e-02**	5.87e-02	8.41e-02	1.08e-01	1.24e-01	1.51e-01	1.72e-01
	G75	3.56e-02	5.93e-02	8.15e-02	**1.05e-01**	**1.21e-01**	1.52e-01	**1.64e-01**
DTLZ2	G	**9.66e-02**	**1.47e-01**	2.04e-01	2.71e-01	3.58e-01	4.41e-01	5.76e-01
	G0	1.05e-01	1.55e-01	2.02e-01	**2.24e-01**	2.89e-01	**3.03e-01**	3.35e-01
	G25	1.01e-01	1.54e-01	2.13e-01	2.41e-01	2.86e-01	3.19e-01	3.34e-01
	G50	1.02e-01	1.55e-01	2.16e-01	2.33e-01	**2.77e-01**	3.06e-01	3.31e-01
	G75	1.03e-01	1.51e-01	**2.00e-01**	2.42e-01	2.81e-01	3.22e-01	**3.23e-01**
DTLZ3	G	1.78e+00	1.74e+00	6.30e+00	2.21e+01	3.35e+01	6.63e+01	1.09e+02
	G0	1.00e-01	1.55e-01	2.02e-01	2.46e-01	2.83e-01	**3.09e-01**	3.54e-01
	G25	1.00e-01	1.50e-01	**1.95e-01**	2.51e-01	2.91e-01	3.34e-01	**3.39e-01**
	G50	**0.00e+00**	1.63e-01	2.05e-01	2.46e-01	2.91e-01	3.10e-01	3.45e-01
	G75	**0.00e+00**	**1.44e-01**	2.95e-01	**0.00e+00**	**2.79e-01**	6.22e-01	6.95e-01
DTLZ4	G	**9.81e-02**	1.48e-01	2.07e-01	2.41e-01	2.82e-01	**3.08e-01**	3.41e-01
	G0	1.04e-01	1.52e-01	2.10e-01	**2.36e-01**	2.80e-01	3.14e-01	3.51e-01
	G25	1.04e-01	1.52e-01	2.00e-01	2.40e-01	2.90e-01	3.12e-01	3.35e-01
	G50	1.00e-01	**1.47e-01**	**1.90e-01**	2.39e-01	2.88e-01	3.13e-01	3.41e-01
	G75	9.97e-02	1.50e-01	1.94e-01	2.38e-01	**2.72e-01**	3.15e-01	**3.30e-01**
DTLZ5	G	**9.75e-02**	2.14e-01	3.31e-01	4.10e-01	4.95e-01	5.28e-01	6.16e-01
	G0	1.01e-01	2.05e-01	**2.85e-01**	3.97e-01	**4.21e-01**	**4.83e-01**	5.45e-01
	G25	1.13e-01	2.17e-01	3.13e-01	**3.81e-01**	4.58e-01	5.15e-01	**5.17e-01**
	G50	1.03e-01	2.19e-01	3.10e-01	4.02e-01	4.48e-01	5.10e-01	5.56e-01
	G75	1.03e-01	**2.05e-01**	3.12e-01	4.20e-01	4.55e-01	5.10e-01	5.46e-01

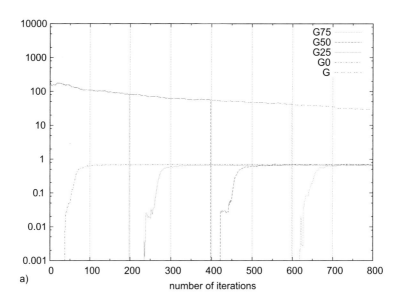

a)

Fig. 1. GD Median for DTLZ1, problem with 8 objectives in GDE3 Algorithm

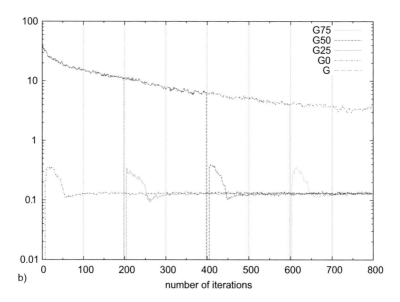

Fig. 2. Spread Median for DTLZ1, problem with 8 objectives in GDE3 Algorithm

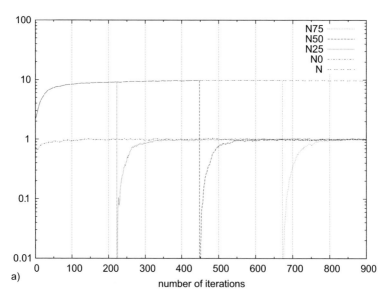

Fig. 3. GD Median for DTLZ2 problem with 9 objectives in NSGA-II Algorithm

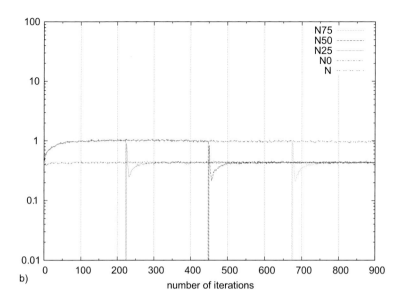

Fig. 4. Spread Median for DTLZ2 problem with 9 objectives in NSGA-II Algorithm

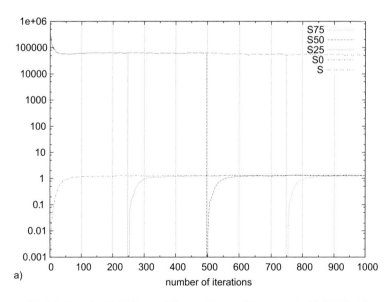

Fig. 5. GD Median for DTLZ3 problem with 10 objectives in SMPSO Algorithm

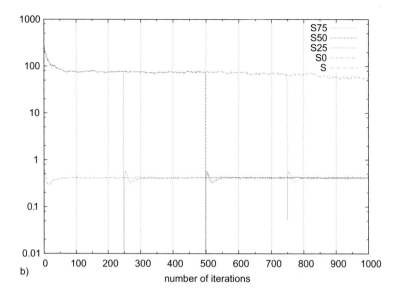

Fig. 6. Spread Median for DTLZ3 problem with 10 objectives in SMPSO Algorithm

The results presented in figures 1,3 and 5 show the convergence improvement achieved with the corner solutions introduction. In some cases after the corner solution insertion the performance appears to be optimal (zero) during a small number of iterations. This is due to the time necessary to the algorithm generate more non-dominated solutions besides the inserted corner solutions. All the presented figures show the performance improvement of the corner solutions introduction when compared to the standard algorithm.

6   Conclusion

A major advantage of evolutionary based multi-objective optimization techniques is to achieve a wide non-dominated set in a single run. However, when applied to many-objective problems, their performance is severely deteriorated. In this paper the idea of introducing non-dominated Pareto front in the beginning of the search procedure proposed by Dasgupta *et al.*[9] is explored. This study addresses the research issue regarding the most appropriate evolutionary stage for inclusion of extreme Pareto front solutions (corners solutions). This technique assumes the availability of corner solutions prior to the many-objective optimization algorithm execution. Corner solutions or approximated solutions can be determined using a specific search algorithm (*e.g.* Nadir point based techniques). Simulation results were presented concerning the corner solutions introduction into the population, in several evolutionary stages. The evolutionary stages considered were 25%, 50% and 75% of a total number of iterations.

The results presented for the considered function test set (DTLZ1-5) and performance metrics allows to conclude that:

- the search efficiency in terms of search convergence rate and diversity is clearly improved relatively to the standard algorithms, as result of corner solutions introduction.
- while the convergence rate is improved, this does not depend on the evolutionary stage in which the informed solutions are introduced. This results are in agreement with the results presented in [9].
- considering the small number of iterations used, the corner solutions insertion showed to be dominant regarding the existing non-dominated archive. This has the effect of significantly reduce the number of non-dominated solutions in the archive until the algorithms find new non-dominated solutions.
- for the DTLZ5 case the corner introduction has a marginal effect in the performance as the standard algorithms are capable of finding the corner solutions despite the number of objectives considered.

It is important to note the small number of algorithm iterations considered in the tests particularly for the problems with more than seven objectives. Future work will be carried on considering the inclusion of other types of informed solutions besides corner solutions.

References

[1] Reyes-Sierra, M., Coello Coello, C.A.: Multi-objective particle swarm optimizers: A survey of the state-of-the-art. International Journal of Computational Intelligence Research 2(3), 287–308 (2006)
[2] Fleming, P.J., Purshouse, R.C., Lygoe, R.J.: Many-Objective Optimization: An Engineering Design Perspective. In: Coello Coello, C.A., Hernández Aguirre, A., Zitzler, E. (eds.) EMO 2005. LNCS, vol. 3410, pp. 14–32. Springer, Heidelberg (2005)
[3] Ishibuchi, H., Tsukamoto, N., Nojima, Y.: Evolutionary many-objective optimization: A short review. In: 2008 IEEE Congress on Evolutionary Computation IEEE World Congress on Computational Intelligence, pp. 2419–2426 (March 2008)
[4] Deb, K., Agrawal, S., Pratap, A., Meyarivan, T.: A fast and elitist multiobjective genetic algorithm: NSGA-II. IEEE Trans. Evolutionary Computation 6(2), 182–197 (2002)
[5] Nebro, A., Durillo, J., Garcia-Nieto, J., Coello Coello, C.A., Luna, F., Alba, E.: SMPSO: A new PSO-based metaheuristic for multi-objective optimization. In: IEEE Symposium on Computational Intelligence in Miulti-criteria Decision-making, MCDM 2009, pp. 66–73 (2009)
[6] Kukkonen, S., Lampinen, J.: GDE3: the third evolution step of generalized differential evolution. In: The 2005 IEEE Congress on Evolutionary Computation, vol. 1, pp. 443–450 (2005)
[7] Adra, S., Fleming, P.: Diversity management in evolutionary many-objective optimization. IEEE Transactions on Evolutionary Computation 15(2), 183–195 (2011)
[8] Deb, K., Jain, H.: Handling many-objective problems using an improved NSGA-II procedure. In: 2012 IEEE Congress on Evolutionary Computation (CEC), pp. 1–8 (2012)

[9] Dasgupta, D., Hernandez, G., Romero, A., Garrett, D., Kaushal, A., Simien, J.: On the use of informed initialization and extreme solutions sub-population in multi-objective evolutionary algorithms. In: IEEE symposium on Computational Intelligence in Miulti-criteria Decision-making, MCDM 2009, pp. 58–65 (2009)

[10] Singh, H.K., Isaacs, A., Ray, T.: A pareto corner search evolutionary algorithm and dimensionality reduction in many-objective optimization problems. IEEE Trans. Evolutionary Computation 15(4), 539–556 (2011)

[11] Bechikh, S., Said, L.B., Ghédira, K.: Searching for knee regions in multi-objective optimization using mobile reference points. In: SAC, pp. 1118–1125 (2010)

[12] Branke, J., Deb, K., Dierolf, H., Osswald, M.: Finding knees in multi-objective optimization. In: Yao, X., et al. (eds.) PPSN 2004. LNCS, vol. 3242, pp. 722–731. Springer, Heidelberg (2004)

[13] Deb, K., Miettinen, K.: Nadir point estimation using evolutionary approaches: Better accuracy and computational speed through focused search. In: Ehrgott, M., Naujoks, B., Stewart, T.J., Walllenius, J. (eds.) Multiple Criteria Decision Making for Sustainable Energy and Transportation Systems. LNEMS, vol. 634, pp. 339–354. Springer, Heidelberg (2010)

[14] Bechikh, S., Ben Said, L., Ghedira, K.: Estimating nadir point in multi-objective optimization using mobile reference points. In: 2010 IEEE Congress on Evolutionary Computation (CEC), pp. 1–9 (2010)

[15] Ishibuchi, H., Tsukamoto, N., Nojima, Y.: Evolutionary many-objective optimization (2008)

[16] Corne, D., Knowles, J.: Techniques for Highly Multiobjective Optimisation: Some Nondominated Points are Better than Others, pp. 773–780 (2007)

[17] Walker, D.J., Everson, R.M., Fieldsend, J.E.: Visualisation and ordering of many-objective populations (2010)

[18] Köppen, M., Yoshida, K.: Substitute Distance Assignments in NSGA-II for Handling Many-Objective Optimization Problems. In: Obayashi, S., Deb, K., Poloni, C., Hiroyasu, T., Murata, T. (eds.) EMO 2007. LNCS, vol. 4403, pp. 727–741. Springer, Heidelberg (2007)

[19] Garza-Fabre, M., Toscano-Pulido, G., Coello Coello, C.A., Rodriguez-Tello, E.: Effective ranking speciation Many-objective optimization (2011)

[20] L'opez, A., Coello Coello, C.A., Oyama, A., Fujii, K.: An alternative preference relation to deal with many-objective optimization problems. In: Purshouse, R.C., Fleming, P.J., Fonseca, C.M., Greco, S., Shaw, J. (eds.) EMO 2013. LNCS, vol. 7811, pp. 291–306. Springer, Heidelberg (2013)

[21] Jaimes, A.L., Coello Coello, C.A., Chakraborty, D.: Objective reduction using a feature selection technique. In: GECCO, pp. 673–680 (2008)

[22] Brockhoff, D., Zitzler, E.: Objective reduction in evolutionary multiobjective optimization: Theory and applications. Evolutionary Comp. 17(2), 135–166 (2009)

[23] Saxena, D.K., Deb, K.: Dimensionality reduction of objectives and constraints in multi-objective optimization problems: A system design perspective. In: IEEE Congress on Evolutionary Computation, pp. 3204–3211 (2008)

[24] Saxena, D.K., Duro, J.A., Tiwari, A., Deb, K., Zhang, Q.: Objective Reduction in Many-Objective Optimization: Linear and Nonlinear Algorithms. IEEE Transactions on Evolutionary Computation 17(1), 77–99 (2013)

[25] Chaudhuri, S., Deb, K.: Applied Soft Computing

[26] Sinha, A., Saxena, D.K., Deb, K., Tiwari, A.: Using objective reduction and interactive procedure to handle many-objective optimization problems. Applied Soft Computing 13(1), 415–427 (2013)

[27] Gutierrez, A.L., Lanza, M., Barriuso, I., Valle, L., Domingo, M., Perez, J.R., Basterrechea, J.: Comparison of different PSO initialization techniques for high dimensional search space problems: A test with FSS and antenna arrays (2011)

[28] Durillo, J.J., Nebro, A.J.: jmetal: A java framework for multi-objective optimization. Advances in Engineering Software 42, 760–771 (2011)

[29] Sierra, M.R., Coello Coello, C.A.: Improving pso-based multi-objective optimization using crowding, mutation and ϵ-dominance. In: Coello Coello, C.A., Hernández Aguirre, A., Zitzler, E. (eds.) EMO 2005. LNCS, vol. 3410, pp. 505–519. Springer, Heidelberg (2005)

[30] Hadka, D., Reed, P., Simpson, T.: Diagnostic assessment of the borg MOEA for many-objective product family design problems. In: 2012 IEEE Congress on Evolutionary Computation (CEC), pp. 1–10 (2012)

[31] Deb, K., Thiele, L., Laumanns, M., Zitzler, E.: Scalable Test Problems for Evolutionary Multi-Objective Optimization, pp. 1–27 (2001)

[32] Veldhuizen, D.A.V., Lamont, G.B.: Evolutionary computation and convergence to a pareto front, Stanford University, pp. 221–228. Morgan Kaufmann (1998)

[33] Schott, J.R.: Fault Tolerant Design Using Single and Multicriteria Genetic Algorithm Optimization. Master's thesis, MIT (May 1995)

Escaping Local Optima via Parallelization and Migration

Vincenzo Cutello, Angelo G. De Michele, and Mario Pavone

Department of Mathematics and Computer Science,
University of Catania,
v.le A. Doria 6 – 95125 Catania, Italy
angelo.demichele@gmail.com,
{cutello,mpavone}@dmi.unict.it

Abstract. We present a new nature-inspired algorithm, $mt-GA$, which is a parallelized version of a simple GA, where subpopulations evolve independently from each other and on different threads. The overall goal is to develop a population-based algorithm capable to escape from local optima. In doing so, we used complex trap functions, and we provide experimental answers to some crucial implementation decision problems. The obtained results show the robustness and efficiency of the proposed algorithm, even when compared to well-known state-of-the art optimization algorithms based on the clonal selection principle.

Keywords: Genetic algorithms, multi-threaded genetic algorithms, trap functions, toy problems, global optimization, optimization.

1 Introduction

Many real-world problems are hard to solve because the parameters that influence the problem structures and dynamics over time are unknown and often impossible to be analytically solved. On such problems Evolutionary Algorithms (EA) seem to be perform quite well, primarily when the solutions are not known *a priori* or they are nonlinear. However, the EA it must be designed in such way to prevent getting trapped into local optima.

One feature that plays a key role on this issue is the diversity of individuals introduced into the population. Population diversity strongly influences, as known, both the *exploration* of the search space, and the *exploitation* of the information gained during the evolutionary process. The aim of this work, therefore, is to develop a population-based algorithm capable to escape from local optima maintaining itself *blind* on the problem's domain, i.e. general purpose and not tailored to the any specific problem. In order to achieve our result, we focused on developing a Genetic Algorithm (GA) based on Multi-Threads, where subsets of individuals evolve on different threads. Moreover, our GA is also equipped with a migration operator which allows pairs of solutions (individuals) to migrate between threads. Migration allows our algorithm to perform a careful and deep

G. Terrazas et al. (eds.), *Nature Inspired Cooperative Strategies for Optimization*
(NICSO 2013), Studies in Computational Intelligence 512,
DOI: 10.1007/978-3-319-01692-4_11, © Springer International Publishing Switzerland 2014

search of the solution space. In what follows, we will denote our algorithm with $mt - GA$.

To check the ability of $mt - GA$ to escape local maxima, we used *Trap Functions,* well-known toy problem, used for understanding the dynamics and search's ability of a generic evolutionary algorithm [10]. We designed two variants of $mt - GA$ (synchronous, and asynchronous threads), and the trap functions are used as testbed in order to analyze and determine which variant is more suitable for our aims. Moreover, what solutions to select for the migration, and in what place, is also subject of this study.

2   Multi-population GA and Migrations

There is a very vast literature on multi-population Genetic Algorithms and the associated concept of migration. A comprehensive analysis of it, is certainly way beyond the scope of our contribution. For sake of completeness, however, we will mention few of the obtained results, especially in relation to the key decision about migration.

The standard GA has a single population which tries to explore the entire search space. The Multi-population approach tries to divide the search space into several parts and then uses a number of small populations to search them separately. The approach can obviously be parallelized and so such separate searches may run either synchronously or asynchronously (we will talk about this in the next sections). If we allow the different populations to "communicate", a key concept is migration.

In [7], the author proposes the random immigrants approach, sociologically inspired by the flux of immigrants that between generations move from one place to another. Technically, some individuals of the current population are replace with random individuals, called random immigrants, from another population at every generation. The choice of individuals to be replaced is usually governed by two main strategies: replacing random individuals or replacing the worst ones. In many ways, random immigrants may act as "genetic mutations" and thus the ratio of number of random immigrants to the the population size, is usually set to a small value.

The effect of the policy used to select migrants and the individuals they replace on the selection pressure in parallel evolutionary algorithms (EAs) with multiple populations is investigated in [1]. In particular, four possible combinations of random and fitness-based emigration and replacement of existing individuals are considered.

In [11] the author investigates a hybrid memory and random immigrants scheme, called memory-based immigrants, and a hybrid elitism and random immigrants scheme, called elitism-based immigrants, for improving the performance of genetic algorithms in dynamic environments.

To underline the importance on modern technical issues of such techniques, we mention the work in [2], where multi population GA's with immigrants schemes are designed for the dynamic dhortest path routing problem in mobile networks.

3 The Trap Functions

The trap functions problem [4, 5] is a toy yet complex problem that simply takes as input the number of $1's$ of bit strings of length ℓ. The fitness function $f(x)$ is defined as a function $\widehat{f}(\cdot)$ of the number of 1-bits, $u(x)$, in the binary input string x

$$f(x) = \widehat{f}(u(x)) = \widehat{f}\left(\sum_{k=1}^{\ell} x_k\right) \tag{1}$$

The definition of the function $\widehat{f}$, which depends on few numerical parameters, gives rise to two different scenarios: *simple trap function* and *complex trap function*. Both scenarios are shown in figure 1.

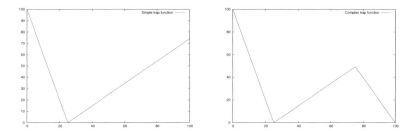

Fig. 1. Simple (left plot) and complex (right plot) trap functions

The simple trap function is characterized by one global optimum (for a bit string of all 0's) and one local optimum (for a bit string of all 1's) that are the complement *bit-wise* of each other (see left plot in figure 1). Its formal definition is given by:

$$\widehat{f}(u(x)) = \begin{cases} \frac{a}{z}(z - u(x)), & \text{if } u(x) \leq z \\ \frac{b}{\ell - z}(u(x) - z), & \text{otherwise} \end{cases} \tag{2}$$

where the 3 parameters a, b, and z are such that [8] $z \approx (1/4)\ell$; $b = \ell - z - 1$; $1.5b \leq a \leq 2b$; a a multiple of z.

The complex trap function, defined using 4 parameters, is instead more difficult to investigate because there are two directions where the algorithm may get trapped (see right plot of figure 1).

It is formally defined as:

$$\widehat{f}(u) = \begin{cases} \frac{a}{z_1}(z_1 - u(x)), & \text{if } u(x) \leq z_1 \\ \frac{b}{\ell - z_1}(u(x) - z_1), & \text{if } z_1 < u(x) \leq z_2 \\ \frac{b(z_2 - z_1)}{z_2}\left(1 - \frac{1}{z_1}(u(x) - z_2)\right) & \text{otherwise.} \end{cases} \tag{3}$$

If z_1, similarly to z in the case of simple trap function, verifies $z \approx (1/4)\ell$, the value of parameter $z_2 > z_1$ could be fixed as $z_2 = \ell - z_1$. We also note that if we fix $z_2 = \ell$ the complex trap function becomes a simple trap one. Summing up, for both kinds of trap functions there are many possible choices for the parameters a, b and z, (with $z_1 = z$ and $z_2 = \ell - z_1$ for the complex trap function) [8]. Some values are shown in table 1 and we will use them for our experiments.

Table 1. Parameter values used by simple and complex trap functions

type	ℓ	z	a	b
I	10	3	12	6
II	20	5	20	14
III	50	10	80	39
IV	75	20	80	54
V	100	25	100	74

The plots in figure 1 were produced using the following parameter values: $\ell = 100, z = 25, a = 100, b = 74$ for the simple trap function, and $\ell = 100, z_1 = 25, z_2 = 75, a = 100, b = 74$ for the complex trap function.

In the next sections, we will focus our discussions on complex trap functions.

4 $mt - GA$: A Multi-Threaded Genetic Algorithm

We began our work with the implementation of a simple, standard GA.

Therefore, we used some classical operators for recombination, mutation, and selection mechanism, namely *2–point crossover*, *bit flip*, and *roulette wheel selection*. Based on this simple algorithm, we studied and tested the impact, and improvements produced by its parallelization where subpopulations evolve separately each on a different thread. The motivation on parallelize a GA via threads is not only the speeding up the running times, but also, and primarily, because it provides a method where different processes are running in cooperation among them, each with specific tasks, and sharing gained information. To strengthen the cooperation produced by the parallelization of GA, we have designed a migration approach, where k individuals migrate from the i-th thread to another. In this way we give the opportunity to exchange among them the gained information. Moreover, such an approach guarantees sufficient introduction of diversity into each subpopulation, which, on the whole, helps $mt - GA$ in escaping from local optima, in according with the aims of this work. Algorithm 1 shows the pseudocode of the designed *Multi-Threaded Genetic Algorithm* ($mt - GA$).

As a classical nature-inspired algorithm, $mt - GA$ starts with a random creation of the initial population ($P^{(t=0)}$), where each chromosome is represented as a bit string of lenght ℓ. Afterwards, the population is divided in subpopulations, as many as the number n of threads, which will evolve independently from each other. The individuals for any subpopulation are selected in sequential order from the overall population ($P^{(t=0)}$); therefore, the

Algorithm 1. Pseudo code of $mt - GA$

$t \leftarrow 0$
$FFE \leftarrow 0$
$P^t \leftarrow$ Create_Initial_Population
for $i = 1$ to n **do**
   $P_i^t \leftarrow$ Assign the individuals $\left((i-1)(\frac{popsize}{n})+1\right), \ldots, i(\frac{popsize}{n})$
end for
while $FFE < T_{max}$ **do**
   for all Thread i $(1 \leq i \leq n)$ **do**
     Compute_Fitness(P_i^t)
     $FFE \leftarrow FFE + \left(\frac{popsize}{n}\right)$
     Select $\left(\frac{popsize}{n}\right)$ individuals via *Roulette Wheel Selection*
     Recombination for generating offsprings via $2-point$ *Crossover* and *Bit_Flip*
     Mutation
     k individuals migrate to the j-th thread $(1 \leq j \leq n,$ and $i \neq j)$
   end for
   $t \leftarrow t + 1$
end while

i-th subpopulation $(P_i^{(t=0)})$, i.e. the i-th thread, will be assigned the individuals $\left(((i-1) \times \frac{popsize}{n})+1\right), \ldots, \left(i \times \frac{posize}{n}\right)$.

From this moment on, all the subpopulations will evolve in independent way. Thus, the description of the subsequent steps, given for the thread i and the population P_i is equivalent for all n threads.

At each timestep t, the fitness function value is computed for each individual. Such an evaluation will, obviously, increase the value of the global variable FFE so to force the termination of the Algorithm within a finite number of steps, as described later. To select individuals for offspring generation, we implemented the classical *Roulette Wheel Selection* mechanism; given a chromosome $x \in P_i^t$ with fitness $f(x)$, the probability p_x that it will be selected is given by

$$p_x = \frac{f(x)}{\sum_{y \in P_i^t} f(y)}.$$

Once two individuals have been chosen for mating, we use the 2-*point crossover* operator to generate two offspring.

As last step, to each offspring is applied a mutation operator, which is basically a bit flip of the selected gene. The mutation is performed with a probability p^i, which accounts for the independent evolution of the population in the thread. $mt-GA$ can can therefore better explore the search space, as well as exploit more efficiently the gained information. In particular, some threads work more on the exploration of wide regions of the landscape; whilst others on the exploitation of the solutions found. Finally, after crossover and mutation operators, k individuals from each thread migrate to other threads. This helps $mt - GA$ in escaping from local optima by introducing diversity into the subpopulations, and brings to a thread information discovered by other threads.

It is obviously crucial to decide which individuals should be chosen from each subpopulation to be migrate, and in which thread they should migrate. Another important issue is synchronicity among threads. If they run synchronized, the cooperation among them is strengthened. If they run in an asynchronous manner, it is possible that one thread does not receive any migrants, because is not running in that time. Thus, we have also designed a third variant of $mt - GA$, which includes a *Birth operator*, which restores the right size of the given subpopulation introducing new elements randomly generated. The three variants of $mt - GA$ (*synchronous*, *asynchronous*, and *asynchronous with birth operator*), have been the subject of our study and are described in section 5.

The algorithm terminates its execution when the fitness function evaluation number (FFE) is great or equal to T_{max}, i.e. the maximum number of allowed objective function evaluations .

5   Experimental Results

In this section we present the study conducted in order to understand the validity of the novelties introduced. In particular, we concentrated on answering the questions:

1. which individuals should be selected as migrants?
2. to which thread should they migrate?
3. which variant of $mt - GA$ shows the best performances?

As anticipated in previous section (sec. 1), the three variants of $mt - GA$ which were implemented are (1) synchronous, (2) asynchronous, and (3) asynchronous with birth.

For the migration thread, we implemented two different protocols: (i) migrate to the next thread (i.e. from i to $i + 1$), and (ii) migrate to a randomly chosen thread.

To choose k individuals for migration, instead, we implemented the following protocols: (a) the best k; (b) the best $k/2$, and the remaining $k/2$ randomly chosen; and (c) k randomly chosen.

A feature that plays also a central role in $mt - GA$, as well as in every evolutionary algorithm, is how to generate the new population for the next iteration. This decision, of course, influences the search ability, and then the overall performances of a generic algorithm.

We implemented three different strategies: *elitism*; *substitution*, and *no preservation*. The first strategy always maintains the best individuals found so far; the second one, instead, replaces the worst offspring with the best of its parents; finally, in the last one we do not preserve any individual.

To make a robust analysis of our study we have tested $mt - GA$ only on the complex trap functions, being the most difficult ones, especially when increasing the size of the search space, and enough challenging to answer the above open

Table 2. The three variants of $mt - GA$ on complex trap functions using *Elitism* approach as preservation strategy

migrants	Trap	SR	AES	best	mean	σ	SR	AES	best	mean	σ
		Synchronous Variant									
			migration place: next thread					migration place: random thread			
best2	C(I)	100	953.04	12.0	12.0	0.0	100	832.16	12.0	12.0	0.0
best2	C(II)	87	45040.92	20.0	18.61	3.59	96	43607.31	20.0	19.57	2.09
best2	C(III)	24	92272.91	80.0	41.86	21.85	58	93684.59	80.0	59.59	24.43
best2	C(IV)	2	135315.5	80.0	35.57	6.7	0	0.0	34.36	34.36	0.05
best2	C(V)	0	0.0	49.33	49.33	0.05	0	0.0	49.33	49.33	0.05
best + random	C(I)	100	905.19	12.0	12.0	0.0	100	1082.91	12.0	12.0	0.0
best + random	C(II)	92	43512.0	20.0	19.15	2.89	98	39015.87	20.0	19.79	1.49
best + random	C(III)	35	84307.17	80.0	47.01	24.21	54	103432.78	80.0	57.46	25.01
best + random	C(IV)	0	0.0	34.36	34.36	0.05	1	111436.0	80.0	34.85	4.54
best + random	C(V)	0	0.0	49.33	49.33	0.05	0	0.0	49.33	49.33	0.05
random	C(I)	100	954.94	12.0	12.0	0.0	100	1049.46	12.0	12.0	0.0
random	C(II)	98	31512.93	20.0	19.79	1.49	97	29823.65	20.0	19.75	1.54
random	C(III)	57	99018.06	80.0	58.21	25.09	57	103747.83	80.0	58.58	24.82
random	C(IV)	1	142402.0	80.0	35.24	6.12	0	0.0	34.36	34.36	0.05
random	C(V)	0	0.0	49.33	49.33	0.05	0	0.0	49.33	49.33	0.05
		Asynchronous Variant									
			migration place: next thread					migration place: random thread			
best2	C(I)	100	1731.06	12.0	12.0	0.0	100	1866.19	12.0	12.0	0.0
best2	C(II)	76	58889.68	20.0	17.44	4.56	88	58618.03	20.0	18.75	3.4
best2	C(III)	27	91646.11	80.0	42.98	22.52	30	107803.47	80.0	44.5	23.24
best2	C(IV)	2	154088.5	80.0	35.28	6.39	1	52527.0	80.0	34.82	4.54
best2	C(V)	0	0.0	49.33	49.33	0.05	0	0.0	49.33	49.33	0.05
best + random	C(I)	100	1649.15	12.0	12.0	0.0	100	1484.57	12.0	12.0	0.0
best + random	C(II)	80	48261.8	20.0	17.87	4.27	90	60251.61	20.0	18.93	3.2
best + random	C(III)	33	87326.87	80.0	46.21	23.79	42	99014.62	80.0	50.57	25.05
best + random	C(IV)	0	0.0	34.36	34.36	0.05	0	0.0	34.36	34.36	0.05
best + random	C(V)	0	0.0	49.33	49.33	0.05	0	0.0	49.33	49.33	0.05
random	C(I)	100	1341.8	12.0	12.0	0.0	100	1665.9	12.0	12.0	0.0
random	C(II)	99	39384.28	20.0	19.89	1.06	99	47915.18	20.0	19.89	1.06
random	C(III)	36	112243.72	80.0	47.65	24.28	34	83118.65	80.0	46.93	24.11
random	C(IV)	1	118636.0	80.0	34.82	4.54	0	0.0	34.36	34.36	0.052
random	C(V)	0	0.0	49.33	49.33	0.05	0	0.0	49.33	49.33	0.05
		Asynchronous Variant with Birth Operator									
			migration place: next thread					migration place: random thread			
best2	C(I)	100	501.11	12.0	12.0	0.0	100	511.01	12.0	12.0	0.0
best2	C(II)	100	27610.78	20.0	20.0	0.0	100	26663.05	20.0	20.0	0.0
best2	C(III)	44	106628.73	80.0	51.98	25.07	40	95407.25	80.0	50.22	24.69
best2	C(IV)	0	0.0	34.36	34.36	0.05	0	0.0	34.36	34.36	0.05
best2	C(V)	0	0.0	49.33	49.33	0.05	0	0.0	49.33	49.33	0.05
best + random	C(I)	100	519.82	12.0	12.0	0.0	100	496.47	12.0	12.0	0.0
best + random	C(II)	100	23491.53	20.0	20.0	0.0	100	25244.08	20.0	20.0	0.0
best + random	C(III)	37	111004.54	80.0	48.14	24.42	46	118486.35	80.0	52.65	25.25
best + random	C(IV)	1	136321.0	80.0	34.82	4.54	1	51949.0	80.0	34.82	4.54
best + random	C(V)	0	0.0	49.33	49.33	0.05	0	0.0	49.33	49.33	0.05
random	C(I)	100	710.88	12.0	12.0	0.0	100	440.12	12.0	12.0	0.0
random	C(II)	100	24147.84	20.0	20.0	0.0	100	22398.65	20.0	20.0	0.0
random	C(III)	39	119943.66	80.0	49.04	24.75	35	112969.88	80.0	47.84	24.21
random	C(IV)	0	0.0	34.36	34.36	0.05	1	191156.0	80.0	34.82	4.54
random	C(V)	0	0.0	49.33	49.33	0.05	0	0.0	49.33	49.33	0.05

questions. After several experiments, and a preliminary investigation on the best parameter tuning, we have fixed ($popsize = 40$) as population size; ($n = 4$) as number of threads; and ($k = 2$) as number of migrants from one thread to another. The low population size is due to both the parallelization and the developed migration strategy, thanks to which $mt-GA$ maintains a good diversity into the subpopulations, performing a proper exploration and exploitation of the landscape.

All the presented experiments were performed on 100 independent runs, and the maximum number of fitness function evaluations allowed has been fixed to 2.5×10^5.

In tables 2, 3, and 4 we show the results obtained by our study in order to understand the right answers to our open questions. For each experiment and each variant, we show the success rate (SR); the average number of fitness

Table 3. The three variants of $mt - GA$ on complex trap functions using *Substitution* approach as preservation strategy

		Synchronous Variant									
migrants	Trap	SR	AES	best	mean	σ	SR	AES	best	mean	σ
		migration place: next thread					migration place: random thread				
best2	C(I)	100	1133.5	12.0	12.0	0.0	100	1208.66	12.0	12.0	0.0
best2	C(II)	87	14040.73	20.0	18.61	3.59	100	18460.4	20.0	20.0	0.0
best2	C(III)	39	62181.23	80.0	49.07	24.73	91	75528.59	80.0	75.81	13.72
best2	C(IV)	1	135149.0	80.0	34.82	4.54	5	155389.8	80.0	36.64	9.95
best2	C(V)	0	0.0	49.33	49.33	0.05	0	0.0	49.33	49.33	0.05
best + random	C(I)	100	987.61	12.0	12.0	0.0	100	1085.56	12.0	12.0	0.0
best + random	C(II)	99	29244.40	20.0	19.89	1.06	100	28607.12	20.0	20.0	0.0
best + random	C(III)	56	108547.59	80.0	57.7	25.16	78	117031.12	80.0	69.56	20.21
best + random	C(IV)	3	146127.0	80.0	35.73	7.78	1	241915.0	80.0	34.82	4.54
best + random	C(V)	0	0.0	49.33	49.33	0.05	0	0.0	49.33	49.33	0.05
random	C(I)	100	1142.13	12.0	12.0	0.0	100	1105.79	12.0	12.0	0.0
random	C(II)	100	22248.69	20.0	20.0	0.0	100	35201.9	20.0	20.0	0.0
random	C(III)	63	117734.73	80.0	62.24	23.84	72	117650.89	80.0	66.11	22.37
random	C(IV)	1	147825.0	80.0	34.82	4.54	2	116774.0	80.0	35.28	6.39
random	C(V)	0	0.0	49.33	49.33	0.05	0	0.0	49.33	49.33	0.05

		Asynchronous Variant									
		migration place: next thread					migration place: random thread				
best2	C(I)	100	2137.36	12.0	12.0	0.0	100	1954.92	12.0	12.0	0.0
best2	C(II)	92	49366.47	20.0	19.15	2.89	97	53139.79	20.0	19.68	1.82
best2	C(III)	47	110869.3	80.0	53.98	25.10	68	116860.8	80.0	64.03	23.33
best2	C(IV)	2	173201.0	80.0	35.28	6.39	2	130577.0	80.0	35.28	6.39
best2	C(V)	0	0.0	49.33	49.33	0.05	0	0.0	49.33	49.33	0.05
best + random	C(I)	100	1512.2	12.0	12.0	0.0	100	2236.88	12.0	12.0	0.0
best + random	C(II)	97	46749.69	20.0	19.68	1.82	97	44977.67	20.0	19.68	1.82
best + random	C(III)	42	100841.66	80.0	50.86	24.84	66	127740.34	80.0	63.81	23.35
best + random	C(IV)	0	0.0	34.36	34.36	0.05	2	157680.5	80.0	35.28	6.39
best + random	C(V)	0	0.0	49.33	49.33	0.05	0	0.0	49.33	49.33	0.05
random	C(I)	100	2223.74	12.0	12.0	0.0	100	1325.07	12.0	12.0	0.0
random	C(II)	98	42429.27	20.0	19.79	1.49	100	41378.41	20.0	20.0	0.0
random	C(III)	49	107010.27	80.0	54.54	25.31	50	111112.34	80.0	54.65	25.35
random	C(IV)	0	0.0	34.36	34.36	0.05	1	128098.0	80.0	34.92	4.63
random	C(V)	0	0.0	49.33	49.33	0.05	0	0.0	49.33	49.33	0.05

		Asynchronous Variant with Birth Operator									
		migration place: next thread					migration place: random thread				
best2	C(I)	100	652.39	12.0	12.0	0.0	100	472.79	12.0	12.0	0.0
best2	C(II)	100	27712.16	20.0	20.0	0.0	100	23326.76	20.0	20.0	0.0
best2	C(III)	38	116789.77	80.0	49.52	4.45	66	121492.37	80.0	63.39	23.33
best2	C(IV)	1	148662.0	80.0	34.82	4.54	0	0.0	34.36	34.36	0.05
best2	C(V)	0	0.0	49.33	49.33	0.05	0	0.0	49.33	49.33	0.05
best + random	C(I)	100	777.71	12.0	12.0	0.0	100	811.5	12.0	12.0	0.0
best + random	C(II)	100	26608.98	20.0	20.0	0.0	100	25715.95	20.0	20.0	0.0
best + random	C(III)	39	128374.16	80.0	49.69	24.52	56	117110.44	80.0	58.93	24.60
best + random	C(IV)	1	123084.0	80.0	34.82	4.54	2	166874.0	80.0	35.28	6.39
best + random	C(V)	0	0.0	49.33	49.33	0.05	0	0.0	49.33	49.33	0.05
random	C(I)	100	742.36	12.0	12.0	0.0	100	633.79	12.0	12.0	0.0
random	C(II)	100	21931.74	20.0	20.0	0.0	100	24375.95	20.0	20.0	0.0
random	C(III)	56	117295.36	80.0	58.04	25.01	50	113501.22	80.0	54.71	25.3
random	C(IV)	0	0.0	36.0	34.38	0.17	1	119316.0	80.0	34.82	4.54
random	C(V)	0	0.0	49.33	49.33	0.05	0	0.0	49.33	49.33	0.05

evaluations to solution (AES); best solution found (*best*); the mean value of the best fitness values for all runs (*mean*); and the relative standard deviation (σ).

In table 2 we show the results obtained by the three variants of $mt - GA$ when the best solutions are always maintained in the new population (elitism approach). By inspecting these results, $mt - GA$ obtains the best performances for both synchronous and asynchronous variants when the $k = 2$ migrants are randomly chosen and the migration place is the next thread; whilst, instead, the last variant (asynchronous with birth operator) works better when the migration place is randomly chosen, and the migrants are selected among both best and random. All in all, this last variant seems to produce better results in all experiments. This can be explained because the introduction of new chromosomes balances the choice to focus the evolution onto the best solutions found so far (feature of any elitism approach).

Table 4. The three variants of $mt-GA$ on complex trap functions using *no preservation* approach as preservation strategy

		Synchronous Variant									
migrants	Trap	SR	AES	best	mean	σ	SR	AES	best	mean	σ
		migration place: next thread					migration place: random thread				
best2	C(I)	100	1451.98	12.0	12.0	0.0	100	1721.94	12.0	12.0	0.0
best2	C(II)	98	70712.0	20.0	19.88	0.89	95	62842.19	20.0	19.47	2.32
best2	C(III)	0	0.0	29.25	29.25	0.0	0	0.0	29.25	29.25	0.0
best2	C(IV)	0	0.0	34.36	34.36	0.05	0	0.0	34.36	34.36	0.05
best2	C(V)	0	0.0	49.33	49.33	0.05	0	0.0	49.33	49.33	0.05
best + random	C(I)	100	1825.64	12.0	12.0	0.0	100	1923.4	12.0	12.0	0.0
best + random	C(II)	100	61141.63	20.0	20.0	0.0	100	45439.82	20.0	20.0	0.0
best + random	C(III)	0	0.0	40.0	29.49	1.53	0	0.0	29.25	29.25	0.0
best + random	C(IV)	0	0.0	34.36	34.36	0.05	0	0.0	34.36	34.36	0.05
best + random	C(V)	0	0.0	49.33	49.33	0.05	0	0.0	49.33	49.33	0.05
random	C(I)	100	1561.16	12.0	12.0	0.0	100	1759.7	12.0	12.0	0.0
random	C(II)	100	41329.28	20.0	20.0	0.0	100	36417.68	20.0	20.0	0.0
random	C(III)	14	202747.08	80.0	40.54	17.84	27	169550.36	80.0	44.72	22.62
random	C(IV)	0	0.0	34.36	34.36	0.05	0	0.0	76.0	34.92	4.35
random	C(V)	0	0.0	49.33	49.33	0.05	0	0.0	49.33	49.33	0.05
		Asynchronous Variant									
		migration place: next thread					migration place: random thread				
best2	C(I)	100	4520.26	12.0	12.0	0.0	100	3726.43	12.0	12.0	0.0
best2	C(II)	90	70742.43	20.0	19.05	2.92	93	70598.37	20.0	19.31	2.55
best2	C(III)	0	0.0	29.25	29.25	0.0	0	0.0	29.25	29.25	0.0
best2	C(IV)	0	0.0	34.36	34.36	0.05	0	0.0	34.36	34.36	0.05
best2	C(V)	0	0.0	49.33	49.33	0.05	0	0.0	49.33	49.33	0.05
best + random	C(I)	100	4060.6	12.0	12.0	0.0	100	3626.43	12.0	12.0	0.0
best + random	C(II)	100	51491.81	20.0	20.0	0.0	100	42154.58	20.0	20.0	0.0
best + random	C(III)	10	187785.3	80.0	37.18	15.07	42	149380.88	80.0	52.31	24.29
best + random	C(IV)	0	0.0	34.36	34.36	0.05	0	0.0	36.0	34.4	0.23
best + random	C(V)	0	0.0	49.33	49.33	0.05	0	0.0	49.33	49.33	0.05
random	C(I)	100	3963.11	12.0	12.0	0.0	100	3542.22	12.0	12.0	0.0
random	C(II)	100	50164.49	20.0	20.0	0.0	100	39202.16	20.0	20.0	0.0
random	C(III)	18	176831.11	80.0	41.68	19.42	76	147446.23	80.0	69.48	19.62
random	C(IV)	0	0.0	34.36	34.36	0.05	4	163231.75	80.0	36.58	9.62
random	C(V)	0	0.0	49.33	49.33	0.05	0	0.0	49.33	49.33	0.05
		Asynchronous Variant with Birth Operator									
		migration place: next thread					migration place: random thread				
best2	C(I)	100	2269.42	12.0	12.0	0.0	100	2041.41	12.0	12.0	0.0
best2	C(II)	100	58625.36	20.0	20.0	0.0	100	40907.38	20.0	20.0	0.0
best2	C(III)	0	0.0	29.25	29.25	0.0	0	0.0	29.25	29.25	0.0
best2	C(IV)	0	0.0	34.36	34.36	0.05	0	0.0	34.36	34.36	0.05
best2	C(V)	0	0.0	49.33	49.33	0.05	0	0.0	49.33	49.33	0.05
best + random	C(I)	100	2012.56	12.0	12.0	0.0	100	1948.07	12.0	12.0	0.0
best + random	C(II)	100	48335.62	20.0	0.0	0.0	100	34830.15	20.0	20.0	0.0
best + random	C(III)	7	150069.58	80.0	34.88	13.67	23	147841.39	80.0	41.4175	21.11
best + random	C(IV)	0	0.0	34.36	34.36	0.05	0	0.0	36.0	34.38	0.17
best + random	C(V)	0	0.0	49.33	49.33	0.05	0	0.0	49.33	49.33	0.05
random	C(I)	100	2355.47	12.0	12.0	0.0	100	1195.44	12.0	12.0	0.0
random	C(II)	100	43478.27	20.0	20.0	0.0	100	28972.43	20.0	20.0	0.0
random	C(III)	12	173612.67	80.0	37.22	16.86	57	151960.47	80.0	59.14	24.54
random	C(IV)	0	0.0	34.36	34.36	0.05	1	232352.0	80.0	35.51	6.49
random	C(V)	0	0.0	49.33	49.33	0.05	0	0.0	49.33	49.33	0.05

Table 3 presents the results of $mt-GA$ when the worst offspring x_1 is replaced with the best one of its parents x_2; of course this is done if $f(x_1) < f(x_2)$. Using this strategy for the generation of the new population, the synchronous variant produces the best overall performances on all experiments conducted, and with respect to all the studied variants. In particular the absolute best results are obtained when the 2 best individuals of each thread become migrants, and the place of migration is randomly selected. Choosing randomly the migration thread helps considerably $mt - GA$ in finding better solutions independently on how the migrants are chosen, because in this way the cooperation between the $n = 4$ threads is improved. Also in this table, for both asynchronous variants the best place where to migrate is randomly chosen; the migrants, instead, are selected randomly for the simple variant, whilst with the birth operator they are chosen by picking the best, and one randomly. Comparing only these two last variants on table 3, is possible to see how the use of the birth operator helps $mt - GA$ to achieve a higher success rate.

For the last experiments, showed in table 4, we obtain a different behavior on both $mt - GA$ variants. In particular, without any preservation of the best solutions during the construction of the new population, the synchronous variant shows its worst performances (random migrants, and random migration place); whilst the asynchronous ones achieve the best results in overall. The best selection for both asynchronous variants is given by random migrants and random migration threads. With respect to previous tables, these experiments are the only one where the simple variant outperforms the one with the birth operator. By inspecting all three tables together, it is possible to claim that the best performances, in order to achieve the aims of this work, are given by the synchronous variant selecting the best two individuals for migrating in a thread randomly chosen.

Table 5. Comparisons between the three best variants of $mt - GA$ and two clonal selection algorithms: CLONALG [5, 6], and $opt - IA$ [3, 4]

Trap	SR	AES	SR	AES	SR	AES	SR	AES	SR	AES
			opt-IA [3], [5]					CLONALG$_1$		
	Inv		Macro		Inv+Macro		$\left(\frac{1}{\rho}\right)e^{(-f)}$		$e^{(-\rho*f)}$	
C(I)	100	371.15	100	737.78	100	388.42	100	272.5	100	251.3
C(II)	100	44079.57	100	27392.18	100	29271.68	100	17526.3	10	191852.7
C(III)	0	-	54	115908.61	24	149006.5	0	-	0	-
C(IV)	0	-	7	179593.29	2	154925	0	-	0	-
C(V)	0	-	2	353579	0	-	0	-	0	-
	Synchronous		Asynchronous		Asynchronous & Birth		CLONALG$_2$			
C(I)	100	1208.66	100	3542.22	100	1195.44	100	254.0	100	218.4
C(II)	100	18460.4	100	39202.16	100	28972.43	29	173992.6	24	172434.2
C(III)	91	75528.59	76	147446.23	57	151960.47	0	-	0	-
C(IV)	5	155389.8	4	163231.75	1	232352.0	0	-	0	-
C(V)	0	-	0	-	0	-	0	-	0	-

In order to better understand the robustness of the performances and the quality of the solutions produced by all $mt - GA$ variants, we have compared the designed algorithm with two well-known clonal selection algorithms: CLONALG [6], and $opt - IA$ [3, 4]). We underline the fact that today $opt - IA$ represents one of the best bio-inspired algorithms for optimization tasks [9]. The showed results for these two algorithms have been taken mainly from [5], and are showed in table 5.

By inspecting this table, it is possible to see how the three variants of $mt - GA$ outperforms all compared algorithms, achieving higher values of success rate (SR), except for the macro version of opt-IA on the traps C(IV) and C(V). From an overall point of view by inspecting the obtained results, it is possible to claim that $mt - GA$, as well as its three variants, are competitive on optimization tasks; are able to get out from local optima; and, finally, they prove to us that the implemented strategies are efficient, and robust.

6   Conclusions

The overall aim of this work is to develop a multi population-based algorithm capable to escape from local optima, which are the main reason why a generic optimization algorithm fails. In order to achieve the fixed aim, it is crucial to answer questions such as: (1) the best performance is obtained by running threads in a synchronous way, which strengthen the cooperation, or in asynchronous form? (2) which individuals should be selected as migrants? and (3) to what thread should they migrate? Our Algorithm, denoted $mt - GA$, addresses those questions and it contains a migration strategy which improves a the cooperation between the subpopulations, yet maintaining a sufficient amount of diversity.

Trap functions are a classical toy problems that represent a really useful tool in order to well understand the main features of a given EA, albeit they are not of immediate scientific interest. In particular, trap functions are mostly used for understanding the dynamics, and search's ability of a generic evolutionary algorithm. Although there exist two different scenarios (simple and complex), we focused our experiments only on complex trap functions, since they are sufficiently challenging to properly evaluate the goodness of $mt - GA$.

Many experiments were conducted on different complex trap functions, from which we tried to provide the right answer the above questions. We have tested three variants of $mt - GA$: (1) synchronous, (2) asynchronous, and (3) asynchronous with birth; this last variant is necessary when one thread does not receive migrants because it may not be running at that specific time. The migration place, and which migrants to select were problems studied as well.

By inspecting all performed experiments, we concluded that the best overall results are obtained by the synchronous variant, where the $k = 2$ migrants correspond to the best two individuals, and the migration place is randomly chosen. The two asynchronous variants instead show a slightly lower performances to the synchronous one, and they both obtain the best performance when the migrants, and migration place are randomly chosen.

Finally, in order to properly evaluate the efficiency, and robustness of $mt-GA$, we compared the best results obtained by each variant with two well-known optimization algorithms based on the clonal selection principle, CLONALG [6] and $opt-IA$ [3, 4] (belonging to Artificial Immune Systems class). In particular, the latter represents today the state-of-the-art for global optimization tasks. From the comparisons, we can see that the three variants of $mt - GA$ are very competitive with the other algorithms. This proves the efficiency and robustness of $mt - GA$, and the high success rates achieved on the most of the instances confirms us that $mt - GA$ is really suitable for the fixed aim.

Acknowledgements. We wish to thank the anonymous referees for their very valuable comments.

References

[1] Cantú-Paz, E.: Migration Policies, Selection Pressure, and Parallel Evolutionary Algorithms. J. Heuristics 7(4), 311–334 (2001)

[2] Cheng, H., Yang, S.: Multi-population Genetic Algorithms with Immigrants Scheme for Dynamic Shortest Path Routing Problems in Mobile Ad Hoc Networks. In: Di Chio, C., et al. (eds.) EvoApplicatons 2010, Part I. LNCS, vol. 6024, pp. 562–571. Springer, Heidelberg (2010)

[3] Cutello, V., Nicosia, G., Pavone, M.: Exploring the Capability of Immune Algorithms: A Characterization of Hypermutation Operators. In: Nicosia, G., Cutello, V., Bentley, P.J., Timmis, J. (eds.) ICARIS 2004. LNCS, vol. 3239, pp. 263–276. Springer, Heidelberg (2004)

[4] Cutello, V., Narzisi, G., Nicosia, G., Pavone, M., Sorace, G.: How to Escape Traps Using Clonal Selection Algorithms. In: 1st International Conference on Informatics in Control, Automation and Robotics (ICINCO), vol. 1, pp. 322–326. INSTICC Press (2004)

[5] Cutello, V., Narzisi, G., Nicosia, G., Pavone, M.: Clonal Selection Algorithms: A Comparative Case Study using Effective Mutation Potentials. In: Jacob, C., Pilat, M.L., Bentley, P.J., Timmis, J.I. (eds.) ICARIS 2005. LNCS, vol. 3627, pp. 13–28. Springer, Heidelberg (2005)

[6] De Castro, L.N., Von Zuben, F.J.: Learning and Optimization using the Clonal Selection Principle. IEEE Transaction on Evolutionary Computation 6(3), 239–251 (2002)

[7] Grefenstette, J.J.: Genetic algorithms for changing environments. In: Proc. 2nd Int. Conf. on Parallel Problem Solving from Nature, pp. 137–144 (1992)

[8] Nijssen, S., Back, T.: An analysis of the behavior of semplified evolutionary algorithms on trap functions. IEEE Transaction on Evolutionary Computation 7(1), 11–22 (2003)

[9] Pavone, M., Narzisi, G., Nicosia, G.: Clonal Selection - An Immunological Algorithm for Global Optimization over Continuous Spaces. Journal of Global Optimization 53(4), 769–808 (2012)

[10] Prugel-Bennett, A., Rogers, A.: Modelling GA Dynamics. In: Theoretical Aspects of Evolutionary Computing, pp. 59–86 (2001)

[11] Yang, S.: Genetic Algorithms with Memory- and Elitism-Based Immigrants in Dynamic Environments. Evolutionary Computation 16(3), 385–416 (2008)

An Improved Genetic Based Keyword Extraction Technique

J. Dafni Rose[1], Divya D. Dev[2], and C.R. Rene Robin[3]

[1] St. Joseph's Institute of Technology, Chennai-119, India
`jdafnirose@yahoo.co.in`
[2] St. Joseph's College of Engineering, Chennai-119, India
`divyaddev@gmail.com`
[3] Jerusalem College of Engineering, Chennai-100, India
`crrenerobin@gmail.com`

Abstract. Keyword extraction plays an increasingly crucial role in several texts related researches. Applications that utilize feature word selection include text mining, web page retrieval, text clustering and text categorization. Current methods for computing the keywords of a document are subject to a series of evolutions. Nevertheless, the methods do not perform well in very high dimensional state spaces. The methods are quite inefficient as they depend greatly on a human form of input. This attribute of the existing keyword extraction methods is not ideal in several applications. This paper presents a technique which will extract keywords without any kind of manual support. Genetic based extraction computes the list of key terms for each document. Irrespective of the text size, the novel method is able to perform the required computation with a higher echelon of performance. Calculations are done with the information taken from a structured document. Then the document is converted into a numerical representation by bestowing the distinct words with a numerical weight. The proposed method uses the knowledge of an iterative computation with a genetic algorithm to discover the optimal key terms. The evolutionary technique is subject to gradual changes that ensure the survival of the fittest. Experiments were done using three different data sets. The proposed method shows a high degree of correlation when the performance was checked against the existing methods of weighted term standard deviation, The Differential Text Categorizer method and the discourse method.

Keywords: Genetic algorithms, Weighted Term Standard Deviation, Genetic based algorithm, mutation, crossover.

1 Introduction

Digitization of information has led to problems of accessibility and discoverability [5]. The existing problems demand a solution which needs a proper classification of the text, to allow easy extraction of the required content. To achieve this, various keyword extraction techniques are used to extract the keywords from a

G. Terrazas et al. (eds.), *Nature Inspired Cooperative Strategies for Optimization (NICSO 2013)*, Studies in Computational Intelligence 512,
DOI: 10.1007/978-3-319-01692-4_12, © Springer International Publishing Switzerland 2014

plain textual document. Various keyword extraction techniques have been used in processes like text classification, text categorization, text summarization, etc [1]. These techniques rely on exploiting various document parameters like the word count, term occurrences and term position. "Ease of use" and the "ascertain temperament" of the text are the two key problems, which have been instantiated as a result of high digitalisation. Understanding a corpus would require navigating through the entire manuscript to discover the foremost terminologies. For an effective level of keyword extraction, information has to be gathered from the document's structure. Word building methods with iterative principles work on all the different aspects of document analysis. This increases the accuracy and precision of the possible outputs.

Each document comprises of a topic. The documents have an appropriate title that is used to describe the content. The title is followed by the body of the document. The body is not a single entity but a collection of many factors. The factors would include the sub paragraphs that elaborate on the topic, the links and other references. The importance of the data varies according to the location of the content within the document [9]. The initial words of a paragraph can be given more priority, than the others that occur later in the document.

In this paper, the initial weight of the words is calculated using the basic equations, as stated by the Weighted Term Standard Deviation method in an extended fashion. The basic weighing function is followed by the Genetic Procedures. The term weighing function works on certain aspects of the document features, by which the word profile is established. The initial weights that are calculated are then passed into the genetic algorithm procedures, which are used to enhance the feature-weighting schemes. The total number of words derived is based on the dimensionality of the content. Several test documents are used to decide the efficiency of the extraction. The end result is compared against a prebuilt set of words and their weights. The similarity values are also computed to ensure the output. In this paper we exploit a technique that does not rely on manpower and can vary outputs across all possibilities of the text parameters. It ensures the accuracy and optimization of the end results.

The rest of the paper is organized as follows: Section 2 presents the literature on feature extraction. Section 3 discusses the design and development of the new method for keyword extraction, and introduces the experimental procedures for the performance measures. Section 4 analyses the data and the results are obtained. Section 5 concludes with the summary, identifying the limitations of the research and proposes suggestions for future study.

2   Literature Review

The fundamental keyword extraction method would be the term frequency (TF) and inverse document frequency (IDF) where the occurrences of the term in the document are considered [7, 8]. The IDF gives the weight with respect to the whole pack of documents. The weight of the characteristic term is divided into two parts [8, 9].

$$tfidf(ti, dj) = tf * \log(N/n)$$

where
tf - term frequency of the ith term in the jth document,
N- Total number of documents in the collection,
n - Document frequency of the ith term.

The weight of a term is directly proportional to the number of times it occurs in the document. This has a positive and negative impact on the final weighing. The advantage of using the term frequency is that, it clearly shows the words which occur repeatedly. Nevertheless, less priority is given for the ones that occur less frequently. To improve the effect of the computation, newer methods take into consideration the relevancy of the terms in the document. The relevance frequency (RF) [8] is given by the following:

$$RF(t) = \log(1 + n(+)/n(-))$$

n(+) - number of positive documents with a particular term,
n(-) - number of negative documents with a particular term.

KEA [3, 16] is a keyword extraction algorithm that was generated in the later stages of the feature selection research. The new KEA++ includes the phrase length of the words and the frequency of the various nodes. KEA was released in several versions. Each had a predominant variation from the other. Various factors like the document type, the structure of the corpus and the context can be used to help the feature calculation [13]. The existing techniques are not reliantly fast when the above considerations are made. KEA is a supervised keyword extraction strategy. The other supervised methodologies include the Support Vector Machine (SVM) [6, 15].

In [11] a statistical equation called Weighted Term Standard Deviation (WTSD) is proposed to represent the dispersion of the concepts in a document. They divided the standard deviation by a maxline (di), to normalize the standard deviation by the size of each document.

$$Vard_i c_j = \sqrt{\frac{\sqrt{x^2 - (\Sigma x)^2 / D_c}}{(D_c - 1)maxline(d_i)}}$$

where
x - position of the ith term in the jth document,
Dc - total number of words in a document "d",
$maxline(d_i)$ - the maximum line of document with a maximal width,
$Vard_i c_j$ - standard deviation of the concept position.

The standard deviation method was used to represent the distribution of the dispersion in a document. The higher the standard deviation of the document is, the larger the dispersion in the distribution would be [11].

The discourse method ought to take place in the initial sections of a document. It is not applicable for the latter portions. The word weight is computed with respect to the frequency. The position of each word in the sentence is taken into consideration in the discourse technique [12]. The discourse is similar to the TF-IDF method. The equation representation is as follows:

$$w_i = tf * \log(s/s_p)$$

where
w_i - weight of word in the ith position of the jth sentence,
tf - term frequency of the ith term in j^{th} document,
s - number of sentences in document "d",
s_p - position of the ith word during its first appearance in document "d".

In [14] various distributional features of a word are used to characterize the importance of a word in the document. The various distributional features include the compactness of the word, the frequency of its appearance and the position of its first appearance. But all these features are dependent on various factors, like the length of the document etc.

From the above, it is evident that the discourse method does not consider the later appearances of a word. The equation takes only the first occurrence of a word into consideration. This is a major drawback of the technique. The procedure would be effective in a smaller document space, but when applied in large corpuses where the terms tend to repeat, the method will result in incorrect results. The discourse is a simple formulation that will work perfectly with documents with a lower number of words. As the number of terms and sentences increases, the overall output accuracy declines.

On the whole, it is evident that the existing methods do not dig into the intrinsic features of a document [2]. The term frequency method relies on the frequency of the keywords. This is not sufficient to come to a conclusion, that the term which keeps repeating is the ideal feature word of the document. The method has to be revised and more parameters need to be included in it. The supervised method is a tedious one, which needs the user to input test data into the system. For effective extraction there has to be an effective training process. KEA is a common keyword extraction algorithm, but it has a lot of scope for improvement.

Our method is an enhancement of the Weighted Term Standard Deviation (WTSD) technique. It has been the base of many feature selection methodologies. The Weighted Term Standard Deviation method does not consider the position of a word with respect to each sentence. This is a feature that can increase the accuracy of a keyword weight computation. The discourse method does not take into consideration the parameters of the words, which occur in the later positions of a document, which in turn, reduces the accuracy of the output. The genetic based algorithm has identified the drawbacks of the existing systems, and modulated a new set of formulas which gives the terms a better weight value. Statistically and experimentally the new algorithm has established an improved performance.

3   Proposed Method

This paper shows how the genetic algorithm can be used to haul out the supreme set of keywords. The computational challenges are analysed for efficient and effective retrieval of the scoring words from both the real and synthetic data sets. This involves the repetitive application of the mutation, crossover, and fitness functions along with the selection operators.

The main objective is to the extract keywords from each individual document. The initial phase of the keyword extraction involves pre processing the document. The terms need to be assigned with a weight that will help in prioritizing them within the population. Once the initial weight is calibrated, the genetic procedures are executed to gain a final keyword population. The terms are chromosomes and the weights are the numeric representation of genes. A simple, modified arithmetic technique is applied for crossover, trialed by the "Expected Number of Elements in the Population" viewpoint to declare the fitness of the engendered populace. Mutation is alleged only if the fitness utility is not contended for cessation.

3.1   Document Pre-processing

The content is initially converted into the appropriate form [5]. The processes for stop word removal and stemming are applied to the text, which is given as the input. Pre-processing involves the elimination of non-textual information like punctuation, HTML tags etc. The reduction of the document dimension improves the performance of the extraction technique. Most of the keywords would not be less than 2 characters; with this assumption the words that confine to the above range are dropped. The space complexity is reduced with the pre-processing methods.

3.2   Initial Weighing of the Keywords

Keyword retrieval systems are based on the weight of terms. Term weighing is divided into two sections. The initial phase gives each term a weight. The second phase involves the execution of several genetic engineering modules. The net output of the combined processes would be a set of terms that describes the document.

The initial weighing equation is once again divided into two parts. When words occur for the first time, an extension of the weighted term standard deviation method as given by [11] was used. The equation that was proposed previously exploits only certain document features. To improve the accuracy the new formulation weighs the terms based on all the parameters, which could identify the terms in the document.

The equation (1) takes into consideration the "Position of a term in a given sentence" along with "Average position in whole document, $Avg(w_x)$". $Avg(w_x)$ is a parameter that has not been used in many keyword extraction mechanisms. $Avg(w_x)$, gives priority to a word which occurs many times and at diverse positions, than a word which occurs only once in the first sentence. Thus, this incorporates the idea of "Survival of the Fittest". Additionally, equation (1) takes into consideration the number of words in each sentence, s_c. This is an important parameter which gives value to each and every sentence. Thus words are weighted with respect to the document and each sentence. This increases the precision and accuracy of the term weights. The very first equation in the genetic based method is as follows.

$$w_i = \frac{s_c/Avg(w_x)}{maxline(d_i)/D_c} \tag{1}$$

where
w_i - weight of word in the ith position of the jth sentence,
s_c - word count in the ith sentence,
$Avg(w_x)$ - Average position of the ith term in a document "d",
$maxline(d_i)$ - the maximum line of document with a maximal width,
D_c - total number of words in a document "d".

When a word starts repeating, the next equation (2) is used. The equation has utilized the concept of term frequency. As a term keeps repeating over and over again, it could have a certain amount of relevance. The idea is made prominent by considering several other document attributes, which reveal the real relevance of each term. When the same word repeats at position "$t + 1$", its weight is calculated using the weight that was being computed at position "t" and this is continued every time a term re-appears. The equations (2) and (3) now deviate from the base equations and navigate into the unique document features. This increases the accuracy of the output.

As a word repeats, an incremental factor Wicr is computed. This increases the weight of the "repeating" term. Thus, as the weight of a term increases, its priority in the document will also enhance. W_{icr} is computed with the term's current parameters. This includes "the number of words in the sentence in which it currently appears" and "the improved average position of the term". These are the factors, which will keep the recurrent term at a higher position. With these factors, the weight of a term is computed with W_{icr} and the "square root of its previous weight". This generalizes the weight and prevents it from deviating away from the "document's term populace".

$$W_{icr} = \frac{1}{s_c^* \sqrt{Avg(w_x)}} \tag{2}$$

$$Vw_i = W_{icr} + \sqrt{Vw_i(t-1)} \tag{3}$$

where W_{icr} - value to be added to the weight of a word when it is encountered more than once, $Vw_i(t-1)$ - Weight of word at position $t-1$, Vw_i - Net Weight of the "i^{th}" term after increment.

This helps with the evaluation, and words with lower weights give a reduced prominence to the document meaning. Words that occur deeper down in the passage may be seen to have less weight-age but the positional factors of the words in the particular sentence neutralises the drop. A balance is made between the position of the word in the document and its occurrences. The combination of the above two equations will produce the best set of keywords to be passed on to the next phase of genetic algorithms. Algorithm 1 gives an overview of the process.

Algorithm 1. Calculating the initial weights of Words

INPUT

A Text document "D" with:

"S" number of sentences

"C" number of words in each sentence

Let "E" be the set of distinct words

Let "Vw_i" be the weight of each distinct word in the document

INITIALISE

The weight Vw_i of each word w_i to 0

EXECUTE

for i from 1 to S **do**

for j from 1 to C **do**

Get the word w_j in position "j"

if sentence s_i contains the word w_j that is not in E

then

add w_i to E

Calculate Vw_i of word w_i according to (1)

else

Calculate Vw_i of word w_i according to (2)

End if

End for

End for

OUTPUT

A set of distinct words with a weight Vw_i

3.3 Application of the Genetic Algorithm

The genetic algorithm helps to obtain improved results than the previous techniques. The genetic algorithm relies on a population referred to as chromosomes, which have been encoded as solutions to emphasize the optimization [4]. The process starts with a randomly selected population, but ends with a stable and coherent resultant. It works with mutual processes like the fitness function and mutation.

The fitness function will evaluate the populations and the next generation is finalised by the process of mutation. The algorithm concludes when, an utmost amount of production has been achieved, or an adequate suitability echelon has been accomplished. The objective is to engender solutions with optimization. Keyword extraction is an idea of pronouncing the terms restating a document the search is optimised with the assistance of the inherent schemas. The results of the initial weighing equations are passed onto the genetic procedures. The input weights are sorted in the descending order, before the methods of crossover, fitness and mutation are applied.

3.4 Probability Crossover

The performance of the genetic algorithm is controlled by the crossover value. The choice of the crossover value should be such as to control the performance of the genetic algorithm in a positive manner. The probability crossover for our procedures is indicated with respect to the word weights. The efficiency of the mechanism is increased since there is no need to convert the chromosomes into the binary type; thus, no loss in precision by discretion from binary to other values. The equations are framed in such way that values which are not likely to occur in the final population will sense the required changes and determine their chance of occurrence thereafter.

The basic principle behind crossover involves the divide and conquer method. The populationis brokeninto two halves where the first segment contains the better half,and the rest holds the weight of the lower probability poulation. W_i indicates the weight of a word in position "i". Pc_i is the probability ratio of the most feasible word with respect to the word that has the highest occurence in the other part of the division. The genetic based algorithm creates a balance between the words with high and low weight values. It divides term weights, which are sorted in descending order into two distinguishable halves. Where the first segment contains the better half and the rest holds the weight of terms in the lower probability population. Thereforth, equations (5) and (6) can be used to measure the probability crossover of terms corresponding to each other in the broken set of words.

For example, if the document term set has 50 different words, the first half as a result of "divide and conquer" will have 25 terms with high weights. While the other half will have words with lower weights. To compute probability crossover, the "first word in the first half" will be paired with the "first word in the second half". This has to be continued till the whole term populace is computed against probability crossover. As a result, words which were thought to be meaningless will be given a "second chance" to prove its priority. Probability Crossover is succeeded by fitness and mutation. Thus, words which had lower weights but a higher likelihood of defining the document will be allowed to come within the final term set. This clearly stresses on the genetic principle the "Survival of the Fittest". Equations (4), (5) and (6) computes in an organised manner, following the principles of simple crossover. It does not select words in a random order. The equations exploited in the crossover mechanism include the following:

$$P_i = W_i$$
$$P_k = W_k$$
$$K = P_i + P_k \tag{4}$$
$$a_i = \frac{P_i}{(K)}$$
$$a_k = \frac{P_k}{(K)}$$
$$Pc_i = (a_i * P_i) + (1 - a_k)P_k \tag{5}$$
$$Pc_k = (a_k * P_k) + (1 - a_i)P_i \tag{6}$$

where
P_i - Weight of a word "i" in the first half of the population,
P_k - Weight of a word "k" in the second half of the population,
a_i - Probability occurrence of a word "i",
a_k - Probability occurrence of a word "k",
Pc_i - modified weight of the word "i",
Pc_k - modified weight of a word "k".

The probability method of "$p = (1-q)$" is used to obtain the likeliness between words. "P" denotes the weight of a highly prioritised word and "q" indicates the words with a lower priority. The above methodology has the ability to push words with less feasibility to a better position. The base of the crossover is to achieve a population by choosing parents that have the ability to generate offspring that can satisfy the fitness calculations. The values of crossover can affect the whole algorithm, as its unhealthy values could lead to an un-optimal local solution that may cease [10].

3.5 Fitness

The robustness of the solution depends on the selection of parents passed from the current process to the next iteration. The fitness function is used to generate a functional assessment of comparative fitness expressions. The survival of an individual term is considered independently. The algorithm flows with the standard of "Expected number of copies of the word "i" in the forthcoming population". The equation representation would be as follows:

$$E(n_i) = \frac{Pc(i)}{avg(Pc)} * D_{max} \tag{7}$$

$$T(i) = E(n_i) * D_{distinct\_terms\_in\_documents} \tag{8}$$

$$if(T(i) < avg(P_c))$$

$$Pm_i = 0$$

where
$E(n_i)$ - Expected copies of the ith term,
$Pc(i)$ - Probability Crossover weight of the ith term,
$Pm(i)$ - Probability Mutation of the ith term,
$avg(Pc)$ - Average weight of "all" words in a document "d",
$D_{distinct\_terms\_in\_documents}$ - Number of distinct words in a document.

The novel genetic based algorithm computes a fitness value for each term. It calculates the likelihood of a term in the final populace, with respect to its probability crossover. This is followed by a test condition, which calibrates each individual term against all the other terms in the document. The term which ceases to fall within the average probability crossover of all the other terms will be eliminated. Similarly, words with higher weights and greater divergence from the normal population of terms will also be eliminated. This allows the final set of terms to be confined with a small state space.

3.6 Mutation

Mutation endorses the amendment of term weights with a probability mutation "pm". Mutation has the capability to reinstate the mislaid genetic material into the populace, thus thwarting the convergence of the solution into a suboptimal region, or its divergence into an infinite loop. $T(i)$ decides whether the mutation process is to be applied or not. When a word does not lie within the fitness condition, the process of mutation is not applied. The fitness value determines the mutation. When two consecutive iterations have a similar weight-age for the terms, the ultimate keyword list is generated. The equational representation of mutation is given by the following formula:

$$Pm_i = \sqrt{Vw_i(i)} + \frac{tf}{Avg[x]} \qquad (9)$$

where $Avg(x)$ - average position of the ith term in the jth document.

Mutation considers the term frequency and the average position of words. This allows the terms to retain its documental parameters. Words with greater weights, positioned at the initial stages of a sentence and has incurred a greater frequency will be allowed to have a greater weight value. This will also prevent the genetic based algorithm from ceasing at a suboptimal region.

4   Computational Results

Three different datasets were used for the study. One is the Reuters-21578, manually processed set of abstracts and the test data set of KEA. The Reuters-21578 is an experimental data collection that appeared on the Reuters newswire of the year 1987. The dataset was obtained from

http://kdd.ics.uci.edu/databases/reuters21578/reuters21578

Nearly 12300 articles from Reuters under different categories were uniquely assigned, and used for evaluation. There is no standard keyword list for this dataset; so the keywords extracted using the TF method, are used as the baseline for our evaluation. The manually processed set of abstracts is obtained from cornell.edu. This contained around 500 documents with a predefined collection of keywords. The keywords were extracted using the test methods WTSD, discourse, text categoriser and genetic based extraction. The output was evaluated against the keywords extracted, using term-frequency method. The KEA dataset is obtained from the KEA website. The dataset had a standard keyword list attached with every document. The dataset that was pre-processed and executed by using KEA was also tested with the WTSD, discourse, text Categoriser and genetic based algorithm, and calibrated to show the level of performance. The keywords that were extracted by these methods are compared with the standard keyword list given in the dataset.

After pre-processing, the set of words that can describe the document is left over. The words are given a document term weight. The initially mentioned weight age equations are applied to gain the numerical representation. Thereafter, the weights are passed to the genetic algorithm that comprises the newly devised formulations for the feature selection.

4.1 Percentage of the Keywords Extracted

The percentage of the keywords extracted is found using the following formula

$$\%words\_found = \frac{(100 * words\_extracted)}{Total\_Number\_of\_keywords}$$

Table 1 shows the percentage of the keywords extracted by the several methods. The results did justify that the new way of extraction is more appropriate than the existing ones.

Table 1. % of the Keywords Extracted

Dataset	No. of keywords	WTSD	Discourse	Text Categoriser	Genetic based method
Reuters	3	56.6	11.1	66.7	66.7
	4	63.3	12.3	75.0	66.7
	5	99.6	23.6	80.0	99.6
	6	64.4	54.9	66.7	76.6
	7	82.3	69.3	14.2	87.4
	8	73.3	66.7	25.0	72.6
	9	33.3	33.5	66.7	83.3
Cornell	4	34.5	18	25	38
	5	42.4	21.2	60.0	44.4
	6	51.6	30.4	66.7	54.7
	7	47.5	29.9	42.8	50.6
	8	47.5	29.5	37.5	56.5
	9	38.5	15	22.2	35.6
	10	38.3	17.6	20.0	38.1
KEA	7	34.4	54.6	57.1	58.5
	8	49.6	30.9	45.6	58.6
	9	37.3	15.8	55.5	37.8
	10	39.6	17.5	38.7	40.3
	11	21.2	18.8	47.6	29.9

4.2 *F*-Measure

The evaluation of the algorithms can be done using the F-Measure. This evaluation method works with the Recall and Precision values. The higher the F-Measure, the better will be the performance of extraction. The F Measure is a standard method that is used to analyze the performance of the keyword extraction methods. These standard measures to address the performance are given by:

$$Recall, R = \frac{Terms\_found\_and\_correct}{Total\_number\_of\_keywords}$$
$$Precision, p = \frac{Terms\_found\_and\_correct}{Keywords\_extracted}$$

F-Measure for a document, D is given by:

$$F(D) = \frac{2PR}{(P+R)}$$

Table 2 shows that the genetic based method have higher values than other methods. This is a clear evidence that the genetic based method is better than other methods.

Table 2. F-Measure values

Dataset	No. of keywords	WTSD	Discourse	Text Categoriser	Genetic based method
Reuters	3	0.27	0.06	0.29	0.30
	4	0.24	0.06	0.14	0.24
	5	0.28	0.08	0.35	0.28
	6	0.18	0.15	0.19	0.21
	7	0.19	0.17	0.09	0.21
	8	0.16	0.13	0.14	0.16
	9	0.07	0.07	0.38	0.16
Cornell	4	0.58	0.31	0.07	0.64
	5	0.57	0.28	0.07	0.59
	6	0.63	0.37	0.12	0.67
	7	0.55	0.34	0.09	0.58
	8	0.50	0.31	0.16	0.59
	9	0.37	0.16	0.12	0.38
	10	0.40	0.18	0.08	0.40
KEA	7	0.21	0.49	0.18	0.37
	8	0.22	0.53	0.15	0.53
	9	0.19	0.64	0.28	0.71
	10	0.33	0.57	0.42	0.75
	11	0.21	0.53	0.60	0.21

4.3   Entropy

The entropy measure can also be used for a homogenous evaluation. The total entropy is evaluated as the sum of the precision. It is necessary to minimize the entropy of the results. The formula for entropy calculation is given below.

$$Entropy = -\sum_{i \in P} P(i) \times \log(P(i))$$

The output is tabulated in Table 3.

Table 3. Entropy Values

Dataset	No. of keywords	WTSD	Discourse	Text Categoriser	Genetic based method
Reuters	4	0.28	0.57	0.30	0.21
	5	0.31	0.59	0.76	0.31
	6	0.27	0.53	0.35	0.20
	7	0.38	0.40	0.56	0.35
	8	0.37	0.39	0.59	0.35
	9	0.40	0.44	0.53	0.41
	10	0.56	0.56	0.05	0.40
Cornell	4	0.28	0.33	0.99	0.08
	5	0.28	0.30	1.11	0.08
	6	0.33	0.36	0.85	0.10
	7	0.30	0.33	0.97	0.10
	8	0.28	0.34	0.57	0.09
	9	0.23	0.25	0.58	0.05
	10	0.26	0.28	0.83	0.08
KEA	7	0.67	0.45	0.39	0.32
	8	0.63	0.18	0.36	0.18
	9	0.71	0.13	0.32	0.19
	10	0.52	0.17	0.18	0.30
	11	0.67	0.67	0.19	0.22

The genetic based method had the least value in the entropy computation. This is clear evidence that the novel method is better than the existing ones. Table 3 shows the numerical values of entropy for the data sets.

5   Conclusion

The paper demonstrates the execution of the keyword extraction using the genetic algorithm techniques. A variety of features are proposed beyond the traditional frequency and position based features extracted from the system generated summaries. We have proposed a looping structure that works with the leverage of relationship between the keyword extraction and summarization. According to the experimental data examination the projected procedure is better than the standard deviation methods by virtually 15% in terms of the F-Measure. The method appears to be more powerful and has given satisfactory levels of achievement, when large data sets were worked with. The fitness function that has been keenly focused on, makes certain that the best word is not lost through the iterations. It also gives room for the lower value words to catch up with the final population. The genetic algorithm bestows a healthier approach in pronouncing the words that depict the document. There is scope for enhancement where the restrictions on the dimensions of the state space can be imposed. This enables reducing the deviated word list from being generated. Upgrading, in terms of dimension and accuracy, can allow further supremacy of the keyword extraction technique.

References

[1] Abdelmalek, A., Zakaria, E., Ladjel, B., Michel, S., Mimoun, M.: Concept - Based Clustering of Textual Documents Using SOM. In: Computer Systems and Applications AICCSA, pp. 156–163 (2008)

[2] Berend, G., Farkas, R.: Feature engineering for keyphrase extraction. In: Proceeding of the 5th International Workshop on Semantic Evaluation, pp. 186–189. ACL, Uppsala (2010)

[3] Bracewell, D.B., Ren, F., Kuriowa, S.: Multilingual single document keyword extraction for information retrieval. In: IEEE International Conference in Natural Language Processing and Knowledge Engineering, pp. 517–522 (2005)

[4] Khalessizadeh, S.M., Zaefarian, R., Nasseri, S.H., Ardil, E.: Genetic Mining: Using Genetic Algorithm for Topic based on Concept Distribution Word. World Academy of Science, Engineering and Technology

[5] Kian, H.H., Zahedi, M.: An efficient approach for keyword selection: improving accessibility of web contents by general search engines. International Journal of Web & Semantic Technology 2(4) (2011)

[6] Zhang, K., Xu, H., Tang, J., Li, J.: Keyword Extraction Using Support Vector Machine, pp. 85–96. Springer, Berlin (2006)

[7] Matsuo, Y., Ishizuka, M.: Keyword Extraction from a single document using word co-occurrence statistical information. Int. J. Artificial Intelligence 13 (2004)

[8] Murugeshan, M.S., Lakshmi, K., Mukerjee, S.: A negative category based approach for Wikipedia document classification. Int J. Knowledge Engineering and Data Mining 1(1), 84–97 (2010)

[9] Salton, G., Buckley, C.: Term-weighting approaches in automatic text retrieval. Information Processing Management 24(5), 513–523 (1988)

[10] Srinivas, M., Patnaik, L.M.: Adaptive Probabilities of Crossover and Mutation in Genetic Algorithm. IEEE Transactions on Systems, Man and Cybernetics 24(4) (1994)

[11] Weng, S.S., Lin, Y.-J.: A Study on searching for document based on multiple concepts and distribution of concepts. Expert Systems with Applications, pp. 355–368. Elsevier (2003)

[12] Kawahara, T., Hasegawa, M., Shitaoka, K., Kitade, T., Nanjo, H.: Automatic indexing of lecture presentations using unsupervised learning of presumed discourse Markers. IEEE Transactions on Speech and Audio Processing 12, 409–419 (2004)

[13] You, W., Fontaine, D., Barthes, J.-P.: An automatic Key phrase extraction system for scientific documents. In: Knowledge Information System. Springer-Verlag London Limited (2012), doi:10.1007/s10115-012-0480-2

[14] Xue, X.-B., Zhou, Z.-H.: Distributional Features for Text Categorization. IEEE Transactions on Knowledge and Data Engineering

[15] Zhang, K., Xu, H., Tang, J., Li, J.: Keyword extraction using support vector machine. In: Yu, J.X., Kitsuregawa, M., Leong, H.-V. (eds.) WAIM 2006. LNCS, vol. 4016, pp. 85–96. Springer, Heidelberg (2006)

[16] Li, Z., Zhou, D., Juan, Y.F., Han, J.: Keyword Extraction for Social Snippets. In: Proceedings of the 19th International Conference on World Wide Web, pp. 1143–1144 (2010)

Part-of-Speech Tagging Using Evolutionary Computation

Ana Paula Silva[1], Arlindo Silva[1], and Irene Rodrigues[2]

[1] Escola Superior de Tecnologia do Instituto Politécnico de Castelo Branco, Portugal
{dorian,arlindo}@ipcb.pt
[2] Universidade de Évora, Portugal
ipr@uevora.pt

Abstract. Part-of-speech tagging is a task of considerable importance in the field of natural language processing. Its purpose is to automatically tag the words of a text with labels that designate the appropriate parts-of-speech. The approach proposed in this paper divides the problem into two tasks: a learning task and an optimization task. Algorithms from the field of evolutionary computation were adopted to tackle each of those tasks. We emphasize the use of swarm intelligence, not only for the good results achieved, but also because it is one of the first applications of such algorithms to this problem. This approach was designed with the aim of being easily extended to other natural language processing tasks that share characteristics with the part-of-speech tagging problem. The results obtained in two different English corpora are among the best published.

Keywords: Part-of-speech Tagging, Disambiguation Rules, Evolutionary Algorithms, Particle Swarm Optimization, Natural Language Processing.

1 Introduction

The words of a language are usually grouped in lexical categories or parts-of-speech (POS). A tagger is a system that should receive a text, made of sentences, and, as output, should return the same text, but with each of its words associated with the correct POS tag. These tags are acronyms for the lexical categories chosen for labeling the words. The process of classifying words into their POS, and labeling them accordingly, is known as POS tagging, or, simply, tagging. In most languages, each word has a set of lexical categories that represent the roles that they can assume in a sentence. When the cardinality of this set is greater than one, we say that the word is ambiguous. The context of a word, i.e., the lexical categories of the surrounding words, is the fundamental piece of information for determining its role in a sentence. For instance, the word *wind* can assume the function of a **verb**, if it follows the word *to*, or can be used as a **noun** if it is preceded by a determiner like *the*. According to this, most taggers take into consideration the context of a word to decide which should be its tag.

G. Terrazas et al. (eds.), *Nature Inspired Cooperative Strategies for Optimization (NICSO 2013)*, Studies in Computational Intelligence 512,
DOI: 10.1007/978-3-319-01692-4_13, © Springer International Publishing Switzerland 2014

However, each of the words belonging to a word's context can also be used in different ways, and that means that, in order to solve the problem, a tagger should have some type of disambiguation mechanism that allows it to choose the proper POS tags for all the words of a sentence.

The methods used for solving the POS tagging problem can be divided into two distinct groups, based on the information they use. In one group, we can gather the approaches that use statistical information about the possible contexts of the various word tagging hypotheses. Most of the stochastic taggers are based on hidden Markov models. In the other group, we find rule based taggers [1–3]. The rules are usually discovered automatically, and its purpose is to correct errors resulting from an initial basic tagging. Brill's tagger [1] is perhaps the most popular tagger based on rules.

More recently, several works following an evolutionary approach have been published. These taggers can also be divided by the type of information they use to solve the problem: statistical information [4, 5], and rule-based information [2]. In the former, an evolutionary algorithm is used to assign the most likely tag to each word of a sentence, based on a training table that basically has the same information that is used in the traditional probabilistic approaches. The later is inspired by Brill's rule based tagger. In this case a genetic algorithm (GA) is used to evolve a set of transformations rules, which will be used to tag a text in much the same way as in Brill's tagger. While in [4, 5], the evolutionary algorithm is used to discover the best sequence of tags for the words of a sentence, using an information model based on statistical data, in [2] the evolutionary algorithm is used to evolve the information model itself, in the form of a set of transformation rules.

Although the POS tagging problem is a task that has had a special attention in the field of natural language processing (NLP), the evolutionary approach deserves, in our opinion, a more thorough study. We believe that this study should include the application of other algorithms from the evolutionary computation field. Moreover, previous work suggest the exploitation of these algorithms on two key aspects of the task: the information gathering and the automatic process to perform the tagging, according to the information collected. In this paper, we present a new evolutionary approach to the POS tagging problem. Our strategy implies a division of the problem into two different tasks: a learning task and an optimization task. These are tackled using not only evolutionary algorithms, but also particle swarm optimization (PSO), resulting, as far as we know, in the first attempt to approach this problem using swarm intelligence. Although focusing mainly on the POS tagging problem, we believe that this work may be the foundation for a new paradigm to solve other NLP tasks. This paradigm is based, however, in two fundamental assumptions:

– With the help of a classification algorithm, it is possible to generalize, from linguistic resources, the information typically used in the probabilistic approach, by learning a set of disambiguation rules. These rules will not play the role of a classifier, instead they will be used as an heuristic to help solve the task in question.

– It is possible to formalize the main problem as a search problem and use the rules discovered in the first phase as an heuristic to guide the search for a solution in the problem state space.

The field of evolutionary computation includes a set of global optimization algorithms that have been applied, with recognized success, to a vast and varied set of problems in areas such as optimization, search and learning. These algorithms are characterized by being easily adapted to different representations and tasks. They are also global optimization algorithms, hence outperforming many of the greedy approaches. Therefore, they present themselves as a suitable tool to integrate the approach we propose here, since they can be used in both phases of the strategy. Also the inherent versatility of these algorithms contributes to strengthen the possibility of applying this approach to other NLP tasks.

2   Rules Discovery Using Evolutionary Computation

It is our belief that the information stored in the training tables of the probabilistic approach can be interpreted as a set of instances. Each of these instances is typically described by a set of measurable attributes related to the tags of the surrounding words, and is associated with a numerical value that identifies the number of times each one occurs in the training corpus. Naturally, this information is specific to the corpus from which it was collected and does not show any degree of generalization, instead it can easily be interpreted as an extensive and comprehensive collection of information. Hence we are convinced that it is admissible to investigate the possibility of generalizing this information using a classification algorithm. From this generalization we expect to be able to reduce the amount of information needed to solve the problem and also to improve the tagging accuracy. The learned rules may be used, in a similar way to the training table, to guide the search of the POS tagging problem state space. They aim not to classify a given word, but rather assess the quality of a particular classification.

Previous experience with classification rules discovery [6, 7], using evolutionary computation, has led us to define the classification algorithm based on a covering algorithm. The outline of the algorithm used is defined in Algorithm 1. As we can see, the set of rules is obtained by executing the search algorithm m times. This algorithm is responsible for determining the best classification rule for the set of training examples it receives as input. At each execution, the rule obtained is stored, along with its quality value, and the set of positive examples is updated by eliminating all the instances covered by the rule. The search algorithm will be executed as many times as necessary, so that all positive examples are covered, i.e., the set of positive examples is the empty set. We divided the problem into n distinct classification problems, n being the number of different tags used in the annotated corpus, from which the rules will be learned and that define the tag set E. Each tag $e \in E$ presented in the corpus determines a classifying object, with possible classes taking values from the discrete set $Y = \{Yes, No\}$. Two different search algorithms were tested: one based on a GA and another based on a PSO. A more detailed description of the implemented algorithms can be found in [8, 9].

Algorithm 1. Covering Algorithm

Require: PosExemples, NegExemples
Ensure: RulesSet
   while PosExemples $\neq \varnothing$ **do**
      $\langle$BestRule, Quality$\rangle \leftarrow$ SearchAlgorithm(PosExemples, NegExemples)
      PosExemples $\leftarrow$ Remove(PosExemples,BestRule)
      RulesSet $\leftarrow$ Add(RulesSet, $\langle$BestRule, Quality$\rangle$)
   end while

2.1 Prediction Attributes and Representation

The use of rules allows, in addition to the grammatical categories of the surrounding words, the consideration of other aspects. Although a word context is perhaps the most determinant piece of information to identify its lexical category, there are also some other aspects that can be helpful. The internal structure of a word may give useful clues as to the word's class [10]. For example, *-ing* is a suffix that is most commonly associated with gerunds, like *walking, talking, thinking, listening*. We also might guess that any word ending in *-ed* is the past participle of a verb, and any word ending with *'s* is a possessive noun. Taking these observations into account, we considered as prediction attributes two distinct groups. The first group includes six attributes related with the context: the lexical categories of the third, second and first words to the left, and the lexical categories of the first, second, and third words to the right of a particular word. The second group comprises the following information about the words: if the word is capitalized, if the word is the first word of the sentence, if the word has numbers or '.' and numbers, and some words' terminations like *ed*, *ing*, *es*, *ould*, *'s*, *s*. The possible values for each of the first group's attributes are the values of the corpus tag set from which the search algorithm will learn the rules. This set will depend on the annotated corpus used, since the tag set will vary for different annotated corpora. The remaining attributes were defined as boolean.

The training sets were built from the Brown corpus. For each word of the corpus, we collected the values of every attribute in the rule's antecedent, creating a specific training example. Next, for each tag of the tag set, we built a training set made by positive and negative examples of that tag. The building process used to define each of the training sets was the following: for each example e_i of the set of examples, with word w and tag t, if w is an ambiguous word, with S the set of all its possible tags, then put e_i in the set of positive examples of tag t, and put e_i in the set of negative examples of all the tags in S, except t.

We used a binary representation for the rules. The attributes related with the context were codified, each one, by six bits. The first bit indicates whether the attribute should or should not be considered, and the following five bits represent the assumed value of the attribute in question. We adopted a table of 20 entries to store the tag set, and used the binary value represented by five bits to index this table. If the value exceeds the number 20, we used the remainder of the division by 20. The remaining attributes were encoded by 18 bits, two bits for each of the nine attributes. In the same way, the first bit indicates if the attribute should or shouldn't be considered, while the second bit, indicates whether the property is, or is not, present. We adopted a Michigan approach, thus, in both implementations of the search algorithm, each particle/individual represents a rule using the codification described. In short, each particle/individual was composed by $6 \times 6 + 2 \times 9 = 54$ bits.

2.2   Search Algorithm

For the PSO based search algorithm we adopted the binary version presented by Kennedy [11]. The genetic algorithm based version follows the classical GA with binary representation [12]. We used, as genetic operators, the two point crossover (with 0.75 probability) and the binary mutation (with 0.01 probability). The selection scheme used was a tournament selection with tournaments of size two and $k = 0.8$.

The formula used to evaluate each rule, and therefore to set its quality, is expressed in Equation 1. This formula penalizes a particle/individual that represents a rule that ignores the first six attributes, which are related with the word's context, forcing it to assume a more desirable form. The others are evaluated by the well known F_β-measure (see Equation 2). The F_β-measure can be interpreted as a weighted average of precision and recall. We used $\beta = 0.09$, which means we put more emphasis on precision than recall.

$$Q(X) = \begin{cases} F_\beta(X) & \text{if } X \text{ tests at least one of the first six attributes} \\ -1 & \text{otherwise} \end{cases} \quad (1)$$

$$F_\beta(X) = (1 + \beta^2) \times \frac{precision(X) \times recall(X)}{\beta^2 \times precison(X) + recall(X)} \quad (2)$$

3   POS-Tagger

By definition, a POS-tagger should receive as input a non annotated sentence, $\mathbf{w}$, made of n words, w_i, and should return the same sentence, but now with all the w_i marked with the appropriate tag. Assuming we know all the possibilities, W_i, of tagging each of the words w_i of the input sentence, the search space of the problem can be defined by the set $W_1 \times W_2 \times \cdots \times W_m$. Therefore the solution can be found by searching the problem state space. We believe that this search can be guided by the disambiguation rules found earlier. We tested two different global search algorithms: a genetic algorithm (GA-Tagger) and a binary particle swarm optimizer (PSO-Tagger).

The taggers developed were designed to receive as inputs a sentence, $\mathbf{w}$, a set of sets of disambiguation rules, D_t, and a dictionary, returning as output the input sentence with each of its words labeled with the correct POS tag. The search algorithm evolves a swarm/population of particles/individuals, that encode, each of them, a sequence of tags for the words of the input sentence. The quality of each particle/individual is measured using the sets of disambiguation rules given as input. Again, a more detailed description of the implemented taggers can be found in [8, 9].

3.1   Representation

The representation used in the two implemented algorithms is slightly different. In the GA-Tagger, we adopted a symbolic representation. An individual is represented by a chromosome $\mathbf{g}$ made of a sequence of genes. The number of genes in a chromosome equals the number of words in the input sentence. Each gene, g_i, proposes a candidate tag for the word, w_i, in the homologous position. The possible alleles for gene g_i, are the elements of the set W_i.

Since we adopted the binary version of the PSO algorithm, we used, in this case, a binary representation. To encode each of the tags belonging to the tag set, we used a string of 5 bits. Therefore, a particle that proposes a tagging for a sentence with n ambiguous words will be represented by $n \times 5$ bits. Each five bits of a particle encode a integer number that indexes a table with as much entries as the possible tags for the correspondent ambiguous word. If the integer number, given by the binary string, exceeds the table size, we use as index the remainder of the division by the table size value.

3.2   Tagging Evaluation

The quality of the overall tagging, $\mathbf{t}$, is given by the sum of the evaluation results of each tag assignment, t_i for each word w_i. A particle/individual representing a sequence of n tags, $\mathbf{t}$, for a sentence with n words will give rise to a set of n pairs $\langle \mathbf{x}_i, t_i \rangle$, with $\mathbf{x}_i$ denoting the correspondent 15-tuple collecting the values of the 15 attributes presented in the antecedent of the disambiguation rule. The quality of each tag assignment, t_i, is measured by assessing the quality of the pair $\langle \mathbf{x}_i, t_i \rangle$, with $\mathbf{x}_i$ using Equation 3.

$$h(\langle \mathbf{x}_i, t_i \rangle) = \begin{cases} q_k & \text{If } \langle r_k, q_k \rangle \in D_{t_i} \text{ and } r_k \text{ covers } \mathbf{x}_i \\ 0 & \text{Otherwise} \end{cases} \tag{3}$$

The quality of a particle/individual is given by Equation 4, with T representing the set of all n pairs $\langle \mathbf{x}_i, t_i \rangle$.

$$Quality(T) = \sum_{j=1}^{n} h(T_j) \tag{4}$$

4   Experimental Results

We developed our system in Python and used the resources available on the NLTK (Natural Language Toolkit) package in our experiences. The NLTK package provides, among others, the Brown corpus and a sample of 10% of the WSJ corpus of the Penn Treebank. It also provides several Python modules to process those corpora. The experimental work was done in two phases. First the disambiguation rules were discovered and, after that, the POS taggers were tested. The results achieved in each phase are presented in the next subsections.

4.1 Disambiguation Rules

As we said before, tagged corpora use many different conventions for tagging words. In order to be able to use the disambiguation rules learned from the Brown corpus to tag text from other corpora, we used the *simplify_tags=True* option of the *tagged_sentence* module of NLTK corpus readers. When this option is set to *True*, NLTK converts the respective tag set of the corpus used to a uniform simplified tag set, composed by 20 tags. This simplified tag set establishes the set of classes we use in our algorithm. We ran the covering algorithm for each one of these classes and built, for each one, the respective sets of positive and negative examples.

We processed 90% of the Brown corpus in order to extract the training examples, and, for each word found, we built the corresponding instance. The total number of examples extracted from the corpus equaled 929286. We used 6 subsets of this set (with different cardinality) to conduct our experiments. We used sets of size: $3E4$, $4E4$, $5E4$, $6E4$, $7E4$ and $8E4$, which we identified with labels A, B, ..., F. For each subset, we built the sets of positive and negative examples for each tag, using the process described in the previous section.

We tested the classification algorithm both with the GA and the PSO implementation of the search algorithm. We ran the classification algorithm two times with each different implementation for each of the training sets. The GA was run with populations of size 200 for a maximum of 80 generations and the PSO with swarms of 20 particles over 200 generations. In Table 1 we present the average number of rules achieved by both algorithms and the correspondent reduction, considering the total number of positive examples (+) adopted.

Although the publications describing previous evolutionary approaches, based on training tables, do not clearly indicate the number of entries of those tables, their size is explicitly mentioned as a sensitive point concerning the algorithm

Table 1. Average number of rules discovered by the classification algorithm

Set	+	GA	Reduction	PSO	Reduction
A	25859	2719	89.49%	2715.5	89.49%
B	33513	3081	90.81%	3124.5	90.68%
C	41080	3358.5	91.82%	3327.0	91.90%
D	48612	3735.5	92.32%	3696.5	92.39%
E	55823	4137	92.59%	4033.0	92.78%
F	63515	4399	93.07%	4288.5	93.25%

(table header: *Average number of rules* spanning GA, Reduction, PSO, Reduction columns)

time execution [4]. While unknowing these values, the total number of positive examples considered from each of the training sets adopted, can give us an idea of the size of these tables, since the information used is similar. However, while the large training set in our case has a total of $8E4$, the previous approaches use sets with typically more than $1.5E5$. As we can see in Table 1, the rules discovered by both algorithms, allowed a significant reduction (around 90%) in the number of positive examples considered. The results also show that there are no significant differences in the number of rules discovered by the GA and the PSO.

In order to evaluate the quality of the heuristic represented by each of the discovered rules sets, we use them as input for the implemented taggers. At this point, our goal was to compare the accuracy results given by the taggers for each of the rules sets, but at the same time confirm the second hypothesis of our approach. We executed 10 times each of the taggers for each of the rules sets with the same test set. We ran the GA-Tagger with 50 individuals during 10 generations and the PSO-Tagger with swarms of 10 particles during 50 generations. The best accuracy results were systematically achieved by the PSO-Tagger and they are presented in Tables 2 and 3. We observed that the best tagging was achieved with the rules discovered from the set F during the first execution of the classification algorithm based on a GA (GA F.1)). The best set of rules discovered by the classification algorithm based on a PSO was achieved from the training examples of set C during the first run (PSO C.1).

Table 2. Tagging accuracy results achieved using the rules discovered by the GA

Set	Number of rules	Average	Best	Standard deviation
GA A.1	2740	0.9655128	0.9659605	$2.3120E - 4$
GA A.2	2698	0.9647239	0.9652956	$3.7227E - 4$
GA B.1	3059	0.9651449	0.9654286	$2.6196E - 4$
GA B.2	3103	0.9644358	0.9649854	$2.4651E - 4$
GA C.1	3355	0.9664569	0.9667583	$2.6329E - 4$
GA C.2	3362	0.9654596	0.9658718	$2.6350E - 4$
GA D.1	3742	0.9664258	0.9667139	$1.7882E - 4$
GA D.2	3362	0.9661023	0.9664480	$3.2624E - 4$
GA E.1	4166	0.9669666	0.9672458	$1.7858E - 4$
GA E.2	4108	0.9666209	0.9669356	$2.4496E - 4$
GA F.1	4440	**0.9672369**	**0.9677334**	$2.3248E - 4$
GA F.2	4358	0.9671128	0.9677334	$2.7878E - 4$

This first set of experiments enable us to identify the best heuristic and also to confirm that it is possible to formalize the POS tagging problem as a search problem and use the disambiguation rules as an heuristic to guide the search for a solution in the state space of the problem. We also concluded that the classification algorithm based on a GA was more successful than the one based on a PSO. Also we can observe, from the results achieved with the rules discovered

Table 3. Tagging accuracy results achieved using the rules discovered by the PSO

Set	Number of rules	Average	Best	Standard deviation
PSO A.1	2695	0.9635227	0.9641876	$2.5759E - 4$
PSO A.2	2736	0.9629022	0.9633011	$2.8265E - 4$
PSO B.1	3148	0.9628668	0.9631682	$2.6163E - 4$
PSO B.2	3101	0.9649366	0.9651183	$1.5132E - 4$
PSO C.1	3385	**0.9669356**	**0.9673345**	$2.3172E - 4$
PSO C.2	3269	0.9650962	0.9654286	$2.3937E - 4$
PSO D.1	3664	0.9655749	0.9660048	$2.3643E - 4$
PSO D.2	3729	0.9661378	0.9664923	$2.4005E - 4$
PSO E.1	3958	0.9650740	0.9654286	$2.0365E - 4$
PSO E.2	4108	0.9654286	0.9658275	$2.4809E - 4$
PSO F.1	4309	0.9655394	0.9658718	$2.1435E - 4$
PSO F.2	4268	0.9656901	0.9660934	$2.0836E - 4$

by the GA, a correlation between the size of the training examples set and increasing accuracy values. This allows us to expect that better heuristics could be learned by the GA using larger training sets.

4.2   POS Tagging Results

We tested the PSO-Tagger and the GA-Tagger on a test set made of 22562 words of the Brown corpus using the best set of rules found (AG F.1). We ran the PSO-Tagger 20 times with swarms of 10 and 20 particles during 50 and 100 generations. The GA-Tagger was also executed 20 times with populations of 50 and 100 individuals during 10 and 20 generations. These values were chosen so that we could test both algorithms with similar computational effort, considering the number of necessary evaluations the effort measure.

The results achieved are shown in Table 4. As we can see, the best average accuracy was achieved with the PSO-Tagger using a swarm of 20 particles evolving during 50 generations. The best accuracy result returned by the GA-Tagger is

Table 4. Tagging accuracy results achieved by both POS-taggers on a test set made of 22562 words of the Brown corpus using as heuristic the set GA F.1

Tagger	Part/Ind	Generations	Average	Best	Standard Deviation
PSO-Tagger	10	50	0.9672658	0.9679550	$2.6534E - 4$
		100	0.9673123	0.9676004	$1.9373E - 4$
	20	**50**	**0.9674896**	**0.9678220**	$1.9158E - 4$
		100	0.9673921	0.9678663	$2.1479E - 4$
GA-Tagger	50	10	0.9672170	0.9675561	$1.9200E - 4$
		20	0.9672968	0.9674231	$1.1707E - 4$
	100	10	0.9672591	0.9675561	$1.4097E - 4$
		20	**0.9672835**	**0.9675117**	$1.0978E - 4$

worst than the best result obtained with the PSO-Tagger and it needs the double number of evaluations required by the PSO-Tagger. However, the accuracy values displayed by the GA-Tagger are still very competitive when compared with others published using similar approaches.

We also tested the taggers on a test set of the WSJ corpus of the Penn Treebank made of 1000 sentences, in a total of 25184 words, using the rules discovered from the Brown corpus (see Table 5). As expected, the results achieved by the two taggers on the WSJ corpus, using as heuristic the disambiguation rules learned from the Brown corpus, are inferior to the ones obtained on the Brown corpus. However, we believe that they allow us to conclude that the discovered rules are sufficiently generic so that they can be used in different corpora. This conviction emerges from comparing the obtained results with those published by other evolutionary approaches (see Table 6). Indeed, we found that the accuracy achieved is comparable with the best published results. It is also important to stress that this values are achieved with no previous training on this corpus. The accuracy values for the WSJ corpus presented in Table 6 were achieved using all the corpus available in the NLTK package, in a total of 100676 words, setting the parameters of the algorithm with the values that provided the best results in the initial set of experiments presented in Table 5.

Table 5. Best tagging accuracy results achieved by both POS-taggers on a test set made of 25184 words of the WSJ corpus using as heuristic the set AG F.1

Tagger	Part/Ind	Generations	Average	Best	Standard Deviation
PSO-Tagger	20	50	**0.9659943**	**0.9668837**	$3.1277E-4$
GA-Tagger	100	20	**0.9660598**	**0.9663675**	$2.4541E-4$

An overview of the accuracy values achieved by the taggers in both English corpora used, is presented in Table 6, along with the results published by works using similar approaches. These results only reveal that the accuracy values obtained by the two taggers are competitive with those of past approaches. We can not directly compare our results with those published since we have no access to the test set used in the experiments made in the cited works. Nevertheless, we may conclude that for comparable size words sets (in the case of the evolutionary approaches), taken from the same corpora, the results obtained in this work are among the best published. The values shown in Table 6 were converted to percentage values and rounded to the second decimal place, so that they could be more easily compared with the ones presented in the publications cited.

Table 6. Results achieved by the two taggers on two english corpora along with the ones published by similar approaches. (Araujo - [4]; Alba, Alba-GA, Alba-PGA, Alba - [5]; Wilson - [2]; Brill - [1])

Corpus	Tagger	Training set	Test set	Best
Brown	PSO-Tagger	80000	22562	**96.78**
	GA-Tagger	80000	22562	96.76
	Araujo	185000	2500	95.40
	Alba-GA	165276	17303	96.67
	Alba-PGA	165276	17303	96.75
WSJ	PSO-Tagger	∅	100676	96.67
	GA-Tagger	∅	100676	96.66
	Wilson	600000	=Training	89.80
	Brill	600000	150000	**97.20**
	Alba	554923	2544	96.63

5   Conclusions

We described a new evolutionary approach to the POS tagging problem, which we tested using two distinct algorithms from the evolutionary computation field: a GA and a PSO. We would like to emphasize the fact that, to the best of our knowledge, this was the first attempt to apply a PSO to solve the POS tagging problem, and that, in general, there are few approaches based on swarm intelligence to solve NLP tasks.

The experimental work carried out in order to study the influence of the algorithms' parameters in the taggers' output, namely the number of generations and the number of particles/individuals, allowed us to conclude that the algorithms can find promising solutions even with reduced resources. In fact, we could not identify a clear trend of improving accuracy with increasing number of evaluations. We also observed that the best tagging accuracy was displayed by the PSO-Tagger, which allows us to conclude that swarm intelligence based algorithms can also show good results when applied to NLP problems. The results displayed by the GA-Tagger also proved to be competitive with the best ones published in previous works following a evolutionary approach.

The experiments made using the WSJ corpus and the disambiguation rules extracted from the Brown corpus gave us an idea of the degree of generalization achieved by the adopted classification algorithm. From those results, we were able to confirm that the rules obtained are sufficiently generic to be applied on different corpora. The attained generalization also reflected a substantial reduction in the information volume needed to solve the problem, while contemplating, besides the typical context information, other aspects related, not to the POS tags, but to the characteristics of the words. Although we did not present any example of the learned rules, we would like to point out the advantages of representing the information in the typical classification rule format, when compared to the numerical values used in the probabilistic approaches. The comprehensibility of the learned rules, which can be represented by predicate logic, allows its easy application in different contexts.

It is our conviction that the presented approach can be viewed as a new paradigm for solving a set of NLP tasks that share some of the features of the POS tagging problem and that are currently mainly solved by probabilistic approaches. Therefore, we are planning to extend this method to other tasks that also need some kind of disambiguation in the resolution process, like noun-phrase chunking, the named-entity recognition problem, sentiment analysis, etc.

References

[1] Brill, E.: Transformation-based error-driven learning and natural language processing: a case study in part-of-speech tagging. Comput. Linguist. 21, 543–565 (1995)
[2] Wilson, G., Heywood, M.: Use of a genetic algorithm in brill's transformation-based part-of-speech tagger. In: Proceedings of the 2005 Conference on Genetic and Evolutionary Computation, GECCO 2005, pp. 2067–2073. ACM, New York (2005)
[3] Nogueira Dos Santos, C., Milidiú, R.L., Rentería, R.P.: Portuguese part-of-speech tagging using entropy guided transformation learning. In: Teixeira, A., de Lima, V.L.S., de Oliveira, L.C., Quaresma, P. (eds.) PROPOR 2008. LNCS (LNAI), vol. 5190, pp. 143–152. Springer, Heidelberg (2008)
[4] Araujo, L.: Part-of-speech tagging with evolutionary algorithms. In: Gelbukh, A. (ed.) CICLing 2002. LNCS, vol. 2276, pp. 230–239. Springer, Heidelberg (2002)
[5] Alba, E., Luque, G., Araujo, L.: Natural language tagging with genetic algorithms. Inf. Process. Lett. 100(5), 173–182 (2006)
[6] Sousa, T., Neves, A., Silva, A.: Swarm optimisation as a new tool for data mining. In: Proceedings of the 17th International Symposium on Parallel and Distributed Processing, IPDPS 2003, p. 144. IEEE Computer Society, Washington, DC (2003)
[7] Sousa, T., Silva, A., Neves, A.: Particle swarm based data mining algorithms for classification tasks. Parallel Computing 30(5), 767–783 (2004)
[8] Silva, A.P., Silva, A., Rodrigues, I.: Biopos: Biologically inspired algorithms for pos tagging. In: Proceedings of the 1st International Workshop on Optimization Techniques for Human Language Technology, OPTHLT/COLING 2012, Mumbai, India, pp. 1–16 (December 2012)
[9] Silva, A.P., Silva, A., Rodrigues, I.: Tagging with disambiguation rules: A new evolutionary approach to the part-of-speech problem. In: Proceedings of the 4th International Conference on Evolutionary Computation Theory and Applications, IJCCI 2012, Barcelona, pp. 5–14 (2012)
[10] Steven Bird, E.K., Loper, E.: Natural Language Processing with Python. O'Reilly Media (2009)
[11] Kennedy, J., Eberhart, R.C.: Swarm intelligence. Morgan Kaufmann Publishers Inc., San Francisco (2001)
[12] Holland, J.: Genetic Algorithms. Scientific American 267(1) (July 1992)

A Cooperative Approach Using Ants and Bees for the Graph Coloring Problem

Malika Bessedik, Asma Daoudi, and Karima Benatchba

Laboratoire des Méthodes de Conception de Systèmes(LMCS),
Ecole nationale Supérieure d'Informatique
BP 68M, 16309, Oued-Smar, Alger, Algérie
{m_bessedik,as_daoudi,k_benatchba}@esi.dz
http://www.esi.dz

Abstract. Recent works in combinatorial optimization shows that the cooperation of activities allows obtaining good results. In this work, we are interested in a parallel cooperation between Ant Colony System (ACS) and Marriage in honey Bees Optimisation (MBO) for the resolution of the graph coloring problem (GCP). We first present two ACS new strategies (ACS1 and ASC2) and an MBO approach (BeesCol) for the GCP, then, we offer several collaboration modes and parallelisation for the proposed methods using a parallel machine simulated on a cluster of PCs. An empirical study is undertaken for each method. Moreover, to test our approach, we have also implemented effective algorithms for the GCP. A comparison between the different algorithms shows that ACS1 (construction strategy) gives best results and is quite fast compared to other methods. Moreover, the parallel implementation of ACS reduces significantly the execution time. Finally, we show that the cooperation between ACS and MBO improves the results obtained separately by each algorithm.

Keywords: Graph coloring problem, Ant Colony System, Marriage in honey Bees Optimisation, cooperation, parallel solution.

1 Introduction

The graph coloring problem (GCP) is one of the most studied NP-hard optimization problems in graph's theory, completeness theory and operational research [15]. Its importance is justified by its diverse and interesting applications such as scheduling [14], register allocation [16], and frequency assignment [17]. The GCP consists in assigning to each graph node, a color different from those of its neighbors while using a minimum of possible colors. In this paper, we present the main steps of a new ACS approach for the GCP, with two strategies: construction and improvement. We compare the ants' approach versus the bees' approach using BeesCol [4]. Then, we propose a parallel implementation of ACS and different modes of collaboration between the proposed ACS and BeesCol.

In section 2, we give the general concept of Ant Colony Optimisation (ACO) and the details of our new ACS approach for the GCP. In section 3, we give the

G. Terrazas et al. (eds.), *Nature Inspired Cooperative Strategies for Optimization*   179
(NICSO 2013), Studies in Computational Intelligence 512,
DOI: 10.1007/978-3-319-01692-4_14, © Springer International Publishing Switzerland 2014

general principle of BeesCol. In section 4, we present a parallel solution for ACS alone and collaboration between the proposed ACS and BeesCol. In section 5, we discuss the tests results and finally, in section 6, we draw some conclusions.

2   Ant Colony Optimisation

Ant Colony Optimisation (ACO) is a proposed metaheuristic approach for solving hard combinatorial optimization problems [9]. It is an evolutionary method inspired by real ants foraging behaviour that enables them to find shortest paths between a food source and their nest. In an ACO algorithm,a complete graph denoted by $G=(C,L)$ whose vertices are the solution components associated to the problem. It is called *"construction graph"*. Moreover, a simple agents called *artificial ants* communicate indirectly to find good solutions for the optimization problem. Informally, the behaviour of ants in ACO algorithms can be summarised as follows: The ants of a colony concurrently and independently move through adjacent states of the problem on the *construction graph*, applying a stochastic local decision. While moving, ants incrementally build solutions to the optimisation problem. Typically, good quality solutions emerge as the result of the collective interaction of the ants, which is obtained via indirect communication *(pheromone trails)*. Indeed, during the construction of the solution, ants evaluate the partial solution and deposit pheromone on components or connections *(online update)*. This information will guide the future ants search. when moving on the construction graph ants make decision based on *pheromone trails* (τ) and an information specific to the problem (η). In many cases, η is the cost, or an estimate of the cost. These values are used by the ant's heuristic rule to make probabilistic decisions on how to move on the construction graph. The probabilities involved in this case are commonly called *transition probabilities*.The first ACO algorithm proposed was Ant System (AS) [11]. Several ant algorithms were developed later on improving AS such as Ant Colony System (ACS)[10][13][5]. In this paper, we will be interested by this last algorithm. In fact, we implemented two strategies for ACS: Construction and improvement strategy.

2.1   Construction Strategy: ACS1

The construction strategy, we propose in this work, does not use in its self adaptation phase, a constructive specific method such as RLF (Recursive Largest First) [12] or DSATUR (Degree of Saturation) [6][5]. Every ant is initially put on a node of the construction graph randomly or according to a well defined criterion. Each one builds a feasible solution iteratively by inserting, at every iteration, a couple (node, color) in the partial solution until obtaining a complete one. Note that the graph built by the ant is the graph to color. The algorithm main steps are summarized below: (1) Initialization. Before applying the algorithm, the following parameters are initialized: The pheromone initial values

($\tau0$), the ants' number, the number of iterations (stopping criterion), the transition rule parameters α,β, q0,$\varphi\in$ [0,1] pheromone trails persistence and finally the candidate list size.

(2) Ants position. Every ant is put on a node randomly.

(3) Selection of the next node to color. The choice of the next node (s) to color is carried out by applying a transition rule in the following manner. Let q be a uniformly distributed random variable on [0, 1], if q$\leq$ q0 then the chosen node (s) is the one for which the product between the pheromone value and the heuristic information value is maximum; ie $s = \arg\max(\tau_{ij})^\alpha(t).(\eta_{ij})^\beta(t)$, If q >q0 , the node is chosen using the following probability form:

$$\frac{(\tau_{ij}(t))^\alpha.(\eta_{ij}(t))^\beta}{\sum_{l\in J_{ik}}(\tau_{il}(t))^\alpha.(\eta_{il}(t))^\beta}. \tag{1}$$

Where τ_{ij} (t) is the pheromone value on the edge (i,j) at iteration t,Values of pheromone are associated to pairs of nonadjacent vertices having the same color. $\eta_{ij}(t)$ is the heuristic information on node j, q0 (0 $\leq$ q0 $\leq$ 1), a parameter of the algorithm which determines the relative importance between the exploitation and the exploration. α and β are two parameters which control the relative influence of the pheromone trails and heuristic information and J_{ik} is the feasible neighborhood of ant k; that is, the set of nodes which ant k has not yet visited. The next node to color (ncur) is selected among those nodes. When the size of the instance is very large, a candidate list is used. It is an intelligent strategy that does not consider the totality of the neighborhood but rather a subset. Nodes in this list can be sorted out in a descending order of their degrees. (n_{cur}) is added to the partial solution and put in the ant's taboo list. This list is used to save the path (solution) built by the ant. It is also used to make sure that a node already colored will not be colored a second time, and consequently it guarantees the feasibility of the built solution. (4) On-line Updated step by step. This stage consists in decreasing the pheromone values associated to pairs of nodes having the same color (nodes which has been just added to the stable in construction) to make them less attracting for the future ants and then avoid a fast convergence to the same solution. Pheromone is either put on nodes (2) or on edges (3):

$$\tau(s) = (1 - \varphi).\tau(s) + \varphi.\tau_0 \tag{2}$$

$$\tau_{ij} = (1 - \varphi).\tau_{ij} + \varphi.\tau_0 \tag{3}$$

where s or j is the chosen node to color. The preceding steps are repeated as long as the ant has not finished building a complete solution. (5) Off-line Update. Extra-pheromone is added to the best solution found in this generation which will help future ants in their search. The Off-line Update is carried out according to the following formula:((4) case of pheromone on nodes and (5) case of pheromone on connexions):

$$\forall s \in S^*, \tau(s) = (1 - \eta) * \tau(s) + \eta * cost(S^*) \tag{4}$$

$$\tau_{ij} = (1 - \eta) * \tau_{ij} + \eta * cost(S^*) \qquad (5)$$

Where S* is the best solution found until now, η is a parameter indicating the pheromone decline and cost(S) is defined as the number of colors used in the solution S. (6) Evaporation. When stagnation (same solution appears during several iterations) is detected, then evaporation is made according to the following equation:

$$\tau(v_i, v_j) = (1 - \varphi) * \tau(v_i, v_j) \qquad (6)$$

where $(1 - \varphi)$ is the evaporation coefficient. This process is repeated as long as the stopping criterion (a maximum number of iterations) is not satisfied.

2.2   Improvement Strategy (ACS2)

In this strategy, the search space is also a graph, where every node is a complete solution. Initially, each ant is on a node. It moves, at each iteration, to a neighbor node to improve its solution. When moving, the ant changes the color (recolor) of a node. The steps of this strategy are summarized as follows: 1) Initialization. Similar to the first strategy's one. 2) Initial solution generation. Each ant generates an initial complete solution using one of these strategies: a. Assign to each node a color randomly, the number of colors nc is fixed beforehand, such as $n_c \leq$ (maximum degree of G) +1. b. The initial solution is built using a constructive algorithm that partitions the nodes set in K stables (all nodes of a stable have the same color), where K is the maximum degree of the graph [7]. 3) Recolor a node. At each iteration, the ant chooses in its ordered candidate list a node to recolor, applying the pseudorandom transition rule. This corresponds to move the chosen node to another partition in order to minimize the conflicts if they exist or to improve the solution. This process is repeated a fixed number of iterations. 4) On-line Delayed Update. After an ant recolors a node, it updates the pheromone table according to the (7) (case of pheromone on nodes) or formula (8) (case of pheromone on connections):

$$\tau_i j = \tau_i j + \varphi/cost(S^*) \qquad (7)$$

$$\tau(s) = \tau(s) + \varphi/cost(S^*) \qquad (8)$$

where S* represents the best solution found until now and the cost is the number of conflicts. Thus, the amount of the added pheromone is proportional to the cost of the ant's solution. 5) Off-line update. It is similar to the first strategy. 6) Evaporation. As in the first strategy, if stagnation occurs, then evaporation is made according to the equation (6). 7) Stopping Criterion. The algorithm stops when a maximum number of iterations pre-defined is reached. Heuristic information The choice of the next node to color is influenced by the heuristic information. In the case of the GCP, this information is tightly bound to the characteristic of the nodes (degree, degsat). In this work, we propose a static and a dynamic heuristic. The static heuristic is represented by a vector of size

$|V|$, which contains the degrees of the nodes to color, computed during the initialisation phase. However, the value of the dynamic heuristic information is recomputed after each node re-coloration (by an ant). It defines the node's saturation degree. Let's recall that the saturation degree of a node v (DSATUR(v)) is the number of colors already assigned to the neighbours of v.

3   Marriage in Honey Bees Optimisation

Bees are social insects living in very organized colonies. Each honey-bees colony consists of one or several queens, drones, workers and broods. Queens specialize in egg laying, workers in brood care and sometimes egg lying, drones are the sires of the colony and broods the children. Marriage in Honey Bees Optimization (MBO) is a swarm approach to combinatorial optimization problems inspired by the bee's reproduction process which can be summarised as followed: In a mating flight, each queen mates with seven to twenty drones. During each mating, sperm reaches the spermatheca and accumulates there to form the genetic pool of the colony. Each time a queen lays fertilized eggs (broods), it retrieves at random a mixture of sperms accumulated in the spermatheca to fertilize the egg [2]. Contrary to most of the swarm intelligence algorithms such as Ant colony optimisation, MBO uses self-organization to mix different heuristics. MBO was first used on the 3-SAT problem [3]. It was also used for the GCP (BeesCol) [4]. BeesCol uses as worker Taboo Search, Local Search or ACS2. The worker (heuristics) intervenes on two levels of the algorithm; it improves the initial solutions at first and then improves the solutions obtained after the crossover. The pseudo code of BeesCol is given in [4].

4   A Cooperative Approach For The GCP

Cooperation among the different implemented methods requires a parallel machine having certain characteristics: parallel execution of different methods, a good communication among the different applications, synchronization and control of the co-operation between these applications. The machine we use consists of a master processor and several slaves supervised and controlled by the master. It was simulated on a cluster of PCs where a particular PC is identified as a master processor. The communication among the different processors is made via the TCP / IP protocol, the choice of this protocol is justified by its reliability. For an execution, the user selects a number of slaves to run and the methods (ACS or MBO) to assign to each one. Moreover, he initialises the parameters of each method. Then, the master processor ensures the communication and the synchronisation among the different slaves during their execution. The exchange of solutions among slaves is done via the master. In this work, we first present a parallel solution of ACS, then, two cooperation modes: (1) BeesCol with ACS2; (2) ACS1with BeesCol and ACS2.

4.1   Parallel Solution of ACS

If a large colony (n) is needed to solve GCP, using a sequential version of ACS can be time consuming. To overcome this limit, we can use the machine defined above with m $(m < n)$ slaves. Each slave executes the ACS algorithm (ACS1 or ACS2) with a number of ants $= \lfloor n/m \rfloor$ (the largest integer less than or equal than the total number of ants n divided by the number of slaves m) . As a result, a large colony is divided into smaller size colonies. This form of parallelization is said to be coarse-grain. Each processor builds its solution and sends the solution as well as the pheromone information periodically (the period is defined by the user) to the master. The latter compares all received solutions, and sends the best solution pheromone information to all slaves. The pheromone trails are then updated by all Slaves.

4.2   Parallel Cooperation

In the first mode (BeesCol and ACS2), there is a total cooperation between all slaves. Each slave executes independently the assigned optimization method with different parameters and sends its solution periodically (defined by the user) to the master. The following figure (Fig. 1) shows an example of cooperation between ACS2 and two different BeesCol algorithms (with different workers).

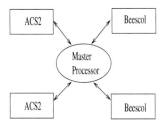

Fig. 1. A cooperation example

In the second mode (ACS1 with BeesCol and ACS2), there is a partial cooperation, as the slave executing ACS, is the only one sending its solution to others, via the master. Indeed, ACS1 is a constructive method contrarily to BeesCol and ACS2. In both modes, the master sends the best received solution to all slaves.

5   Tests Results

In this section, we present a summary of the results obtained by different implemented methods applied to the GCP. The first part is devoted to the ACS behavior study. Then, a comparison is carried out on the implemented methods and some ants' algorithms, namely: Ant Colony System where an ant is implemented as a constructive method (RLF or Dsatur) [5] and Ant System [8] (well known algorithm). Finally, we conclude with a synthesis study of the various cooperation modes among the implemented methods. All tests are performed either on generated random graphs used in [8], or on well known benchmarks taken from [1]. A random graph n_d is a graph with n vertices and such that there exists an edge with a probability p=0.d between any pair of vertices. Random graphs are deeply studied with respect to coloring, especially for $p \in \{0.4, 0.5, 0.6\}$. The execution times are written in the format "Seconds: mili-seconds". This time covers only the effective algorithms' treatment excluding the input and output display. In all tables presented below, each line gives the best, and the average solution of 50 executions.

5.1   ACS Results

We have performed several tests to adjust the different parameters of ACS. These experiments are carried out on random generated graph 100_4 (100 nodes, 1941 edges and a density of 0.4) [1]. In the following, we have used these abbreviations: Cs: colony size; It: iterations number; Gn: generations number; Cd: size of the candidate list; τ_0: initial value of pheromone; φ : Evaporation rate of pheromone; α : pheromone influence; β : heuristic information influence; q_0: intensification/ diversification parameter; St : pheromone strategy (N: pheromone on nodes; C: pheromone on connections); H: heuristics (S: static; D: Dynamics); G: Generation where best solution found; In: generation strategy of initial population (0: for random; 1: a constructive algorithm); time: time to obtain the best solution in seconds and cost: Cost of the best solution. After preliminar tests, we found that the best values for Cd and q0 are respectively 3 and 0.8. These values will be used in all coming tests.

ACS1 Results.   The obtained results for benchmarks 100_4 , described above, are summarized in tables 1. Note that after preliminar tests, we use pheromone on nodes as it is the strategy that gives better results. In the first and second lines of table 1, we use respectively the static and the dynamic heuristic. Same importance is given to both heuristic and pheromone information ($\alpha = \beta$). A closer look at the results shows that even though dynamic heuristic gives better results, it slows down the search process (line 1, 2, 6 and 7). When we cancel the pheromone influence ($\alpha = 0$) and give more importance to heuristic information($\beta = 1$), we notice that a local optimum is reached only at the 6th iteration (line 4). This can be explained by the fact that last ants do not take advantage of the experience of preceding ones. We conclude that ants' learning process is very important as pheromone trails improve the search through the generations. The comparison of lines 5 and 7 results shows that the use of a bigger colony size, even with a reduced number of generations, helps reaching the optimum at generation 3; thus emphasizing the importance of diversification in the solutions space.

The   configuration   that   gives   the   best   results   is   the   following   one: $BestConfig = \tau_0 = 0.1, \varphi = 0.1, \alpha = 0.01, \beta = 1, Cs = 37, Gn = 3$, use of the dynamic heuristic and pheromone on nodes.

Table 1. ACS1 results for the benchmark 100_4

Cs	Gn	St	H	τ_0	ρ	α	β	Best cost	aver	G	time
5	10	N	S	0.1	0.1	0.50	0.50	15	15.24	4	01:05
5	10	N	D	0.1	0.1	0.50	0.50	14	14.50	5	01:75
5	10	N	S	0.1	0.1	0.50	0.50	15	15.44	6	01:30
5	10	N	D	0.1	0.1	0	1	15	15.24	6	02:05
5	10	N	D	0.1	0.1	0.01	1	15	15.50	6	02:12
50	3	N	S	0.1	0.1	0.01	1	15	14.20	3	00:93
37	3	N	D	0.1	0.1	0.01	1	14	14.30	3	01:46

ACS2 Results. In table 2, we give the most significant results of the tests performed with the improvement strategy on benchmark 100_4. Contrary to ACS1, the optimum (14) is never reached. The cost of the best solution found is (15), obtained with the following parameters: $\tau_0 = 0.1, \varphi = 0.1, \alpha = 0.01, and \beta = 1$. Thus, the construction strategy seems to be more suitable for the GCP. This is due to the fact that ACS1 takes into account the nature of the problem, therefore its constraints.

Table 2. ACS2 results for benchmark 100_4

Gn	St	H	τ_0	ρ	α	β	cst	G	time
10	N	S	0.1	0.1	0.01	1	16	1	10:30
10	N	D	0.1	0.1	0.01	1	15	2	13:48

5.2 MBO Results

The test results obtained for MBO shows that the optimum is reached for all the workers but with different execution times. The use of Tabu Search (TS) reduces the execution time when compared to the use of the two other workers (Local Search and ACS2). Results obtained for the first benchmark are close to the optimum (14). The details of the execution of MBO are given in [4].

5.3 Comparative Study among Optimization Methods

We have done tests on several benchmarks which chromatic number is known, to make a comparative study among the results of ACS1, ACS2, BeesCol and the methods presented in [8] and [5]. The following table gives the optimum found for each benchmark. The notations below are used in table 3: AS_D (Ant System where an ant is implemented as DSATUR), AS_F (Ant System where an ant is implemented as RLF), ACS1_D (ant implemented as DSATUR, and ACS1_R (ant implemented as RLF).

Table 3. Comparison among various methods

Graph	opt	BeesCol	ACS1	ACS2	AS_D	AS_R	ACS_D	ACS_R
queen6.6	7	8	7	8	8	8	8	8
queen7.7	7	7	7	8	10	7	8	9
queen8.8	9	10	9	10	10	10	9	9
miles1500	73	73	73	73	75	74	73	74
mulsol.i.4	31	31	31	31	31	32	32	31
zeroin.i.2	30	30	30	30	31	31	30	31
100-4	14	13	13	14	14	13	13	14
100-5	16	16	16	17	17	16	17	18
100-6	19	20	19	21	19	19	19	20
300-4	28	29	28	29	31	29	28	30
300-5	35	35	35	35	40	36	35	35
300-6	42	44	42	45	44	46	42	43

From this table, we note that ACS1 reaches the optima in most cases. Concerning ACS2 and BeesCol, the obtained results are satisfactory compared to those of AS_D, AS_R.

Compared to all other methods, ACS1 gives better results. Indeed, in BeesCol, the crossover can deteriorate the quality of the solutions and sometimes create conflicts, which moves away research from the interesting zones and slowed down the algorithm. Whereas, the ACS approach by construction, manages at the same time the constraints posed by the problem, the exploitation of the research space and the exploration of the promising areas in this space.

5.4   Cooperation Results

Recall from section 4 that we propose in this paper two types of parallel solutions for the GCP: a parallel solution of ACS and a parallel cooperation of several methods. In what follows we present the results obtained.

Parallel ACS Results. We have chosen to run two ACS1 sub colonies (25 ants each) in parallel on benchmark 100_4. Each colony is assigned to a slave. The user initializes the parameters as well as of the periodicity of the pheromone table and solution exchange via the master. Note that the parameters can be the similar or different. In this case, both slaves are initialized with the values of Best config and several executions varying the exchange periodicity (1 generation to 5) are run. Both slaves reached the optimum, but with different execution times. For a periodicity of 1, there is no gain in the execution time as most time is devoted to the exchange. However for a periodicity of 2, we notice that the execution time is cut by 25%. For a periodicity over 3, this parallelization is not efficient in terms of time performance compared to the sequential execution of ACS1.However, these results are promising for a large scale parallelization. Concerning ACS2, even when we execute two sub colonies (25 ants each) in parallel on benchmark

100_4 with the best configuration, the optimum could not be reached. The cost of the best solution found through all runs, for all slaves, is 15. The ACS2 parallel version does not improve the results of the sequential version; however, it reduces the execution time.

Parallel Cooperation Results. We propose two modes of cooperation. In the first one, there is a total cooperation among all initialized slaves. In this execution, we use two slaves to which we assign BeesCol and ACS2. In the second mode, there is a partial cooperation where the slave executing ACS1 sends periodically its best solutions to the other slaves.

Total Cooperation Results. Each method exploits the solutions sent by the other. The best parameters deduced from previous tests are used for each method. The results are summarized in table 4.

Table 4. Total cooperation results for benchmark 100_4

processor	method	average cost	average time
P1	ACS2	14.83	11:562
P2	MBO	14.25	8:589

Recall that ACS2 does not reach the optimum when executed alone (Table 2). However, when ACS2 collaborates with BeesCol, it reaches the optimum after two exchanges (on average). Moreover ACS2 helps BeesCol improving its average cost if we refer to table 4.

Partial Cooperation Results. In this type of cooperation, ACS1 sends periodically its best solutions to the other two slaves (BeesCol, ACS2). Table 5 gives a summary of the results.

Table 5. Partial cooperation results for the benchmark 100_4

processor	method	cost	time
P1	ACS1	14	01:27
P2	ACS2	14	7:33
P3	MBO	14	13:254

One notices that ACS1 reaches the optimum the first (second generation) while the two others methods find the optimum starting from the 6th generation. ACS1 "helps" the other methods to overcome the local optimum.

6   Conclusion

In this paper, we were interested in a parallel cooperation between ACS and MBO for the resolution of the GCP. We, first, presented two new strategies for ACS to solve the GCP. The construction strategy, we proposed in this work, do not use in its self adaptation phase constructive specifics methods such as RLF or DSATUR and thus, was faster than ACS-R and ACS-D proposed in [3]. We gave briefly the general principle of BeesCol, the GCP MBO approach. We offered several parallelisation and collaboration modes between proposed methods using a parallel machine simulated on a cluster of PCs. An empirical study was done for each method. To test our algorithms, we implemented known algorithms for the GCP: Ant System and another approach of ACS. The preliminary results obtained on some well studied DIMACS graphs showed that ACS1 (construction strategy) gave best results and was quite fast compared to other methods. It approached the results of best coloring on some large instances of famous Dimacs benchmarks. Parallel implementation of ACS reduced significantly the execution time and computed quickly the best parameters (pheromone trails). Our application offered partial and total cooperation modes of different optimization methods to improve the solution and reduce the execution time. On one hand, we showed that cooperation between ACS and MBO improved the results obtained separately by each algorithm. Indeed, total cooperation allowed ACS2 to reach the optimum and BeesCol to improve its average cost. On the other hand, in the partial cooperation, ACS1 helped BeesCol and ACS2 to overcome the local optimum.

References

[1] http://mat.gsia.cmu.edu/color02
[2] Abbass, H.A.: Marriage in honey bees optimization a haplometrosi polygynous swarming approach. In: The IEEE Congress on Evolutionary computation (CEC 2001), Seoul, Korea (2001)
[3] Abbass, H.A.: An agent based approach to 3-sat using marriage in honey-bees optimisation, vol. 6, pp. 1–8 (2002)
[4] Bessedik, M., Bouakline, T., Drias, H.: How can bees color graphs. International Journal of Bio-Inspired Computing 3(1), 67–76 (2011)
[5] Bessedik, M., Laib, M., Boulmerka, R., Drias, H.: Ant colony system for the graph coloring problem. In: International Conference on Computational Modeling Control and Modeling, CIMCA, pp. 786–791 (2005)
[6] Brélaz, D.: New methods to color vertices of a graph. In Communications of ACM 22, 251–256 (1979)
[7] Costa, D.: Méhodes de résolution constructives, séquentielles et évolutives pour des problémes d'affectation sous contraintes. PhD thesis Num. 1411, Lausanne, EPFL (1995)
[8] Costa, D., Hertz, A.: Ants can color graphs. The Journal of the Operational Research Society 48(3), 295–305 (2007)
[9] Dorigo, M., Di Caro, G., Gambardella, L.M.: Ant algorithms for discrete optimization. Artificial Life 5(2), 137–172 (1999)

[10] Dorigo, M., Gambardella, L.M.: Ant colony system: A cooperation learning approach to the travelling salesman problem. IEEE Trans. Evol. Comp. 1(1), 53–66 (1997)

[11] Dorigo, M., Maniezzo, V., Colorni, A.: The ant system: Optimization by a colony of cooperating agent. Proceedings of the IEEE Transactions on Systems, Man, and Cybernetics, Part B 26(1), 29–41 (1996)

[12] Leighton, F.T.: A graph coloring algorithm for large scheduling problems. Journal of Research of the National Bureau of Standards 84, 489–503 (1979)

[13] Mahmoudi, A., Bessedik, M., Drias, H.: Two hybrid ant algorithms for the general t-coloring problem. International Journal of Bio-Inspired Computing 2(5), 353–362 (2007)

[14] Marx, D.: Graph coloring and their applications in scheduling. Periodica Polytechnica Ser. El. ENG. 48(1), 11–16 (2004)

[15] Papadimitriou, C.H., Steiglitz, K.: Combinatorial optimization- algorithms and complexity. Prentice Hall (1982)

[16] Smith, M.D., Ramsey, N., Hollowey, G.: A generalized algorithm for graph coloring register allocation. SIGPLAN 39(6), 277–288 (2004)

[17] Zufferey, N.: Heuristiques pour les problémes de coloration des sommets d'un graphe et d'affectation de fréquences avec polarité. PhD Thesis, Ecole Polytechnique Fdrale Lausane Switzerland (2002)

Artificial Bee Colony Training
of Neural Networks

John A. Bullinaria and Khulood AlYahya

School of Computer Science, University of Birmingham
Birmingham, B15 2TT, UK
{j.a.bullinaria,kya020}@cs.bham.ac.uk

Abstract. The Artificial Bee Colony (ABC) is a recently introduced swarm intelligence algorithm for optimization, that has previously been applied successfully to the training of neural networks. This paper explores more carefully the performance of the ABC algorithm for optimizing the connection weights of feed-forward neural networks for classification tasks, and presents a more rigorous comparison with the traditional Back-Propagation (BP) training algorithm. The empirical results show that using the standard "stopping early" approach with optimized learning parameters leads to improved BP performance over the previous comparative study, and that a simple variation of the ABC approach provides improved ABC performance too. With both improvements applied, we conclude that the ABC approach does perform very well on small problems, but the generalization performances achieved are only significantly better than standard BP on one out of six datasets, and the training times increase rapidly as the size of the problem grows.

1 Introduction

Recently, the study of different insect behaviours, animal colonies and swarms has led to the introduction of many nature inspired optimization algorithms [6]. Such swarm intelligence algorithms typically involve a group of simple agents that cooperate with each other locally, either directly or indirectly, and these simple interactions lead to the emergence of complex intelligent global behaviour for solving problems. The best known examples are Particle Swarm Optimization (PSO), inspired by the social behaviour of flocks of birds, and Ant Colony Optimization (ACO), inspired by the foraging behaviour of ants.

A more recent, and less well studied, swarm intelligence algorithm is the Artificial Bee Colony (ABC) originally proposed by Karaboga [7], and inspired by the foraging behaviour of honeybees. There are many possible applications of ABC, but this paper will concentrate on their use in optimizing the weights of artificial Neural Networks (NNs). Of course, there already exist many hybrid neural network learning algorithms that aim to improve upon standard gradient descent algorithms such as Back-Propagation (BP), but the advantages of those approaches are debatable. In particular, Cantu-Paz and Kamath [4] have shown that most combinations of Evolutionary Algorithms (EAs) and neural

G. Terrazas et al. (eds.), *Nature Inspired Cooperative Strategies for Optimization*   191
(NICSO 2013), Studies in Computational Intelligence 512,
DOI: 10.1007/978-3-319-01692-4_15, © Springer International Publishing Switzerland 2014

networks performed no better than simple BP on the classification tasks they tested. Karaboga and colleagues [9, 11], however, have previously applied ABC to neural network learning and claimed some success. The aim of this paper is to explore more carefully how effective ABC really is for training feed-forward neural networks to perform classification tasks.

In the following sections, we first describe the ABC algorithm and how it can be applied to neural network training. Then we describe and present results from a series of computational experiments that explore the power of standard and improved ABC for neural network applications in comparison with appropriately optimized BP. The paper ends with our conclusions and a discussion of the implications.

2   The Standard Artificial Bee Colony Algorithm

The ABC algorithm is a stochastic optimization algorithm inspired by the foraging behaviour of honeybees [7]. The algorithm represents solutions in the given multi-dimensional search space as food sources (nectar), and maintains a population of three types of bee (employed, onlooker, and scout) to search for the best food source. Comparative studies [8, 10] have indicated that ABC performance is competitive with other population-based algorithms such as PSO, Genetic Algorithms (GA) and Differential Evolution (DE).

The general idea of the ABC is that it starts with random solutions and repeatedly attempts to find better solutions by searching the neighbourhoods of the current best solutions and abandoning the poor solutions. The current problem solutions are represented as food sources that are each associated with an employed bee. An equal number of onlooker bees each choose one of those food sources to be exploited based on their quality or fitness, using standard roulette wheel selection [6]. Both onlooker and employed bees continuously try to locate better food sources in the neighbourhood of their current food source by changing a randomly chosen dimension of their food source position (i.e., a randomly chosen parameter of their solution) by a random amount in the direction of another randomly chosen food source. Specifically, at each stage, a randomly chosen parameter x_i of food source i is updated by $r.(x_i - x_j)$ where r is a random number drawn uniformly from the range $[-1, 1]$, and x_j is the corresponding parameter of a different randomly chosen food source j [11]. If that update results in a better solution, the existing food source is replaced by the one at the updated position. Meanwhile, scout bees carry out global exploration of the search space by randomly choosing new food sources to initialize the algorithm, and to replace food sources that have been deemed exhausted because they have failed too many times to lead to improvements.

It is clear from the above description that the standard ABC algorithm has only three control parameters that need to be set appropriately for the given problem. First, the bee colony size, equal to twice the number of food sources, and effectively equivalent to an EA population size. Second, the local search abandoning limit. Third, the maximum number of search cycles, that is equivalent to an EA number of generations, and can be defined indirectly by a suitably chosen fitness-based termination criterion.

3   Training Neural Networks Using ABC

Applying the ABC algorithm to training neural networks is relatively straight-forward. The multi-dimensional search space is the space of network connection weights and neuron thresholds, and the fitness is a standard measure of network output performance (such as sum-squared error or cross entropy) on the training data. However, the main objective here is for the trained network to generalize to perform well on previously unseen testing data, and it is well known that learning the training data too precisely can lead to "over-fitting" and unnecessarily poor generalization performance [1]. With gradient descent training, such as BP, that is typically avoided by "stopping training early", or by adding a regularization term to the cost function (such as "weight decay"), and optimizing those with reference to an independent validation dataset [1]. In principle, similar approaches can be applied to optimize the ABC training, though that does not appear to have been done in the previous studies.

Karaboga and Ozturk [11] have already tested the ABC approach on nine PROBEN1 benchmark classification problems [13], and compared their results with those they obtained using two traditional neural network learning algorithms (BP and Levenberg-Marquardt) and three population based algorithms (PSO, GA and DE). Overall, the ABC achieved good results. More recently, further improved results have been obtained with hybrid learning algorithms involving ABC combined with traditional neural network training algorithms [12, 14]. The key question to be addressed in this paper is: how can these good ABC results be reconciled with the earlier negative results that Cantu-Paz and Kamath obtained for the closely related population-based EAs [4]?

For the purpose of fair comparison, we shall follow as closely as possible the approaches used in the previous studies in this area. As with the earlier comparative study of using EAs for NN training [4], the ABC algorithm will be compared here with standard BP. Following the earlier study of using ABC for NN training [11], we shall concentrate on standard fully connected feed-forward classification neural networks with one hidden layer and use sigmoidal hidden and output activation functions. Sum squared error will again be used as the training cost function, a simple winner-take-all approach will be used to determine the predicted output classes during testing, and performance will be given as percentage correct scores.

An important issue when comparing learning algorithms is that many of the standard benchmark datasets in the UCI Machine Learning Repository [2] are actually trivial in the sense that even the simplest low complexity $O(nd)$ algorithms do not perform significantly worse on them than more sophisticated algorithms [5]. In fact, four of the nine datasets used in the Karaboga and Ozturk study [11] are trivial in that sense (Cancer, Card, Diabetes and Glass) [5], so we shall not consider them any further. They will be replaced by the more challenging Optical Digits dataset that has 64 inputs representing pixelated images and

Table 1. Neural network architectures, numbers of weights, and training, validation and testing dataset sizes

Dataset	Architecture	Weights	Train	Valid.	Test
Thyroid	21-6-3	153	3600	1800	1800
Heart	35-6-2	230	460	230	230
Horse	58-6-3	375	182	91	91
Soybean	82-6-9	631	342	171	170
Gene	120-6-3	747	1588	794	793
Digits	64-40-10	3010	3058	765	1797

10 output classes for the digits 0 to 9, with 3823 training patterns and 1797 for testing [2]. The same network architectures were used as in the Karaboga and Ozturk study [11] for their five remaining datasets. For the new Optical Digits set, 6 hidden units was nowhere near enough, so 40 were used. Table 1 summarizes the properties of the six datasets studied, showing the corresponding network architectures, numbers of weights, and dataset sizes.

Throughout this study we shall use standard unpaired two-tailed t tests to determine the statistical significances of any performance differences found. Such tests on the Karaboga and Ozturk [11] results (repeated in Table 2) for each of their five datasets indicate that BP is significantly better ($p < 0.001$) than ABC on one (Gene), significantly worse ($p < 0.001$) on three (Heart, Soybean, Thyroid), and not significantly different ($p > 0.1$) on one (Horse). A potential problem with these results, however, is that the performance of both algorithms appear surprisingly poor, particularly for the Thyroid and Soybean datasets, so the following sections will attempt to optimize the performance of each algorithm, and repeat the comparisons using the improved results.

4   Training Neural Networks Using Optimized BP

A common problem with all comparisons against BP is that it is very easy for BP to perform poorly on the chosen datasets if its learning parameters are not optimized well, and that can be difficult to do by hand, because the parameters are not independent, and the best values depend on the properties of the given dataset. The study of Karaboga and Ozturk [11] simply used the same learning parameters for all nine datasets, and it is likely that they were far from optimal for at least some of them. One solution would be to use an evolutionary algorithm to optimize the key BP learning parameters, such as the random initial weight range $[-\rho, \rho]$ and the learning rate η. With a fixed, sufficiently large, number of training epochs for each problem, the evolved learning rate is then able to implement a form of early stopping and avoid over-fitting, and that consistently leads to improved performances [3]. However, this evolutionary approach tends to be rather computationally intensive, and might be regarded as giving BP an unfair advantage over ABC. Instead, we can abstract a consistent emergent property of the evolutionary approach, namely that very small initial weight ranges and

Table 2. Mean neural network Classification Error Percentages (CEP) and standard deviations (s.d.) for the six datasets using: BP from [11], ABC from [11], Optimized BP, Optimized ABC, and Optimized Unconstrained ABC

Dataset		BP [11]	ABC [11]	Opt. BP	Opt. ABC	Opt. UABC
Thyroid	CEP	7.26	6.95	2.06	6.14	1.87
	s.d.	0.00	0.01	0.21	0.07	0.14
Heart	CEP	21.44	19.48	19.43	19.13	19.49
	s.d.	0.55	1.41	0.54	1.34	0.57
Horse	CEP	27.84	28.63	28.43	27.69	27.14
	s.d.	2.12	2.61	2.70	1.23	1.69
Soybean	CEP	61.16	38.63	10.08	13.93	9.91
	s.d.	19.18	3.18	1.98	1.13	1.04
Gene	CEP	11.37	29.50	13.23	19.55	12.22
	s.d.	1.15	1.88	0.57	0.71	0.52
Digits	CEP	-	-	4.32	6.29	4.27
	s.d.	-	-	0.27	0.18	0.34

very slow learning rates tend to work best, and use a standard stopping early approach to set the number of epochs. The details of the experimental set-up and analysis were then chosen to provide the closest possible match with the ABC approach discussed in the next section.

The datasets were each split into standard training, validation and testing sub-sets (as indicated in Table 1), with the validation set performance used to determine the optimal stopping point for the training on the training set. For each training run, for each dataset, the initial network weights were drawn uniformly from the range [-0.03,0.03] and a maximum of one million epochs of BP training were applied. Clearly, a learning rate for each training dataset was required that consistently resulted in achieving the maximum validation set performance in the allowed number of epochs. These were found by initially trying a learning rate of 0.000001 in each case, and then increasing that by factors of ten till it was large enough, giving 0.000001 for Gene, 0.00001 for Heart and Digits, 0.0001 for Horse, 0.001 for Soybean, and 0.01 for Thyroid. These large differences serve to emphasise again how important it is to set the learning parameters differently and appropriately for each dataset. It is quite likely that the learning could be speeded up in some cases (by using fewer epochs and larger learning rates), but determining by how much would potentially require more computational effort overall for no improvement in performance.

As always, the random factors result in fluctuating performances within and across runs, so there are often no clear optimal stopping points for the training, and it is not obvious that all runs should be selected for use in computing the average test set performances. A number of valid model selection approaches were possible, but it made best sense to choose an approach to averaging that most closely matched the natural averaging approach for the ABC. An average test set performance was therefore determined using the network weights corresponding to the top ten validation set performances from five BP runs. This was then repeated ten times to give an indication of the variance of the results. These results are presented in the "Opt. BP" column of Table 2 for comparison with the corresponding results from the earlier study [11]. With the optimized parameter values, BP is now significantly better ($p < 0.001$) than ABC on three of the datasets (Thyroid, Soybean, Gene), and not significantly different ($p > 0.1$) on the other two (Heart, Horse), despite the fact that BP has been trained on less data (i.e., not on the subset of the full training data set that was kept aside to be the validation set). So, at this stage, the empirical results show that ABC is significantly worse than BP for training neural networks.

5   Training Neural Networks Using Optimized ABC

In the same way that non-optimized learning parameter values resulted in misleadingly poor BP results, it may be that better optimization of the ABC parameters can bring that approach back up to, or even beyond, the performance levels of BP. This is the issue that we address next.

We proceed by investigating how the ABC performance depends on its parameters, and thus determine the best values that will enable a fair comparison against BP. A preliminary investigation revealed that the bee colony size and abandoning limit had very little effect on the results achieved, but the number of search cycles was extremely important. This is not surprising, since the ABC will obviously be prone to under- and over-fitting in exactly the same way as gradient descent algorithms such as BP, and stopping training early (at an optimum point determined by performance on a validation set) can be expected to lead to improved generalization performance on the test set. The way to get the best generalization results is therefore to apply the ABC algorithm for enough cycles that over-fitting has clearly begun, and then go back and take the solutions (i.e. network weights) corresponding to the best validation set performances to be the ones to represent the Optimized ABC. As with the above averaging approach for BP, we take the average test set performance over the ten sets of weights corresponding to the ten best validation performances from each ABC run, and repeat that ten times to estimate the variances. The use of five BP runs to give the ten best sets of BP weights can now be seen as providing a reasonable approximation to picking the best weights from whole bee colonies.

For neural network training using ABC, there is another crucial parameter that can have a big effect on the results, namely the size of the search space, which here corresponds to the limit on the network weights. It is known that optimizing the initial random weight range for BP can have a big effect on the generalization performance [3], so it is not surprising that it also has a big effect for ABC too. The obvious way to proceed is to start with the default ABC colony size of 30 and abandoning limit of 1000 used by Karaboga and Ozturk [11], but to train for a range of search space limits to find the best for each dataset.

Figure 1 shows how the performance varies with search space size, i.e. the weight range $[-\rho, \rho]$ used to generate the initial solutions and to limit the weights throughout training. There is inevitable problem dependence, but if the range is

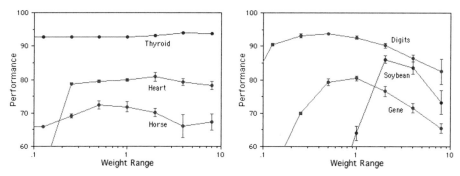

Fig. 1. Generalization performance as a function of weight range for the ABC training algorithm with limited random initial weight range, and the same limited weight range throughout training

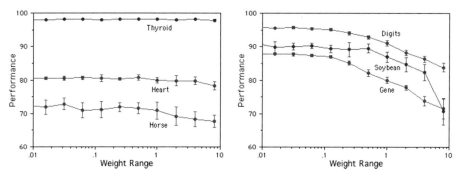

Fig. 2. Generalization performance as a function of initial weight range for the ABC training algorithm with limited random initial weight range, but unconstrained weights at later stages of training

too small or too large, the generalization performance deteriorates in each case. The study of Karaboga and Ozturk [11] simply used the same range of $[-2, 2]$ for all the datsets, but that is significantly worse than optimal for four of the six datasets (Thyroid, Horse, Gene, Digits), and not significantly different for the other two (Heart, Soybean). The performances of the optimal data points from Figure 1 are shown in the "Opt. ABC" column of Table 2, and despite the reduced amount of training data caused by excluding the validation set, no datasets have reduced performance compared with the original study. However, even with the optimized weight ranges, ABC is still significantly worse ($p < 0.01$) than BP on four data sets (Thyroid, Soybean, Gene, Digits), and not significantly different ($p > 0.1$) on the other two (Heart, Horse).

The general pattern found for BP initial weight ranges is that smaller values tend to result in better generalization until a point is reached when any further reductions make little difference. The problem the ABC approach has is that smaller values will lead to an over-restricted search space if the weights are

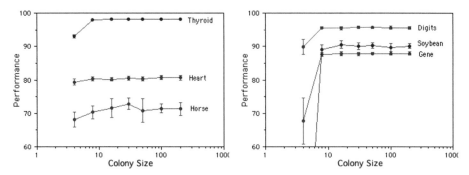

Fig. 3. Generalization performance as a function of the bee colony size for the UABC training algorithm with optimal initial weight range and abandoning limit of 1000

constrained to stay within that range throughout training. However, it is a simple variation of the standard ABC (that we shall call Unconstrained ABC, or UABC) to define an initial weight range, but allow the ABC algorithm to take the weights outside that range. Doing that leads to the improved pattern of performances shown in Figure 2. Now the performance is fairly level for small weight ranges, and the range $[-0.03, 0.03]$, that we used for the BP runs, is small enough to work well for all the datasets. Smaller values tend to increase the number of training cycles without significant performance improvement, so there is no point in using a smaller range. The optimized performances using this approach and initial weight range are given in the "Opt. UABC" column of Table 2. This shows significant performance improvement ($p < 0.01$) over the restricted weight range approach (in the "Opt. ABC" column) for four of the datasets (Thyroid, Soybean, Gene, Digits), and no significant difference ($p > 0.1$) for the other two (Heart, Horse). Comparing the optimized UABC results with the optimized BP results shows no significant difference ($p > 0.1$) for five of the six datasets (Thyroid, Heart, Horse, Soybean, Digits), but UABC is now significantly better ($p < 0.01$) than BP for the Gene dataset.

It was stated above that the bee colony size and abandoning limit had little effect on the results obtained by ABC for neural network training, but we now need to check that claim more carefully, in case their optimization can lead to further improvements in performance. First, Figure 3 confirms that, as long as the colony size does not fall below about 10, it makes no significant difference to the final performance what the colony size is. Obviously, larger colonies will inevitably result in longer compute times per cycle, and that tends to not be fully compensated by a reduction in the number of cycles required, so there is an overall advantage in keeping the colony size reasonably low. The default size of 30 used above is well within the range of good values, but not so high as to have serious adverse computational resource implications.

The effect of varying the abandoning limit is shown in Figure 4. As long as it is not below about 30, it makes no significant difference what the limit is. In fact, for the default limit of 1000, or more, the scout bees are virtually never employed, and that has no adverse effect on performance.

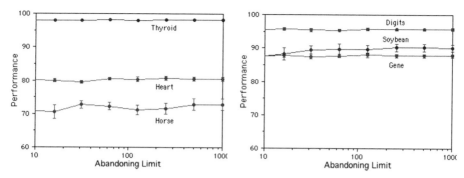

Fig. 4. Generalization performance as a function of the abandoning limit for the UABC training algorithm with optimal initial weight range and bee colony size of 30

Thus, we have now fully optimized the ABC algorithm parameters, and the results shown in Table 2 are the best possible without further modification of the algorithm itself. The ABC has achieved neural network generalization performance significantly better than BP on the Gene dataset, but the results for the other five datasets studied are not significantly different to those obtained using standard BP with appropriate learning parameter values.

6   Conclusions and Discussion

This paper has investigated the use of the ABC algorithm for training neural networks, and shown how it can be optimized to give better results than those found in previous studies. However, in most cases, the best ABC generalization performance levels obtained are not significantly different to standard BP that has been properly optimized for the given problems.

One could argue that ABC algorithms are relatively minor extensions of standard EAs: they both involve populations of solutions, the generation of new solutions based on existing solutions, and the discovery of better solutions by iteratively using fitness based selection to determine which "offspring" should replace which existing solutions. The obvious question to ask, then, is whether the offspring generation and selection inspired by bees perform any better on the application of interest (i.e. neural network training) than those inspired by evolution by natural selection. We have seen that the scout bee component of the ABC algorithm is redundant in this case, in that no decrease in performance results from setting the abandoning limit to values so high that the scout bees never become involved after the initial solution set generation. Thus there is effectively no further wide-scale random exploration of the search space during training. This means that all the offspring are generated by changing the value of a single randomly chosen parameter (i.e. network weight) by an amount that depends on the difference between that value and the corresponding value of another individual. That is exactly how a basic EA cross-over and mutation would optimize its genotype [3], so it is not surprising that we have come to a similar conclusion to that of the earlier study of Cantu-Paz and Kamath [4] which showed that weight optimization using EAs gave results that were not significantly better than standard BP.

This paper has shown that the optimized ABC and BP results are not significantly different for five of the six datasets studied, but we are still left with the question of how the ABC performs significantly better than BP on the Gene dataset. With BP learning, the weight update sizes depend on the back-propagated output errors and the chosen value of the learning rate parameter. With ABC optimization, the potential weight update sizes depend on the weight differences between individuals, and that means the algorithm can effectively generate its own learning rates for each weight during training [11]. For example, about half way through training on the Gene dataset, the mean standard deviation across individuals of the input to hidden unit weights is around 0.05, while that of the hidden to output weights is around 3.3. This means that there will be something like a factor of 66 difference in the average effective learning rates for the two sets of weights. Similar large differences in BP learning rates across network components have been found in evolutionary neural network studies to lead to significant improvements in performance for some datasets [3], so it is a reasonable hypothesis that this is why ABC is performing better here than BP with a single learning rate throughout the network.

Another important issue is the increased computational cost of using ABC rather than BP. With ABC updating random network weights one at a time, by amounts involving a random factor, it will inevitably become less computationally efficient as the network sizes increase. Of course, BP also becomes more computationally costly as the network size grows, but to a much lesser extent than ABC. This differing dependence on network size makes fair comparisons of the two approaches difficult, because past empirical results have shown that the generalization performance usually improves with more hidden units, as long as appropriate regularization (such as stopping early) is used (e.g., [3]), and using much larger networks than the current study will not only pose problems with getting the experiments completed in a reasonable time, but will also put ABC at a considerable compute time disadvantage compared with BP. If equal fixed maximum compute times were enforced for both algorithms, it is quite likely that BP would end up being able to use significantly more hidden units, and thus achieve better generalization performances than the ABC in that way.

There clearly remains considerable scope for future work in this area, but, unfortunately, most of it will be extremely computationally expensive. First, of course, the investigation of a wider range of datasets, with many more runs per dataset, will provide a more reliable indication of the patterns of results that can be expected more generally. Then, the application of an evolutionary approach to the optimization of the BP learning parameters, including the evolution of different learning rates for different layers of weights, should allow closer to optimal BP performance than is feasible with parameters set "by hand" [3], and also allow testing of the above hypothesis concerning the superior ABC performance on the Gene dataset. Finally, testing how the generalization performances and run times depend on the number of neural network hidden units will address the computational cost issue noted above. Ultimately, it will be the results of this future work that will determine whether the ABC is a worthwhile algorithm for training neural networks.

References

[1] Bishop, C.M.: Neural Networks for Pattern Recognition. Oxford University Press, Oxford (1995)

[2] Blake, C.L., Merz, C.J.: UCI Repository of Machine Learning Databases. University of California (1998), http://www.ics.uci.edu/~mlearn/MLRepository.html

[3] Bullinaria, J.A.: Using evolution to improve neural network learning: Pitfalls and solutions. Neural Computing and Applications 16, 209–226 (2007)

[4] Cantu-Paz, E., Kamath, C.: An empirical comparison of combinations of evolutionary algorithms and neural networks for classification problems. IEEE Transactions on Systems, Man, and Cybernetics-Part B: Cybernetics 35, 915–927 (2005)

[5] Duch, W., Maszczyk, T., Jankowski, N.: Make it cheap: learning with O(nd) complexity. In: Proceedings of the World Congress on Computational Intelligence, pp. 132–135 (2012)

[6] Engelbrecht, A.P.: Computational Intelligence: An Introduction. Wiley, Sussex (2007)

[7] Karaboga, D.: An idea based on honey bee swarm for numerical optimization. Technical Report TR06, Computer Engineering Department, Erciyes University, Turkey (2005)

[8] Karaboga, D., Akay, B.: A comparative study of Artificial Bee Colony algorithm. Applied Mathematics and Computation 214, 108–132 (2009)

[9] Karaboga, D., Akay, B., Ozturk, C.: Artificial Bee Colony (ABC) optimization algorithm for training feed-forward neural networks. In: Torra, V., Narukawa, Y., Yoshida, Y. (eds.) MDAI 2007. LNCS (LNAI), vol. 4617, pp. 318–329. Springer, Heidelberg (2007)

[10] Karaboga, D., Basturk, B.: On the performance of Artificial Bee Colony (ABC) algorithm. Applied Soft Computing 8, 687–697 (2008)

[11] Karaboga, D., Ozturk, C.: Neural networks training by Artificial Bee Colony algorithm on pattern classification. Neural Network World 19, 279–292 (2009)

[12] Ozturk, C., Karaboga, D.: Hybrid Artificial Bee Colony algorithm for neural network training. In: Proceedings of the IEEE Congress on Evolutionary Computation, pp. 84–88 (2011)

[13] Prechelt, L.: PROBEN1 – A set of benchmarks and benchmarking rules for neural network training algorithms. Technical Report 21/94, Universitat Karlsruhe, Fakult at fur Informatik, Germany (1994)

[14] Qiongshuai, L., Shiqing, W.: A hybrid model of neural network and classification in wine. In: Proceedings of the 3rd International Conference on Computer Research and Development, pp. 58–61 (2011)

Nonlinear Optimization in Landscapes with Planar Regions

Eddy Mesa, Juan David Velásquez, and Gloria Patricia Jaramillo

Universidad Nacional de Colombia, Carrera 80 No. 65-223, Medellín, Colombia
{ejmesad,jdvelasq,gpjarami}@unal.edu.co

Abstract. Metaheuristics are used successfully in several global optimization functions. Problems arise when functions have large flat regions since information, given by the slope, necessary to guide the search is insufficient. In such case, a common solution can be a change in the metaheuristic's parameters in order to attain a optimal balance between the exploration and exploitation. In this paper, we propose a criterion to determine when a flat region can be problematic. It is validated with a very simple hybrid algorithm based on the use of PSO technique for optimizing non-flat regions and Monte Carlo sampling for searching the global optimum in large flat regions. The proposed criterion switches the both algorithms to provide more exploitation for descendent functions and more exploration for planar functions. Preliminary results show that the proposed hybrid algorithm finds better results than PSO and Monte Carlo techniques in isolation for ten well-known test functions.

Keywords: Global Optimization, Particle Swarm Optimization, Monte-Carlo optimization, metaheuristics.

1 Introduction

Heuristic optimization in presence of large flat regions on the landscape of the objective function is an important and difficult challenge. First, several metaheuristics, as Differential Evolution or Particle Swarm Optimization (PSO), are well suited for optimizing non-flat (descendent or ascendant) functions where the gradient provides information about the possible location of the global optimum [1–3]. However, for functions with large flat regions, which means gradients near or equal to zero, there is not information to guide the search [1–3].

Second, each type of objective function has a unique topology. Then, search needs a different balance between exploration and exploitation for algorithm used. That means a new set of parameters for the metaheuristic with the aim to improve the quality of the solution, reducing the computational effort and increasing the robustness of the method [4, 5]. In this sense, the problem of searching in flat regions can be solved by setting the appropriate values for the parameters of the particular metaheuristic [6]. However, the presence of large flat regions is not detectable in advance because of there is not available information about function topology before starting the optimization. The first objective of

G. Terrazas et al. (eds.), *Nature Inspired Cooperative Strategies for Optimization*   203
(NICSO 2013), Studies in Computational Intelligence 512,
DOI: 10.1007/978-3-319-01692-4_16, © Springer International Publishing Switzerland 2014

this work is present a criterion to determine when there is a problematic flat region in the objective function for algorithms that use the information of the search like gradient.

Third, several metaheuristics have problems in presence of large flat regions due to the absence of good candidate solutions for being used in the exploitation; in this case, simple random search, as the Monte Carlo approach, has a better performance for that kind of functions, because of random search does not need previous information to find the new solutions for the search [7, 8]. The second objective of this work is to present the validation of the proposed criterion; to do it, we proposed a simple hybrid optimization algorithm that uses the developed criterion for switching between PSO technique and Monte Carlo sampling for optimizing nonlinear functions.

This paper is organized as follows: in Section II, we describe the problem of planar regions. In Section III, we present the developed of a criterion to determine when we are in the presence of a problematic planar region. In Section IV, we describe a hybrid optimization methodology that switches, between both algorithms used (PSO and Monte Carlo), using the proposed criterion as parameter control. Also, the test and results are discussed.

2   The Problem with Planar Regions

In this section, we present a mathematical description of planar regions, discuss why they are a problem in global optimization and, finally, exemplify the problem using benchmark functions.

2.1   Definitions

Before explaining why metaheuristics have a poor performance in planar regions, we define some basic concepts of global optimization.

Definition 1. *A set A is called a search region if $A = \{\mathbf{x} = (x_1, \ldots, x_n) \, | \, l_1 \leq x_1 \leq u_1, \ldots, l_n \leq x_n \leq u_n\}$ where l_i and $u_i, i = 1, \ldots, n$, are the lower and upper bounds of the variable i [9].*

Definition 2. *A local minimum for the function $f : \mathbb{R}^n \to \mathbb{R}$ in the subregion $N \subseteq A$, with $N \neq \emptyset$, is a solution $\mathbf{x}$, such that $f(\mathbf{x}^*) \leq f(\mathbf{x}), \forall \mathbf{x} \in N$ and $\mathbf{x} \neq \mathbf{x}^*$ [5]. For example, the Figure 1 has three local minimums for the subregions B, C y D.*

Definition 3. *A global minimum for the function $f : \mathbb{R}^n \to \mathbb{R}$ in the region A, with $A \neq \emptyset$ is a solution $\mathbf{x}$, such that $f(\mathbf{x}^*) \leq f(\mathbf{x}), \forall \mathbf{x} \in A$ and $\mathbf{x} \neq \mathbf{x}^*$ [9].*

Definition 4. *A local search algorithm is an iterative procedure that searches the point $\mathbf{x}_k$ in the neighborhood of $\mathbf{x}_{(k-1)}$ such that $f(\mathbf{x}_k) \leq f(\mathbf{x}_{(k-1)})$. The sequence of points $\{\mathbf{x}_k\}$ for $k = 0$ to $k = \infty$ is called trajectory of points [10].*

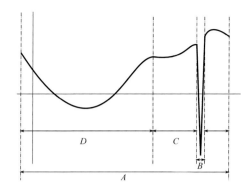

Fig. 1. Search Area A for a objective function $f(\mathbf{x})$

Definition 5. *The attracting region L for the point $\mathbf{x}^*$ where occurs a local minimum for the function $f : \mathbb{R}^n \to \mathbb{R}$ is defined as the region surrounding $\mathbf{x}^*$ such that $\forall \mathbf{x}_0 \in L$, chosen randomly, a local search algorithm starting at $\mathbf{x}_0$ arrives to $\mathbf{x}^*$ in a finite number of iterations [5, 10, 11]. In other words, L is the region that provides information about the location of $\mathbf{x}^*$. (See Figure 1 region B)*

Definition 6. *The point $\mathbf{x}^*$ s a strong minimum of $f(\cdot)$ when $f(\mathbf{x}^*) < f(\mathbf{x})$, $\forall \mathbf{x} \in L$ [8]. In other words, there is only one minimum point in the attraction region.*

Definition 7. *The point $\mathbf{x}^*$ s a weak minimum when $\mathbf{x}^*$ is not a strong minimum [8]. In other words, there is more than one minimum point in the attraction region L.*

2.2   The Problem with Planar Regions

Strictly, a planar region is a region $L \in A$, with $L \neq \emptyset$, such that $\forall \mathbf{x} \in L$, $f(\mathbf{x}) = c \in \mathbb{R}$, and all points $\mathbf{x} \in L$ are weak minimums. This implies that $\triangledown f(\mathbf{x}) = 0$, $\mathbf{x} \in L$. However, in this article we are interested in a definition with relaxed conditions; we consider that the attracting region L for the point $\mathbf{x}^*$ is a flat or planar region when $f(\mathbf{x}^*) \leq f(\mathbf{x})$, $\forall \mathbf{x} \in L$ and $\|f(\mathbf{x}) - f(\mathbf{x}^*)\| \leq \delta$ where δ is an arbitrary quantity near to zero; δ is a small constant that representes the possible amount between complete plans, gradient=0, and a region that could be problematic because of small gradients. This implies that $\triangledown f(\mathbf{x}) \approx 0$ inside of the attracting region and $f(\cdot)$ is a smooth function with slopes near to zero or zero, where the attraction region of the minimum has little information about its location or there are several minimums in the neighborhood. We refer the initial definition of planar region as a pure planar region and the second definition as a relaxed or non-pure planar region.

Classical local optimization based on gradients are not able to optimize functions with optimal point located in flat regions because of gradient algorithm has not direction to follow. Usually, direct methods based on population have similarities with gradient methods, because all of them use the information of the descendent slope like the water's drop that needs the slope to flow; when the slope information is not available, the drop will be stagnated in one point like the algorithms actually does.

Several metaheuristics have problems for finding the global optimum of a function when such point is located inside of a large flat region, due to: first, there is insufficient information for determining descent directions that are required for guiding search of the algorithm. Second, the local search algorithm only finds points with $f\left(\mathbf{x}_k\right) \approx f\left(\mathbf{x}_{(k-1)}\right)$ such that the trajectory of visited points seems to be erratic or random and there is not convergence [4]. Third, the absence of good candidate solutions for being used in the exploitation phase of the used metaheuristic. Thus, metaheuristics presents questionable advantages over pure random optimization when the optimized function has the global optimum inside of a large planar region.

Rarely, we will have a complete planar region for the objective function. Usually, we have a mixture between planar regions and non-planar regions. There is not a unique mathematical criterion to determine when this mixture could be problematic for an algorithm, but we have benchmark functions that are problematic and can be consider as function with large planar regions.

2.3 Benchmark Functions That Present Planar Regions

Commonly benchmark functions are classified in terms of roughness, smoothness and quantity of minimums, but there is not a classification based on the presence or absence of problematic planar regions. However, several well-known benchmark functions are difficult to optimize due to the presence of large flat regions [12–14]. In Table 1 and Figure 2, we present five functions extracted from most relevant literature.

Table 1. Benchmark functions: name, equation, minimum, and search region

Name	Function	Search Area	minimum
Sphere	$f_1(x) = \sum_{i=1}^{D} x_i^2$	$[-500, 500]^{30}$	0
Rosenbrock	$f_2(x) = \sum_{i=1}^{D-1} [100(x_{i+1})] x_i^2$	$[-5.12, 5.12]^2$	0
Rastrigin	$f_3(x) = \sum_{i=1}^{30} [x_i^2 - 10\cos(2\pi x_i + 10)]$	$[-100, 100]^{30}$	0
Foxholes	$f_4(x) = \left[\frac{1}{500} + \sum_{j=1}^{25} \frac{1}{j+\sum_{i=1}^{D}(x_i - a_{i,j})^6}\right]$	$[-65.5, 65.5]^2$	≈ 1
Branin	$f_5(x) = \left(x_2 - \frac{5.1}{(4\pi^2)}x_1^2 + \frac{5}{\pi}x_1 - 6\right)^2$ $+10\left(1 - \frac{1}{8\pi}\right)\cos x_1 + 10$	$[-5, 10][0, 15]$	0.398

Function f_1 and f_3 are descendent and the others present planar regions (see Figure 2). In Figure 2, we presented the three-dimensional plots and two-dimensional contours of these functions; where we can see the planar regions

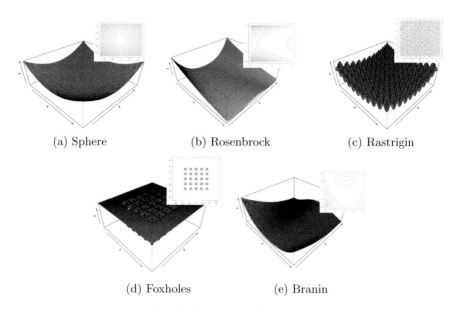

(a) Sphere   (b) Rosenbrock   (c) Rastrigin

(d) Foxholes   (e) Branin

Fig. 2. Benchmark Functions

like bananas for functions f_2 and f_5 and like a foxhole for f_4. Notice that there is not a complete planar in any function, there is a mixture between types of regions with high level of information about location of minimum and zones with a pretty poor clues about minimum location where algorithms, generally, could become struck. All these five function are described in the literature like problematic functions [12–14]. In addition, we use other five functions from CEC 2005 benchmark set (Functions 11 to 15) to validate the criterion; these functions are complex and planar regions are not evident.

3   A Criterion to Determine Problematic Planar Regions

Metaheuristics methods based on population are often initialized by random sampling. Afterwards, population is improve by different strategies using the information given by the last iteration. In real-world problems, the geometrical configuration of the objective function is unknown and to determine when exists a planar regions is not trivial problem. The information given by population can be used for the fine tunning of the parameters of the metaheuristic with the aim of changing the exploration-exploitation balance, or, inclusively, for selecting an alternative optimization method best suited for random search.

In this Section, we propose an empirical criterion for determining when the objective function has a planar region that could be a problem for the optimizer. The proposed criterion is calculated using sampled points of the objective function.

3.1   Proposed Empirical Criterion

Usually, metaheuristic methods use the known values of objective function to direct the search towards regions where it is possible to find the optimum; e.g. evolutionary algorithm uses the values of objective function to calculate the fitness of the population for finding a next generation, which is improved thanks to the knowledge of previous generation. In particle swarm optimization (PSO), the values of objective function calculated for current position of the particles in the swarm are used to calculate the next direction and magnitude of the displacement of each particle; thus, PSO uses an empirical descendent direction in a similar way of the gradient based techniques [9, 15]. In these cases, the similarity with gradient search imposes the same weakness: problems with the performance in presence of large planar regions as explained above.

However, the information provided by the points already visited for the meta-heuristic is not used to extract important features of the objective function as smoothness, roughness and slopes. For example, when $f(\mathbf{x}_i) = c$, with $c \in \mathbb{R}$, for all $\mathbf{x}_i$ belongs to the current population of a population-based optimization algorithm, the standard deviation, mean and median of the values $f(\mathbf{x}_i)$ will be zero and c respectively; this information can be used to determine when the population is located inside a pure planar region.

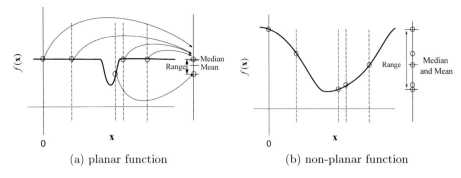

(a) planar function   (b) non-planar function

Fig. 3. Scatter of objective function values versus kinds of objective functions: planar regions with descendent areas and purely descendent functions

But, what will happen with a scatter measures of objective function values when the region has a large (non-pure) planar region with small descendent areas? Figure 3(a) shows the results. First, the range among the optimum, the mean and the median is bigger that in the case of purely planar region mentioned. Second, the most part of the sample of objective function values are found near to the true median of the function; only a small amount of calculated function values will have a smaller value than the true median. Moreover, the median of objective function will be near to constant value of planar region and the mean will be affected by the smaller values. Therefore, there will be a difference between both, mean and median, values.

Accordingly, what will happen when the function is strongly descendent? In Figure 3(b), a descendent function is plotted. In this case, we could see that the difference between the mean and median are zero or could be near to zero. That means, we could identify a large planar region from the scatter measures sampled using the current population of the optimization algorithm.

We propose two measure of variability as indicators of problematic planar regions. For a set of solutions, we calculate the range R_g, higher values indicate that there are a large variation in the values of the samples. Thus, we divide R_g between the dimension of search space to have an average of the change by one unit of this space. We define R_g as:

$$R_g = \frac{f(\mathbf{x})_{max} - f(\mathbf{x})_{min}}{\|\mathbf{ub} - \mathbf{lb}\|} \tag{1}$$

Afterwards, for the same, we calculate the difference, T_c between mean and the median and observe how the sampled are distributed. Low values indicate a well-distributed sample, which implies homogeneity, and it is agree with descendent functions (See Figure 3(a) and 3(b)). We define T_c as:

$$T_C = \frac{f(\mathbf{x})_{mean} - f(\mathbf{x})_{median}}{f(\mathbf{x})_{max} - f(\mathbf{x})_{min}} \cdot 100 \tag{2}$$

We divide the difference between mean and median by the estimated range of the objective function with the aim of normalized it as a percentege to compare with a specific value. Table 2 shows the possible values for the criterion and its meaning in terms of the optimization problem.

Table 2. Criterion meaning

Rg	Tc	Meaning
$Rg \approx 0$	$Tc \approx 0$	Weak minimum zone.
$Rg \gg 0$	$Tc \approx 0$	Descend function.
$Rg \gg 0$	$Tc \gg 0$	Mixture between weak and strong minimum zones.

The criterion presented above is still useless because we have not values for distinguishing when R_g and T_c are near to or differ from zero. To use the criterion first we need to review that:

- The criterion identify between descendent region with small gradient and planar regions.
- The criterion could identify planar regions for high number of dimension.
- The criterion is not sensitive to the sample size used to calculate the criterion it means, the population size.
- The values of R_g and T_c that must be consider problematic for the algorithm, in this case PSO, and if it is the same for all problems and all dimensions.

3.2   Behavior and Sensibility of the Test for Known Functions

The criterion presented is a new approach for solution to the problem of identify planar regions. With the aim of analyzing the behavior of the proposed measures of variability, we conduce a test two tailored functions, a parabola

$$f_d = \alpha x^2 \tag{3}$$

and a piecewise function with a planar region (See Figure 4), defined as:

$$f_p = \begin{cases} \alpha \mathbf{x}^2 & [-2, 2] \\ 4\alpha & (-70, 2) \cup (2, 70) \\ \alpha \mathbf{x}^2 - 140\alpha \mathbf{x} + 4904\alpha & [-100, -70] \cup [70, 100] \end{cases} \tag{4}$$

In both cases, α, varies between $[0.001, 1]$ and controls the gradient of the function. In other words, the parabola and the piecewise function with $\alpha = 0.001$ will be flatter than the both functions with $\alpha = 1$. And $\mathbf{x}$ vary between $[-100, 100]^D$ where D is the dimensions. With the first experiment, we expect to determine the ranges of values for both measures, if the criterion can identify between a planar region and descendent region with small gradients.

In Table 3 we present calculated values for R_g and T_c for one-dimensional functions for different values of α. The sample used consists in 50 points obtained from the Sobol low discrepancy sequence(LDS) [16]. For both one-dimensional functions we found that: first, R_g is directly proportional to α and T_c is independent of the gradient, but changes between both functions; this means that the criterion it is not sensitive to small values for gradient of descendent functions. And, second, the value of T_c for the function f_p is bigger than the value of T_c for the function f_d; this means that f_p is more planar than f_d, which is agree with construction of the functions.

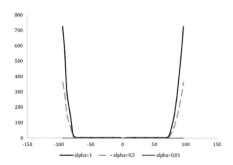

Fig. 4. piece-wise function f_p

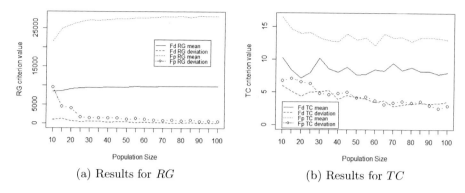

(a) Results for RG (b) Results for TC

Fig. 5. Values for the criterion using k numbers to evaluate the function. The numbers are random with uniform distribution.

Table 3. Results for one-dimensional functions

α	f_d		f_p	
	Rg	Tc	Rg	Tc
1	9384.76	7.85%	26818.06	12.6%
0.5	4692.38	7.85%	13409.03	12.6%
0.1	938.48	7.85%	2681.80	12.6%
0.01	93.647	7.85%	268.18	12.6%
0.001	9.39	7.85%	26.81	12.6%

Table 4. Results for f_d and f_p as a multi-dimensional functions

D	f_d		f_p	
	R_g	T_c	R_g	T_c
1	9384.76	7.85%	26818.06	12.6%
2	15312.50	0.59%	39498.16	13.2%
10	51640.63	1.58%	109868.12	19.8%
30	139824.21	0.26%	298131.17	22.9%

Next, we analyze whether the obtained results for one-dimensional functions are sensitive to sample's size and randomness. For this, we calculate R_g and T_c for different sample sizes and random points. We consider sample sizes from 10 to 100 points every 10 points. For each sample size, we realize 50 independent runs where the points used for calculating R_g and T_c are drawn from a uniform distribution. Figure 5 shows the values of R_g and T_c for both test functions in one-dimension; the mean of the value for criterion and the correspondent standard deviation are shown. We found that R_g is more sensitive to small sample sizes than T_c and both criteria tend to be constant for large sample sizes. Figure 5(b) shows a pretty clear difference between T_c for f_d which is smaller than T_c for f_p. So, the minimal number of points recommended to calculate the criterion is greater than 30 because of the standard deviation found for T_c.

Following, we analyze the behavior of the criterion when the number of dimensions of the search region changes. In Table 4, the values for the proposed measures of variability are calculated for both test functions and for 1, 2, 10 and 30 dimensions. As in the first experiment, we use a Sobol low discrepancy sequence for obtaining 50 samples of the test functions. Results show that the values given by the criterion are sensitive to the number of dimensions; which was expected because with dimensions the difference between plans augment. Nevertheless, the sensitive found is positive because the difference T_c is increase with number of dimensions. In other words, for high number of dimensions the difference identified by the criterion between planar regions and non planar regions will be stronger. However, the relationship between criterion values and dimension is unknown. Based on the results, we did a polynomial regression and obtain that:

$$T_c \geq 7.675(D)^{-0.588} \tag{5}$$

We use the equation 5 to determine when it is a problematic planar region.

3.3   Behavior of the Test for Benchmark Functions

To test the criterion, we choice three benchmark functions that present weak minimums and two the descendent functions as control functions. These functions, and correspondent, dimensions, search areas, and minimums are presented in Table 1 and Figure 2. Table 5 shows the results; functions f_1 and f_3 have not problematic planar region and the value of T_c are smaller than the other functions. R_g in both cases is greater than zero. For the other function T_c is greater than the limit 7.6. The criterion identify different function as planar o non planar regions. In this case, we do not use the other five function because we do not know if they are or not planar.

Table 5. Criterion results for benchmark function

	f_1	f_2	f_3	f_4	f_5
Rg	3495605.47	60301.24	775.55	495.02	204.07
TC	0.26%	13.12%	0.67%	8.65%	12.06%

4   Hybrid Methodology

In this Section, we present the algorithm implemented to validate the criterion proposed to determine the planar regions. The aim of this algorithm is to show how this criterion can uses in making decisions about the optimization strategy. In this case, the criterion will switch between two algorithms, PSO and Monte Carlo sampling. PSO is metaheuristic developed by Kennedy and Eberhart in the middle of 90's [15] and have been applied successful to several problems [4]. PSO has great advantages: high level of convergence, robustness and computational cost. It is pretty known and had been used in hybrids [3, 17, 18]. Monte Carlo is a purely stochastic method; it is used for different applications, it has not a sophisticated method inside. It uses the random sampling to find the optimums, but it has been proved to be superior when the information about localization of minimum is poor [8]. In this approach, the proposed criterion is used to decide when use PSO or Monte Carlo sampling. The sequence of the algorithm is shown in the Pseudocode 1.

Pseudocode 1. Hybrid proposed PSO-MC

1: $S_k = random x_1, ..., x_k$
2: Evaluation of $f(x)$ for S_k
3: Evaluation of Rg and Tc
4: **if** $TC \geq 7.675\,(dimensions)^{-0.588}$ and $Rg > 0$ **then**
5:     Use monte carlo algorithm
6: **else**
7:     use PSO algorithm
8: **end if**

Table 6. Results of hybrid proposed for benchmark functions

Function	Criterion		Type of function	Monte Carlo		PSO	
	Rg	Tc			Mean Desviation		Mean desviation
f_1	3495605.47	0.26%	Non Planar	4,23E-01	3,85E-01	**2,96E-08**	**5,60E-08**
f_2	60301.24	13.12%	Planar	**9,87E-04**	**1,25E-03**	1,64E-03	3,18E-03
f_3	775.55	0.67%	Non Planar	6,53E-01	5,06E-01	**6,70E-06**	**2,12E-05**
f_4	495.02	8.65%	Planar	**9,98E-01**	**4,03E-06**	1,28E+00	5,17E-01
f_5	204.07	12.06%	Planar	1,00E+01	2,40E-09	1,00E+01	5,88E-07
f_6	4.45	0.41%	Non Planar	90.29	0,85E-00	**90.01**	**2.44E-02**
f_7	4662.31	7.5%	Planar	**-459.87**	**8,25E-02**	458.67	3,83E-01
f_8	1.11E05	104%	Planar	**-127,59**	**2.3E-01**	-125.67	2,12E00
f_9	0.018	1.8%	Planar	**298.99**	**5,4E-01**	289.99	5,47E01
f_{10}	195.4	0.74%	Non Planar	194.73	5.5E01	**123.12**	**2.19E00**

4.1   Hybrid Results

In this test, we used the PSO implementation developed by Claus Bendtsen for R language [19] and our own implementation of the most simple version of the Monte Carlo methodology [8]. The proposed optimization methodology is used for optimizing the five functions presented in Table 1. In addition, we optimized other five functions from CEC'05 benchmark test, correspondent to functions f_{11} to f_{15} for details see [20]. The additional functions are for two dimensions because less dimensions is more critical for the criterion because difference bewteen planar and non planar regions are closer. In step 3, the proposed criterion is evaluated using 50 random points drawn from a uniform distribution. In step 4, we sample 12.000 random points drawn from a uniform distribution and use as a result the best point found. In step 5, we run PSO with 300 iterations and 40 individuals; we use the default parameters of the package [19]. Table 6 present our results found by each algorithm (mean and standard deviation for 50 independent runs) and the values for the criterion. The bold results are the ones taken by the hybrid proposed. The results for functions f_2 and f_5 are pretty similar, but the deviation is better in both cases. In terms of computational effort PSO has 12000 calls to objective function it is 300 iterations and population size is 40, but it has internal operations, and Monte Carlo has 12000 calls to the objective function without internal operations. The time used both methods are similar, but it is longer for monte carlo around 0.5 seconds each independent run in 30 dimensions (2.9 Ghz intel i7 processor).

For the complete set of function, we found six functions with problematic planar areas, and four with non planar areas. For the first five functions, we are sure that there are two function described as planar; but for the second part of the tests the functions are hybrids and rotated, which makes difficult to determine the topology. Nevertheless, criterion shows the problematic functions and let use the best strategy for the function except for function 5, in this case, the results for both algorithms are so similar.

5 Conclusions

We conclude that the criterion proposed works successfully as a strategy to identify problematic planar regions of functions with different dimensions, configurations and populations. As a strategy of control, it levers the advantages for each algorithm agrees with the topology of the function. It improves the individual behavior without any parameter change for the algorithm proposed to validate the criterion.

In this case, to obtain the best result compared with other algorithms was not the target, but results for the small benchmark test show criterion's potential as an general strategy to get the relevant information of the function with the aimed to parameters control. The computational effort of the criterion proposed is small and use the information given by the population without other search and calculations, which are desirable. Nevertheless, this is the first approach to the idea then it is immature yet. But now the criterion shows potential as a successful strategy for hybrids and can be used for control and tuning parameters for different metaheuristics.

In future works, the proposed criterion can be used to control parameters inside search with the aim to develop metaheuristics with self-adaptive parameters. Also, control of algorithms for metaevolutionary optimizers.

References

[1] Coello Coello, C.A.: Recent Trends in Evolutionary Multiobjective Optimization. In: Evolutionary Multiobjective Optimization Theoretical Advances and Applications, pp. 7–32. Springer (2005)

[2] Törn, A., Ali, M.M., Viitanen, S.: Stochastic global optimization: Problem classes and solution techniques. Journal of Global Optimization 14(4), 437–447 (1999)

[3] Pant, M., Et, A.: DE-PSO: a new hybrid meta-heuristic for solving global optimization problems. New Mathematics and Natural Computation 7(3), 363–381 (2011)

[4] Talbi, E.G.: Metaheuristics: from design to implementation. Wiley (2009)

[5] Törn, A., Žilinskas, A.: Global Optimization. LNCS, vol. 350. Springer, Heidelberg (1989); Goos, G., Hartmanis, J. (eds.)

[6] Kramer, O.: Self-adaptative Heuristics for Evolutionary Computation, 1st edn. Springer, Berlin (2008)

[7] Mersmann, O., Preuss, M., Trautmann, H.: Benchmarking evolutionary algorithms: Towards exploratory landscape analysis. In: Schaefer, R., Cotta, C., Kołodziej, J., Rudolph, G. (eds.) PPSN XI. LNCS, vol. 6238, pp. 73–82. Springer, Heidelberg (2010)

[8] Yang, X.S.: Engineering Optimization An Introduction with Metaheuristics Applications. Wiley, Hoboken (2010)

[9] Bäck, T.: Evolutionary Algorithms in Theory and Practice. Oxford University Press (1996)

[10] Mohd, I.B.: Identification of region of attraction for global optimization problem using interval symmetric operator. Applied Mathematics and Computation 110, 121–131 (2000)

[11] Hendrix, E., Toth, B.: Introduction to nonlinear and global optimization. Springer, London (2010)

[12] Shang, Y.W., Qiu, Y.H.: A note on the extended rosenbrock function. Evolutionary Computation 14(1), 119–126 (2006)

[13] De Jong, K.: Analysis of the behavior of a class of genetic adaptative systems. PhD thesis, The University of Michigan (1975)

[14] Schwefel, H.P.: Evolution and Optimun Seeking. Wiley, N.Y. (1995)

[15] Eberhart, R., Kennedy, J.: A new optimizer using particle swarm theory. In: Proceedings of the Sixth International Symposium on Micro Machine and Human Science, MHS 1995, pp. 39–43. IEEE (1995)

[16] Gentle, J.E.: Random number generation and monte carlo methods. In: Chambers, J., Eddy, W., Härdle, W., Sheater, S., Tierney, L. (eds.) Statistics and Computing. Statistics and Computing, vol. xvi, p. 381. Springer (2003)

[17] Mirjalili, S., Hashim, S.: A new hybrid PSOGSA algorithm for function optimization. In: 2010 International Conference on Computer and Information Application (ICCIA), pp. 374–377 (2010)

[18] Chuang, L.Y., Yang, C.H., Yang, C.H.: Tabu search and binary particle swarm optimization for feature selection using microarray data. Journal of Computational Biology: A Journal of Computational Molecular Cell Biology 16(12), 1689–1703 (2009)

[19] Bendtsen, C.: PSO, R package version 1.0.3 (2012)

[20] Suganthan, P.N., Hansen, N., Liang, J.J., Deb, K., Chen, Y.P., Auger, A., Tiwari, S.: Problem definitions and evaluation criteria for the cec 2005 special session on real-parameter optimization. Technical Report KanGAL N2005005, Nanyang Technological University, Singapure (May 2005)

Optimizing Neighbourhood Distances for a Variant of Fully-Informed Particle Swarm Algorithm

Vytautas Jančauskas

Faculty of Mathematics and Informatics, Vilnius University, Lithuania
vytautas.jancauskas@mif.vu.lt

Abstract. Most global optimization algorithms offer a trade-off in that they solve one class of problems better for the price of solving another class of problems worse. This is to be expected in light of theoretical results like No free lunch theorem. It is desirable, therefore, to have an automatic method of constructing algorithms tuned for solving specific problems and classes of problems. We offer a variant of Fully-Informed Particle Swarm Optimization algorithm that is highly tunable. We show how to use meta-optimization to optimize it's neighbourhood space and influence function to adjust it to solving various test problems. The optimized neighbourhood configurations and influence functions also give insights in to what it takes for a Particle Swarm Optimization algorithm to successfully solve a problem. These configurations are often contrary to what people would design using their intuitions. This means that meta-optimization in this case can be used as a tool for scientific exploration as well as for practical utility.

1 Introduction

Particle Swarm Optimization is a global optimization method that optimizes a problem by iteratively searching for a solution. It is based on social interactions in a swarm of idealized agents (particles) that share information to adjust their velocity and position in the solution space. It is from a family of metaheuristics that contain such methods as Random Search, Genetic Algorithms, Simulated Annealing and others. It is primarily used for continuous problems, although variants for solving combinatorial problems exist.

There are many studies comparing various different PSO algorithms to other PSO algorithms. It is usually the case that when someone develops a new variant they compare it to several different variants of the PSO algorithm. Among larger and more methodological studies is the one done by Marco A. Montes de Oca [1]. In studies such as these it is almost always the case that certain variants solve certain problems better while other variants solve other problems better. This observation is in line with theoretical results like no free lunch theorems as developed by David Wolpert and William G. Macready [2]. It would be advantageous to have a way of automatically obtaining a variant of PSO algorithm that

G. Terrazas et al. (eds.), *Nature Inspired Cooperative Strategies for Optimization (NICSO 2013)*, Studies in Computational Intelligence 512,
DOI: 10.1007/978-3-319-01692-4_17, © Springer International Publishing Switzerland 2014

would be particularly well suited to solving a specified class of problems. One way to achieve this is meta-optimization — fine tuning of the algorithms parameters using another optimization algorithm. One such study was done by Magnus Erik Hvass Pedersen [3]. Another attempt to meta-optimize the parameters of standard PSO was done by Michael Meissner et al. [4] with the intended use of training artificial neural networks. Another method of tuning a PSO algorithm for a specific class of problems is genetic programming as explored by Riccardo Poli et al. [5].

We propose an algorithm that is purpose built to be meta-optimized. Usually PSO algorithms use a graph based population structure. If two particles are connected by an edge it means that they are neighbours. When two particles are neighbours they directly exchange information about good solutions. In our case we position particles on a two dimensional plane and the Euclidean distance in it represents how much two particles influence each others search process. This means that each particle gets two parameters that can be optimized, namely x and y coordinates in the neighbourhood plane. Additionally, we use an influence function $W_i(j)$ which calculates the amount of influence particle j has on particle i. This function is a third order polynomial of two variables, namely distance between the particles in the neighbourhood plane and the best solution found so far. This polynomial has 10 coefficients that are also optimized in the meta-optimization process. As a result we get an algorithm with $2n + 10$ parameters. We optimize these parameters in the hope that this will adapt the algorithm to specific classes of problems (functions with similar properties).

In the second section we describe various variants of the particle swarm optimization (PSO) algorithms used in this study. Also other relevant theoretical information is presented there, such as the concept of PSO topology. In the third section we describe the proposed changes to the Fully-Informed Particle Swarm Optimization (FIPSO) algorithm that allow it to be meta-optimized. We describe the function $W_i(j)$ that is used to evaluate the influence particle with index j has on particle with index i. The value of this function describes numerically how much particle i is attracted or repulsed from particle j. In the fourth section we describe the meta-optimization procedure we used to optimize function $W_i(j)$ and particle positions in neighbourhood space. We also describe the test functions we used and the experimental procedure that was used to arrive at numerical results presented in this paper. In the fifth section we present results obtained via the experimental procedures, we discuss these results, present and discuss interesting cases, we also sum up the results and present interesting directions for future research.

2   Particle Swarm Optimization

Particle Swarm Optimization is a global optimization metaheuristic designed for continuous problems. It was first proposed by James Kennedy and Russell C. Eberhart in 1995 [6]. The idea is to have a swarm of particles (points in multi-dimensional space) each having some other particles as neighbours and exchanging information to find optimal solutions. Particles move in solution space of some function by adjusting their velocities to move towards the best solutions they found so far and towards the best solutions found by their neighbours. These two attractors are further randomly weighted to allow more diversity in the search process. The idea behind this algorithm are observations from societies acting in nature. For example one can imagine a flock of birds looking for food by flying towards other birds who are signaling a potential food source as well as by remembering where this particular bird itself has seen food before and scouting areas nearby. It can also be viewed as modeling the way we ourselves solve problems - by imitating people we know, who we see as particularly successful, but also by learning on our own. Thus problem solving is being guided by our own experience and by the experiences of people we know to have solved similar problems particularly well. The original algorithm is not presented here since it is very rarely used today and we go straight to more modern implementations.

2.1   Canonical Particle Swarm

Proposed by Maurice Clerc et al. [7] it is a variant of the original PSO algorithm. It guarantees convergence through the use of the constricting factor χ. It also has the advantage of not having any parameters, except for ϕ_1 and ϕ_2 which represent the influence of the personal best solution and the best solution of particles neighbours on the trajectory of that particle. Both of these parameters are usually set to 2.05 as per suggestion in the original paper. Moving the particle in solution space is done by adding the velocity vector to the old position vector as illustrated in (1) equation.

$$\boldsymbol{x}_i \leftarrow \boldsymbol{x}_i + \boldsymbol{v}_i \tag{1}$$

Updating velocity involves taking current velocity and adjusting it so that it will point the particle more in the direction of its personal best and the best of its most successful neighbour. It is laid out in (2) formula.

$$\boldsymbol{v}_i \leftarrow \chi \left(\boldsymbol{v}_i + \boldsymbol{\rho_1} \otimes (\boldsymbol{p}_i - \boldsymbol{x}_i) + \boldsymbol{\rho_2} \otimes (\boldsymbol{g}_i - \boldsymbol{x}_i) \right) \tag{2}$$

where

$$\boldsymbol{\rho}_i = \boldsymbol{U}(0, \phi_i) \tag{3}$$

$$\chi = \frac{2}{\phi - 2 + \sqrt{\phi^2 - 4\phi}} \tag{4}$$

and where $\phi = \phi_1 + \phi_2$ with ϕ_1 and ϕ_2 set to 2.05, $\boldsymbol{U}(a, b)$ is a vector of random numbers from the uniform distribution ranging from a to b in value. Here $\boldsymbol{p}_i$ is the best personal solution of particle i and $\boldsymbol{g}_i$ is the best solution found by a neighbour of particle i. Which particle is a neighbour of which other particle is set in advance.

Algorithm 1. Canonical PSO algorithm

Data: Function f of d dimensions to be optimized.
Result: The best value of function f found and the vector $\boldsymbol{x}$ in solution
        space which evaluates to that value.

1 **for** $j \leftarrow 1$ *to* n **do**
2     $\boldsymbol{x}_j \leftarrow \boldsymbol{U}(a,b)$;
3     $\boldsymbol{v}_j \leftarrow \boldsymbol{0}$;
4 **for** $i \leftarrow 1$ *to* k **do**
5   **for** $j \leftarrow 1$ *to* n **do**
6         $\boldsymbol{x}_j \leftarrow \boldsymbol{x}_j + \boldsymbol{v}_j$;
7   **if** $f(\boldsymbol{x}_j) < f(\boldsymbol{p}_j)$ **then**
8             $\boldsymbol{p}_j \leftarrow \boldsymbol{x}_j$;
9   Update $\boldsymbol{v}_j$ according to (2) formula;

The canonical variant of the PSO algorithm is given in Algorithm 1 and can be explained in plain words as follows: for each particle with index j from n particles in the swarm, initialize the position vector $\boldsymbol{x}_j$ to random values from the range specific to function f and initialize the velocity vector to the zero vector, for k iterations (set by the user) update the position vector according to formula (1) and update velocity according to formula (2), recording best positions found for each particle.

    Particle Swarm topology is a graph that defines how the particles interact. Particles sharing edges are called neighbours. We used two topologies: `lbest` where particles are connected in a ring, each particle connected to 4 closest particles in that ring and `grid` where particles are connected in a von Neumann neighbourhood.

2.2 Fully-Informed Particle Swarm

The original Fully-Informed PSO algorithm was described by Rui Mendes et al. [8] and the original procedure for velocity update is given in (5) formula. The difference between standard PSO and Fully-Informed PSO is that the velocity update formula takes in to account all of particles neighbours, instead of only the one with the best solution found so far. The update formulas are given in the original notation.

$$\boldsymbol{v} \leftarrow \chi \left(\boldsymbol{v} + \phi \left(\boldsymbol{P_m} - \boldsymbol{X} \right) \right) \tag{5}$$

where

$$\boldsymbol{P_m} = \frac{\sum_{k \in N} W(k) \boldsymbol{\phi_k} \otimes \boldsymbol{P_k}}{\sum_{k \in N} W(k) \boldsymbol{\phi_k}} \tag{6}$$

where

$$\boldsymbol{\phi_k} = \boldsymbol{U} \left(0, \frac{\phi_{max}}{|N|} \right) \tag{7}$$

and where N is the set of particles neighbours. As in the canonical PSO algorithm, particles position in solution space is updated by adding velocity vector to position vector $\boldsymbol{x}$. This variant extends the work done by Maurice Clerc et al. [7] on the value of χ - the constriction coefficient in formula (5), which is essential to swarm convergence. This work was done on a particle swarm variant that only takes in to account its own personal best and the personal best of its most successful neighbour, but was extended to use data from all the neighbours.

Riccardo Poli et al. [9] give a slightly different update rule, where $W(k) = 1$, it is shown in (8) formula.

$$\boldsymbol{v}_i \leftarrow \chi \left(\frac{1}{K_i} \sum_{n=1}^{K_i} \boldsymbol{U}(0, \phi) \otimes (\boldsymbol{p}_{nbr_n} - \boldsymbol{x}_i) \right) \tag{8}$$

Here K_i is the number of neighbours of particle with index i and $\boldsymbol{p}_{nbr_n}$ is the best solution found by the n-th neighbour of particle i.

3   Proposed Changes to FIPSO

Our proposed algorithm is identical in all ways to the standard FIPSO algorithm except for the fact that we use all the particles in the swarm to update the velocity and not only the neighbours of each particle. In place of neighbourhood topology (a graph that defines which particles are neighbours of which other particles) we use Euclidean distances in two dimensional space. Each particle gets coordinates in this space and the distances define how socially close particles are to each other. These coordinates are configurable and are considered to be parameters of our algorithm. As such they are subject to meta-optimization procedure. This is shown graphically in Figure 1, where there are 5 particles labeled a, b, c, d and e. If we consider the influence other particles have on particle a, we can arrange them in order of decreasing influence as b, c, d and e. Influence particles have on particle a is proportional to the length of the line connecting them. If we are calculating the influence particle with index j has on particle with index i we will use two variables in the influence function: y_j and d_{ij} which are calculated as follows: $y_j = 1 - \frac{y_j^* - min(\boldsymbol{g})}{max(\boldsymbol{g}) - min(\boldsymbol{g})}$ and $d_{ij} = 1 - \frac{d_{ij}^* - min(\boldsymbol{d}_i)}{max(\boldsymbol{d}_i) - min(\boldsymbol{d}_i)}$, where y_j^* is the best value of the optimized function achieved by particle j, $\boldsymbol{g}$ is the vector of best values achieved by all the particles in the swarm, d_{ij}^* is Euclidean distance in neighbourhood space from particle i to particle j and $\boldsymbol{d}_i$ is the vector of distance from particle i to every other particle in the swarm.

The velocity update function then becomes as shown in (9) formula.

$$\boldsymbol{v}_i \leftarrow \chi \left(\frac{1}{\sum_{j=1}^n W_i(j)} \sum_{j=1}^n W_i(j) \boldsymbol{U}\left(0, \frac{\phi}{n}\right) \otimes (\boldsymbol{p}_j - \boldsymbol{x}_i) \right) \tag{9}$$

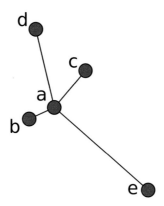

Fig. 1. Particles in neighbourhood plane

Here n is the number of particles in the swarm, p_j is the best solution found by particle with index j and $W_i(j)$ is the influence function described below. All the other parts of the formula mean the same as in the canonical PSO. We also define the influence function $W_i(j)$ as shown in (10) formula.

$$
\begin{aligned}
W_i(j) = w_1 y_j^3 + w_2 d_{ij}^3 + w_3 y_j^2 d_{ij} + w_4 y_j d_{ij}^2 + \\
+ w_5 y_j^2 + w_6 d_{ij}^2 + w_7 y_j d_{ij} + \\
+ w_8 y_j + w_9 d_{ij} + \\
+ w_{10}
\end{aligned}
\tag{10}
$$

We end up with $n \times 2 + 10$ parameters for a swarm that has n particles. This is because each particle gets two coordinates for the neighbourhood plane and additionally there are 10 coefficients for the 3rd order polynomial. We will use meta-optimization to find good values for these parameters for each test problem.

3.1   Meta-optimization

The goal of the meta-optimization procedure is to calculate good values for position of the particle in the 2-dimensional neighbourhood space and polynomial coefficients both of which are used in the influence function for our variant of FIPSO algorithm. The function that was used during meta-optimization process is presented in Algorithm 2.

Algorithm 2. Function used to evaluate the performance of a particle swarm with specific particle positions in neighbourhood plane and influence polynomial coefficients.

Data: A 60-dimensional vector x where first 50 coordinates are the coordinates of particles in a 2-dimensional neighbourhood space and the last 10 coordinates are coefficients for a 3rd order polynomial of the influence function.

Result: Mean value of variables y_1 to y_{50}, holding best function values found by the swarms.

1 **for** $i \leftarrow 1$ *to* 50 **do**
2       $s_i \leftarrow$ a new swarm, where particle j gets the position (x_{2j-1}, x_{2j}) and the coefficients for polynomial $W_j(k)$ are the last 10 coordinates of vector x;
3 **for** $i \leftarrow 1$ *to* 50 **do**
4   **for** *iteration* $\leftarrow 1$ *to* 300 **do**
5   Update position and velocity for each particle in s_i in accordance to formulas (1) and (9) respectively;
6       $y_i \leftarrow$ best result found by s_i;

We used 50 swarms to evaluate given particle positions in neighbourhood plane and polynomial coefficients. We used the average best value found by these 50 swarms. This was done to make sure we get a fairly accurate evaluation for the given parameter set. Each swarm ran for 300 iterations. This function was then optimized for 1000 iterations using Canonical PSO with 36 particles connected in a `grid` topology. The results (particle positions and the coefficients for the polynomial) were recorded. Initial values were chosen so that $x_i \in (-10, 10)$, where $1 \leq i \leq 60$, they were chosen randomly with uniform distribution.

4 Experimental Setup

Two experiments were performed while doing this study. At first we used meta-optimization to find optimal solutions for particle positions in neighbourhood space and coefficients for the polynomials used as influence functions $W_i(j)$. This procedure was performed on every test function we used. After this we compared the results obtained by these tuned versions to results obtained by popular PSO variants. Namely Canonical and FIPSO with $W_j(k) = 1$ as described by Riccardo Poli et al. [9] with `grid` and `lbest` topologies. In this experiment we did not use results obtained by the meta-optimization procedure for our variant. This is because meta-optimization will tend to favor good results thus resulting in bias towards lower values. Instead new swarms were created using coordinates and polynomial coefficients obtained by meta-optimization and these were then tested. When comparing algorithms we used 500 swarm iterations and 100 runs for each swarm/function combination. The results of these runs were averaged and recorded.

4.1   Test Functions

We used 10 test functions defined in Table 1, all functions were used in their 20-dimensional forms so $n = 20$ in all these formulas. These functions can be found in many publications and are often used to test novel optimization methods. Some of them were described by Thomas Bäck [10], others can be found in, for example, [11]. Particles got their initial positions in the solution space from different intervals for each of these functions. They are as follows: Sphere - $(-100, 100)$ for each coordinate, Griewangk - $(-600, 600)$, Rastrigin - $(-5.12, 5.12)$, Rosenbrock - $(-10, 10)$, Ackley - $(-30, 30)$, Michalewitz - $(0, \pi)$, Pathological - $(-100, 100)$, SinEnv - $(-100, 100)$, Schubert - $(-10, 10)$ and $(-10, 10)$ for Stretched-V function.

Table 1. Test Functions

Name	Formula
Sphere	$\sum_{i=1}^{n} x_i^2$
Griewangk	$\frac{1}{4000} \sum_{i=1}^{n} (x_i - 100)^2 - \prod_{i=1}^{n} cos\left(\frac{x_i - 100}{\sqrt{i+1}}\right) + 1$
Rastrigin	$\sum_{i=1}^{n} \left(x_i^2 - 10cos(2\pi x_i) + 10\right)$
Rosenbrock	$\sum_{i=2}^{n} \left(100(x_i - x_{i-1}^2)^2 + (1 - x_{i-1})^2\right)$
Ackley	$-20e^{-0.2\sqrt{\frac{1}{n}\sum_{i=1}^{n} x_i^2}} - e^{\frac{1}{n}\sum_{i=1}^{n} cos(2\pi x_i)} + 20 + e$
Michalewitz	$-\sum_{i=1}^{n} \left(sin(x_i)sin^{20}\left(\frac{(i-1)x_i^2}{\pi}\right)\right)$
Pathological	$\sum_{i=1}^{n-1} \left(sin^2\left(\sqrt{100x_{i+1}^2 + x_i^2} - 0.5\right) \Big/ \left(0.001\left(x_i - x_{i+1}\right)^4 + 0.5\right)\right)$
SinEnv	$\sum_{i=1}^{n-1} \left(sin^2\left(\sqrt{x_{i+1}^2 + x_i^2} - 0.5\right) \Big/ \left(0.001\left(x_{i+1}^2 + x_i^2\right) + 1.0\right)^2 + 0.5\right)$
Schubert	$\sum_{i=1}^{n} \sum_{j=0}^{j<5}(j + 1)sin((j + 2)x_i + (j + 1))$
Stretched-V	$\sum_{i=1}^{n-1} (x_{i+1}^2 + x_i^2)^{0.25} sin^2(50(x_{i+1}^2 + x_i^2)^{0.1}) + 1)$

5   Results

Results of algorithm comparison are shown in Table 2 with best results emphasized in bold font. As can be seen our algorithm gives the best result in 5 out of 10 cases. In cases where it does not give the best result, however, it comes closer than the other of the two algorithms. For example in case of Rosenbrock function FIPSO did best with 18.2477, Canonical PSO only achieved 30.5920 while our variant got 19.3589 - very close to the best result. In the opposite direction Canonical PSO got the best result of -214.7149, while FIPSO only got -106.0894, while our algorithm got -178.9168 which is much closer to the best result. This, we believe, shows the algorithms ability to tune itself to a problem. Two problems where our algorithm was a lot better than any of others are Rastrigin and Michalewitz functions.

Table 2. Algorithm performance comparison

	Canonical		FIPSO		Our Variant
	grid	lbest	grid	lbest	
Sphere	4.2×10^{-4}	7.0×10^{-4}	1.6×10^{-6}	5.6×10^{-7}	$\mathbf{3.4 \times 10^{-8}}$
Griewangk	1.7×10^{-2}	2.0×10^{-2}	3.0×10^{-3}	2.7×10^{-3}	$\mathbf{9.4 \times 10^{-4}}$
Rastrigin	31.0148	31.7726	19.0379	31.4604	**11.9119**
Rosenbrock	42.1298	30.5920	**18.2477**	19.1382	19.3589
Ackley	6.8×10^{-3}	1.0×10^{-2}	3.7×10^{-4}	$\mathbf{2.1 \times 10^{-4}}$	9.5784×10^{-4}
Michalewitz	-15.1072	-15.1069	-11.3147	-11.8751	**-17.0844**
Pathological	-4.8556	**-5.08256**	-3.8177	-3.7988	-4.5861
SinEnv	-22.6788	-22.4641	-25.0864	-24.5566	**-25.8629**
Schubert	-213.9197	**-214.7149**	-97.7311	-106.0894	-178.9168
Stretched-V	2.8070	4.2907	0.2787	**0.2574**	0.4340

Visualizations of optimized swarm parameters are given in Figures 2 to 11. The picture on the left side is a scatter plot of particle positions in two dimensional neighbourhood space, while the picture on the right is a plot of the two variable polynomial that is used as $W_i(j)$ - the influence function. From the scatter plots it seems that particles tend to settle in to a more or less normally distributed configuration. There seems to be some clusterization in the case of Michalewitz, SinEnv and Stretched-V functions, although a larger number of particles would have to be used to make sure. In all three cases there seem to be 3 clusters of particles. Optimized polynomials usually emphasize either distance in neighbourhood space or the best value found by that particle or both. Usually there is a large area where the value of $W_i(j)$ is negative, which means that particle i is repulsed from particle with index j, it is an important point since repulsion is mostly ignored in existing PSO algorithms.

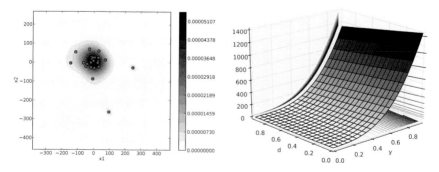

Fig. 2. Optimized swarm parameters for the Sphere problem

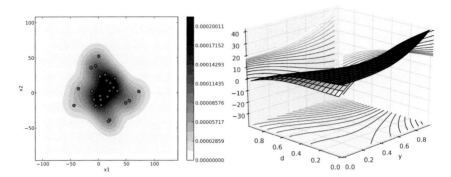

Fig. 3. Optimized swarm parameters for the Griewangk problem

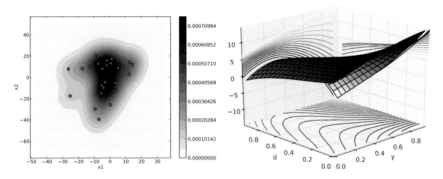

Fig. 4. Optimized swarm parameters for the Rastrigin problem

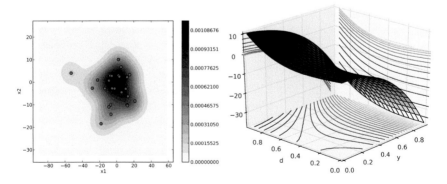

Fig. 5. Optimized swarm parameters for the Rosenbrock problem

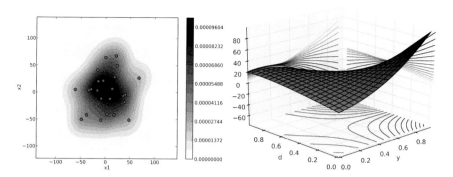

Fig. 6. Optimized swarm parameters for the Ackley problem

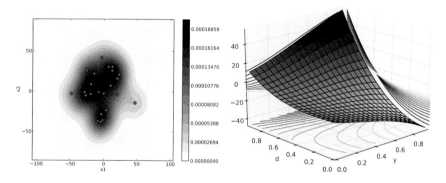

Fig. 7. Optimized swarm parameters for the Michalewitz problem

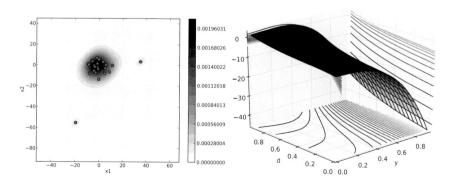

Fig. 8. Optimized swarm parameters for the Pathological problem

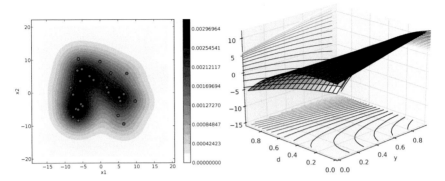

Fig. 9. Optimized swarm parameters for the SinEnv problem

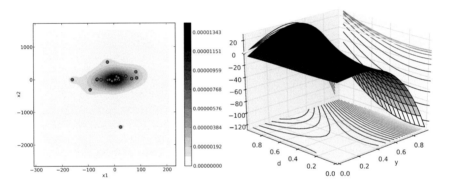

Fig. 10. Optimized swarm parameters for the Schubert problem

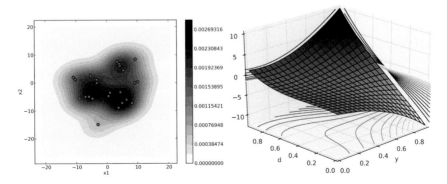

Fig. 11. Optimized swarm parameters for the Stretched-V problem

6   Conclusions

We think we have shown reasons to believe that our algorithm or one like it is capable to adapt to problems and could be used to solve classes of problems more successfully than more generic ones. This avoids the problem of an algorithm being excellent at optimizing some functions yet very bad at optimizing others. There are no reasons why several functions could not be used in the meta-optimization process. There may be some improvements to the method that might make the process more successful, like using higher order polynomials. It would also be interesting to see what properties of test functions result in what polynomials and patterns in neighbourhood space. A more involved study would have the meta-optimization routine run over sets of functions with certain properties and would have us map function properties to properties of neighbourhood plane configurations and polynomial coefficients.

Acknowledgement. This research is supported by the Research Council of Lithuania under Grant No. MIP-063/2012.

References

[1] de Oca, M.A.M., Stützle, T., Birattari, M., Dorigo, M.: A comparison of particle swarm optimization algorithms based on run-length distributions. In: Dorigo, M., Gambardella, L.M., Birattari, M., Martinoli, A., Poli, R., Stützle, T. (eds.) ANTS 2006. LNCS, vol. 4150, pp. 1–12. Springer, Heidelberg (2006)

[2] Wolpert, D., Macready, W.: No free lunch theorems for optimization. IEEE Transactions on Evolutionary Computation 1(1), 67–82 (1997)

[3] Pedersen, M.E.H.: Good parameters for particle swarm optimization. Technical Report HL1001, Hvass Laboratories (2010)

[4] Meissner, M., Schmuker, M., Schneider, G.: Optimized particle swarm optimization (opso) and its application to artificial neural network training. BMC Bioinformatics 7, 125 (2006)

[5] Poli, R., Langdon, W.B., Holland, O.: Extending particle swarm optimisation via genetic programming. In: Keijzer, M., Tettamanzi, A.G.B., Collet, P., van Hemert, J., Tomassini, M. (eds.) EuroGP 2005. LNCS, vol. 3447, pp. 291–300. Springer, Heidelberg (2005)

[6] James Kennedy, R.C.E.: Particle swarm optimization. In: Proceedings of the IEEE International Conference on Neural Networks, vol. 4, pp. 1942–1948 (1995)

[7] Clerc, M., Kennedy, J.: The particle swarm - explosion, stability, and convergence in a multidimensional complex space. IEEE Transactions on Evolutionary Computation 6(1), 58–73 (2002)

[8] Mendes, R., Kennedy, J., Neves, J.: The fully informed particle swarm: simpler, maybe better. IEEE Transactions on Evolutionary Computation 8(3), 204–210 (2004)

[9] Poli, R., Kennedy, J., Blackwell, T.: Particle swarm optimization. Swarm Intelligence 1(1), 33–57 (2007)

[10] Bäck, T.: Evolutionary algorithms in theory and practice: evolution strategies, evolutionary programming, genetic algorithms. Oxford University Press, Oxford (1996)

[11] Oldenhuis, R.: Many test functions for global optimizers (February 2009)

Meta Morphic Particle Swarm Optimization
Simultaneous Optimization of Solution Classes and Their Continuous Parameters

Jesse van den Kieboom, Soha Pouya, and Auke Jan Ijspeert

Biorobotics Laboratory, Institute of Bioengineering, School of Engineering,
Ecole Polytechnique Fédérale de Lausanne (EPFL), Switzerland
{jesse.vandenkieboom,soha.pouya,auke.ijspeert}@epfl.ch

Abstract. Particle Swarm Optimization is a simple and elegant optimization algorithm used to solve a large variety of different real-valued problems. When it comes to solving combinations of continuous and discrete problems however, PSO by itself is not very well suited for the task. There have been previous works addressing the issue of solving solely discrete problems with PSO, but solving problems involving both discrete and continuous parameters at the same time with a PSO-like algorithm has not yet been fully explored. In this paper we provide a novel PSO-based algorithm, called Meta Morphic Particle Swarm Optimization, which looks at solving a particular class of problems for which there exists a discrete set of possible ways to solve the problem where each possibility uses a different subset of a continuous, real-valued parameter space. We introduce a two-layered approach, a PSO in the inner layer for the continuous space, and an outer layer, guided migration scheme using probabilities to choose between the different possible solution sets. We analyze the performance and characteristics of this new algorithm and show how it can be used for real-world applications.

1 Introduction

Since its original inception, the Particle Swarm Optimization algorithm (or PSO) [7] has seen a considerable amount of attention in the evolutionary computation community. Partly due to its simplicity and elegance, since then many new varieties of PSO have been developed by researchers in the community trying either to address some of its shortcomings (such as stagnation [6, 20], diversity [11] or niching [1, 12]) or to improve its performance tailored towards specific sets of problems (such as multiple objectives or constraints [10, 14, 17]).

In this work we look at solving such a particular set of problems. Namely the optimization of a problem for which a discrete set of solution classes exists , each with a (possibly overlapping) subset of continuous, parameters taken from the total parameter set. The optimization then needs to take into account the discrete problem, as well as optimizing the continuous parameters used for this particular solution class. This might seem like an abstract problem, but indeed, many real problems are formulated this way. Often we choose to optimize each

G. Terrazas et al. (eds.), *Nature Inspired Cooperative Strategies for Optimization*
(NICSO 2013), Studies in Computational Intelligence 512,
DOI: 10.1007/978-3-319-01692-4_18, © Springer International Publishing Switzerland 2014

solution class of the problem independently or even manually and compare the results later. This however is 1) not practical for problems for which a large set of different solutions exist, 2) inefficient for problems with a large *possible* solution set, but a *small* probable solution set, and 3) it can be biased by human intervention especially for the cases for which human intervention is not sufficient. To solve these kinds of problems we require an algorithm which 1) makes informed, discrete decisions about which classes of solutions to explore and 2) finds *optimal* parameter values for these classes of solutions.

The first contribution to solving discrete binary problems using PSO came not long after the original PSO algorithm was published. In [8], the original author of PSO details a version of PSO which uses probabilities of a discrete value switching from 0 to 1 (or the other way around) instead of the actual values as the parameters being optimized. More recently, this approach was generalized in [4] in which the definitions and operations of the PSO (position, velocity, subtraction, external multiplication and move) are redefined for the discrete domain. An extension of the original binary discrete PSO algorithm was presented in [16] where discrete multi-valued problems are solved by adding a probability for each possible value that the discrete variable can take. Here we take inspiration from this work and use a similar approach to making discrete choices by using probabilities. However, unlike in previous approaches we will define probabilities related to exploration and exploitation similar to those used in PSO to search the set of discrete solution classes while at the same time solving the continuous problem in each solution class. We believe that this allows for more control of the way the problem is solved while at the same time reusing concepts from the continuous domain, which have worked well in general, to the discrete domain. Although Genetic Algorithms and Genetic Programming could be used in a similar way to provide the discrete part of the optimization, we are interested here in reusing the collaborative/cooperative nature of the PSO in the discrete part, instead of competition. The novel algorithm that we designed for this particular problem is called Meta Morphic Particle Swarm Optimization. In this paper we focus primarily on explaining in detail the design, and properties of this algorithm and provide a working example implementation of the algorithm.

The following sections first shortly describe the base PSO algorithm. We then continue to describe the main MMPSO algorithm in detail. After this description we show some of the MMPSO properties on an example problem. Finally, we discuss some of the applications of the algorithm and future work.

2   Base PSO Algorithm

In this section we first briefly explain the original PSO algorithm. We then continue in the next section to describe how we change the basic algorithm resulting in Meta Morphic PSO.

Particle Swarm Optimization (from here on referred to as PSO) is a population based, stochastic optimization algorithm which has been a popular alternative to Genetic Algorithms since it was first introduced in [7]. In its essence, PSO is a very simple algorithm, consisting only of two simple equations which govern its dynamics. Conceptually, the PSO is a cooperative algorithm where the individual particles share information about known solutions of the particular problem being solved. For this work, we use the PSO algorithm as described in [18] as the basis algorithm. The two equations describing the whole algorithm are given in equation 1.

$$
\begin{aligned}
v_i(t+1) &= w \cdot v_i(t) + r_{i1} \cdot c_1 \cdot (X_i - x_i(t)) + r_{i2} \cdot c_2 \cdot (X_g - x_i(t)) \\
x_i(t+1) &= x_i(t) + v_i(t+1)
\end{aligned}
\tag{1}
$$

Here x_i is the current *position* of particle i in the parameter space. It is thus the vector of real-valued parameter values representing a particular solution to the problem begin solved. v_i is the current *velocity* of particle i, in change of parameter value per iteration. Furthermore, r_{i1} and r_{i2} are two random numbers uniformly distributed between 0 and 1, X_i is the best solution as found by particle i (its personal best) and X_g is the global best known solution. The constants c_1 and c_2 determine the importance of respectively local versus global search. Compared to the original algorithm as described in [7], an additional term is introduced, the so called inertia factor w [18]. The purpose of w is to improve the convergence and has been generally found to improve the performance of the PSO.

The algorithm as presented thus has three parameters, the inertia factor w and the two constants c_1 and c_2. For the remainder of this contribution, these parameters have been set to the values 1.494 for both c_1 and c_2, and 0.729 for w which can be shown to guarantee convergence of the algorithm [2, 3, 5].

Finally, to perform the optimization, an initial population of particles is generated each with an initial position vector x_i and initial velocity vector v_i. Both of these vectors are usually initialized such that they are randomly, uniformly distributed in a bounded parameter space (we will not look at unconstrained parameter spaces in this work). We limit the maximum value of each dimension of the velocity vector to the distance from the minimum parameter boundary to the maximum parameter boundary [5]. This has been shown to give good results in general as there is more exploration (in particular in the beginning of the optimization). After the population has been initialized, at each iteration the fitness of each particle for the parameters x_i is calculated and X_i and X_g are updated accordingly. Then, for each particle, the particle's velocity v_i and x_i are updated using equation 1. The stopping condition is often chosen to be a fixed number of iterations or some measure of convergence.

We should note that we do not focus on the particular values of w, c_1, c_2 and V_{max} in this work and have instead taken sensible values for each from literature which have been proven to work well in the general case.

Now that we have briefly described the base PSO algorithm, the next section will describe the extensions that make up the Meta Morphic PSO Algorithm.

3   Meta Morphic PSO Algorithm

We found PSO to be a well performing and easy to understand algorithm for a wide variety of optimization problems. It often outperforms algorithms such as Genetic Algorithms [9, 13] in a variety of different domains it has been applied to. The elegance of the algorithm, the small number of parameters (c_1, c_2 and w) to tune and the general performance are arguably some of its most prominent features.

The base PSO algorithm as described in the previous section works on continuous parameters. What we are interested in, in this work, however is a combination of a discrete set of parameter subspaces and a simultaneous optimization of each of these parameter subspaces (in which the parameters are continuous). We have coined our algorithm Meta Morphic due to the fact that it performs a sort of **meta** optimization on the possible solution subspaces by **morphing** particles from one subspace to another, reconfiguring its parameter space.

We briefly describe a concrete robotics problem (explained in more detail in section 5) to illustrate for which type of problems MMPSO was designed. Assume a certain robotic structure with K degrees of freedom, for which we want to find control laws for locomotion. Furthermore, assume that we can control each of these DOFs with three different modes of control, namely 1) Oscillation, 2) Continuous Rotation or 3) a Locked constant offset. We now have three choices of control modes to make for each of the K degrees of freedom. Instead of making these choices manually, we designed MMPSO to explore combinations of control modes for each DOF automatically. We will occasionally refer to this application of MMPSO in explaining certain concepts of the algorithm.

3.1   Concepts and Terminology

The Meta Morphic PSO Algorithm (hereafter referred to with MMPSO) has been specifically designed for the type of problem described above. Still staying in the abstract domain, consider the following problem containing 9 parameters to optimize as shown in figure 1.

This schematic representation of the parameter space consists of three entities (**A**, **B** and **C**) which we call parameter *pools*. A parameter *pool* in MMPSO is something which defines a distinct number of possible parameters groupings active at a single given time. Thus, referring to figure 1, in *pool* **A** only either parameter (1) or parameters (2, 3) are active. In the context of MMPSO, we call these different parameter *groups* and in the text we indicate a *group* within parentheses (). The *groups* within each *pool* are mutually exclusive. Although the *groups* are mutually exclusive, the parameters in each *group* need not be. Indeed, shown in pool **B** in figure 1, parameter 5 is active both in *group* 1 and in *group* 2. Similarly, parameter 6 is active in both *group* 2 and in *group* 3 (*groups* are indicated by a superscript in each box).

We have until now only explained the concepts of *pools* and *groups*. We still need to outline the concept of parameter *subspaces*. Given the definitions of the *pool* and *group* above, a parameter *subspace* is one particular, valid combination

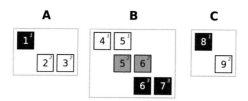

Fig. 1. Example parameter configuration of a single particle. Each of the parameter *pools* **A**, **B** and **C** depict a discrete number of parameter *groups*. The group number is indicated in the superscript of each box as well as by the background shading for clarity. In each *pool*, only one *group* can be active and optimized at a given time. Parameters can overlap between different *groups* as can be seen in *pool* **B**, where a valid set of parameters is either (4, 5), (5, 6) or (6, 7). One complete *subspace* is composed of selecting one *group* for each *pool*, for example {(1), (4, 5), (9)}. There are a total number of 9 parameters in this example.

of *groups* chosen from each *pool*. In the text we will indicate *subspaces* with braces {}. In figure 1 possible *subspaces* are {(1), (6, 7), (8)} or {(1), (4, 5), (9)}. The total number of possible *subspaces* results from simple combinatorics on the *groups* in each *pool*. In our example, the total number of *subspaces* would thus be $2 \times 3 \times 2 = 12$.

To relate the MMPSO parlance to our concrete robotics example, each DOF is represented by a *pool* and each control mode is represented by a *group*. Thus each *pool* contains three *groups* (Oscillation, Rotation, Locked) to choose from. A particular *subspace* is then a specific combination of control modes for each DOF.

The goal of MMPSO is to efficiently search for solutions within these *subspaces*, dividing effort spent in each subspace based on a similar principle of collaboration as used by the base PSO algorithm. To accomplish this we separate the algorithm in two layers.

3.2   The Inner Layer

The *inner* layer is defined as one instance of a *subspace* (i.e. there are 12 distinct *inner* layers in our abstract example). Each *inner* layer runs an **independent** base PSO algorithm. Particles initially are equally distributed over the different *subspaces* (note that there can be more *subspaces* than particles in which case some *subspaces* remain initially unpopulated). Although we use the base PSO algorithm as defined in section 2, it should be noted that *any* extension or variant of PSO could be run without modification in the *inner* layer. The main contribution of MMPSO is the way particles are transferred between *subspaces* in what we call the *outer* layer.

3.3   The Outer Layer

The *outer* layer is a separate algorithm outside the *inner* layers responsible for migrating particles from one *subspace* to another. Figure 2 shows a schematic

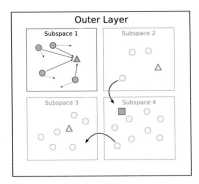

Fig. 2. Schematic overview of the two-layered algorithm. Each of the *subspaces* contains an independent PSO with a population set to the particles which are currently in the *subspace*. The green (triangle) particle represents the best known solution for each *subspace* which we call X_s (this is equivalent to X_g in the base PSO). The blue (rectangle) particle represents the globally best known solution taken over all the *subspaces* and is only known only to the outer layer algorithm. We call this solution X_g.

representation of the two-layered system. Each *subspace* contains a separate PSO and the *outer* layer migrates particles between *subspaces*. A *subspace* best solution, X_s is maintained in each *subspace* and is the equivalent of the globally best known solution X_g in the base PSO algorithm. We also introduce a new X_g which represents a new globally best known solution known only to the *outer* layer algorithm.

To transfer particles between *subspaces* we borrow the concept of the mutation operation from Genetic Algorithms. The basic idea is to *migrate* a particle from one *subspace* to another *subspace* based on migration probabilities. However, unlike in GA where a beneficial mutation is automatically propagated to the next generation, we do not have such a concept in our PSO. Simply moving particles from one *subspace* to another randomly chosen *subspace* will not provide appropriate pressure to explore *subspaces* that have a higher overall fitness more than other *subspaces*, since the particles are moved (and re-moved) randomly.

To address this issue, we take inspiration from the concepts of local versus global and exploration versus exploitation from PSO and introduce three migration probabilities. The *exploration migration probability* P_e, defines the probability of a random migration of the active *group* within each *pool*. The *local migration probability*, P_l, defines the probability of migrating to the *group* within each *pool* which is contained in the best solution known to that particle. Similarly, the *global migration probability*, P_g, defines the same type of migration probability as the *local migration probability*, but towards the globally best known solution X_g over all *subspaces*.

Together, these three migration probabilities will govern the search of the different *subspaces* in a collaborative manner similar to how PSO tries to optimize parameters *within* a *subspace*. We can now define the probability

$P(s_c \rightarrow s_j | s_c \neq s_j)$ for each particle, of migrating the current *group* (*c*) of a *pool* (*s*) to a group (*j*) different from *c* as given in equation 2.

$$P(s_c \rightarrow s_j | s_c \neq s_j) = 1 - (1 - \frac{P_e}{N-1}) \cdot (1 - P_l | s_j = s_l) \cdot (1 - P_g | s_j = s_g)$$

$$P(s_c \rightarrow s_j | s_c = s_j) = 1 - \sum_k^N P(s_c \rightarrow s_k | s_c \neq s_k)$$

$$P_e + P_l + P_g \leq 1 \tag{2}$$

Here the notation $P(a \rightarrow b | a \neq b)$ is used to mean the probability of a transitioning to b given that b is different from a, thus the probability of a particle to migrate a particular group to a different group. This probability is calculated from the probabilities P_e, P_l and P_g as defined above and N the number of different parameter *groups* in the *pool* s. Furthermore, s_l is the parameter *group* l of *pool* s in which the locally best known solution of the particle has been found and s_g is the parameter *group* g of *pool* s in which the globally best known solution (over all parameter *subspaces*) has been found.

Equation 2 proceeds to calculate first the probability of *not* migrating, which is given by the product of the probabilities of *not* migrating due to respectively exploration (P_e), local migration (P_l) and global migration (P_g). The probability of *not* exploring is given by 1 minus the probability to migrate according to P_e to any *other* group, of which there are $N - 1$. Secondly, the probability of *not* migrating towards the locally known best *group* can be calculated by 1 minus P_l, given that the group to be transitioned to (s_j) is the locally best known group (s_l). We have adopted the notation $P_l | s_j = s_l$ here to evaluate to P_l when $s_j = s_l$, or 0 otherwise. The probability of *not* migrating towards the globally best known *group* is calculated in the same way. Finally, the resulting probability $P(s_c \rightarrow s_j | s_c \neq s_j)$ is then given by 1 minus the total probability of not migrating.

For completeness, the second equation provides the probability of staying in the same *group* (i.e. not migrating at all). This probability is simply 1 minus the total probability of migrating to any of the N other *groups*. In practice, only the first equation is used to calculate whether a *group* needs to be migrated. Finally, to guarantee proper probabilities, the sum of P_e, P_l and P_g must be smaller or equal to one.

The probabilities as described in equation 2 are *proper* probabilities in the sense that the sum of all the probabilities equals to 1 (this can be easily seen since the probability of **not** migrating is defined as 1 minus the sum of probabilities of migrating to a different *group*). They are also defined correctly such that setting for example $P_e = 0.5$ will cause on average one particle per two iterations to migrate each *pool* randomly.

As an example, figure 3 shows schematically the migration probabilities involved for a given state of the *pool* $\mathbf{B} \in s$ for a particular particle (as shown before in figure 1). The figure portrays the case where the current *group* of $\mathbf{B}$ (B_c) is *group* 1, or parameters (4, 5). The locally best known *group* (B_l) is *group* 2, or parameters (5, 6) and the globally best known *group* (B_g) is *group* 3. There are

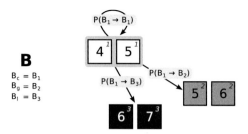

Fig. 3. An example of the migration probabilities involved in migrating the *pool* $\mathbf{B} \in s$ from one particular current *group* (B_c) to each possible *group* of $\mathbf{B}$. The probabilities $P(B_1 \rightarrow B_1)$, $P(B_1 \rightarrow B_2)$ and $P(B_1 \rightarrow B_3)$ can be calculated using equation 2. The resulting probabilities (as functions of P_e, P_l and P_g) are given in equation 3.

then three probabilities $P(B_1|B_1 = B_c)$, $P(B_2|B_2 \neq B_c)$ and $P(B_3|B_3 \neq B_c)$ which respectively represent the migration probability of 1) not changing the current *group*, 2) changing the current *group* from B_1 to B_2 and finally changing the current *group* from B_1 to B_3. Using equation 2, these probabilities then become as shown in equation 3. We will discuss ways to choose P_e, P_l and P_g to design certain behaviors of the algorithm in section 4.

$$P(B_1 \rightarrow B_2) = 1 - (1 - \frac{P_e}{2}) \cdot (1 - P_l)$$

$$P(B_1 \rightarrow B_3) = 1 - (1 - \frac{P_e}{2}) \cdot (1 - P_g)$$

$$P(B_1 \rightarrow B_1) = 1 - P(B_1 \rightarrow B_2) - P(B_1 \rightarrow B_3) \tag{3}$$

3.4   Pseudo Code

A very short and concise pseudo code listing for the algorithm is given in algorithm listing 1[1]. In short, at each iteration, a base PSO is run for each currently non-empty *subspace*. After this, the best local and global *group* (s_l and s_g) for each *pool* are updated according to the fitness of each particle. Finally, particles are migrated from one *subspace* to another by changing the *group* in each *pool* according to the probabilities P_e, P_l and P_g.

4   Properties

There is only one set of parameters left for the user of MMPSO to be chosen. These parameters are the mutation probabilities P_e, P_l and P_g. The values of these parameters are important since they will completely govern the behavior of the *outer* layer algorithm. As such, they need to be chosen carefully.

---

[1] A fully working example of the MMPSO algorithm implemented in matlab is available at: http://biorob2.epfl.ch/~jvanden/mmpso/mmpso_code_nicso_2013.zip

Algorithm 1. MMPSO

Require: Subspaces: the set of all *subspaces*
Require: Pools: the set of all *pools*
Require: P: probability function of $s_c \rightarrow s_i$ with P_e, P_l, P_g
  1: **Particles** $\leftarrow$ initializePopulation
  2: **while** stopping condition not met **do**
  3:　　**for** $u \in$ **Subspaces**, $u \neq \emptyset$ **do**
  4:　　　　PSO(**Particles** $\cup$ u) {base PSO on particles in u}
  5:　　**end for**
  6:　　**for** $s \in$ **Pools do**
  7:　　　　$\{s_l, s_g\} \leftarrow$ updatePoolBest(s) {update s_l and s_g}
  8:　　**end for**
  9:　　**for** $p \in$ **Particles do**
 10:　　　　**for** $s \in$ **Pools do**
 11:　　　　　　migratePool(p, s, $P(s_c \rightarrow s_i | s_c, s_l, s_g)$) {migrate $s_c \rightarrow s_i$ using P}
 12:　　　　**end for**
 13:　　**end for**
 14: **end while**

In general we would normally like to stimulate exploration early in the optimization, so the various *subspaces* are explored sufficiently and general (sub)optima can be located. As the optimization progresses, particles should start to focus more on their locally best known *subspaces* to explore these in more detail. Finally, particles should start to converge on the globally best known *subspace* to maximally optimize for that particular space during the late phases of the optimization process.

To get this kind of behavior, we can design the mutation probabilities using probability curves as functions of the number of iterations. Note that we assume here a stopping criterion based on the maximum number of iterations. If a measurable convergence criterion is used, then the probability curves can be a function of the convergence instead. Figure 4 shows one particular choice of the probability curves. Here we used sigmoid shaped functions for the exploration and global exploitation probabilities, and a Gaussian shaped curve for the local exploitation probability.

The exact choice of where to switch from exploration to exploitation will choosing the shapes of the curves similarly to the ones shown in figure 4 generally works well.

4.1　Example

In this section we will briefly show some characteristics of the MMPSO on the most simple numerical problem. Although the example is a trivial one, it makes it equally trivial to analyze its behavior. More complex examples are currently outside of the scope of this paper (see also the discussion in section 6). In this simple example we are going to consider only two parameters, x and y both

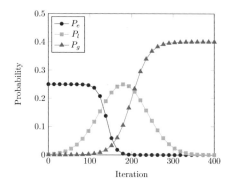

Fig. 4. Mutation probability characteristics for the exploration probability P_e, local exploitation probability P_l and global exploitation probability P_g, emphasizing early exploration and late convergence

bounded in $[0, 1]$. We define one *pool* containing two *groups*. The first *group* is (x) and the second is (x, y). Thus a particle either optimizes for only x or both x and y. We further define two objective functions. The first is evaluated for particles optimizing $\{(x)\}$ and the objective is simply x itself, with a maximum value of 1. The second objective function is evaluated for $\{(x, y)\}$ and is given by $2 - (|x - 0.5| + |y - 0.5|)$, which has a maximum value of 2 at $x = 0.5$ and $y = 0.5$. These objectives were chosen such that the maxima in both *subspaces* are at different values of x, to illustrate the ability of the algorithm to find both.

The population size in this example is set to 40 particles and the optimization lasted for 70 iterations. The probabilities P_e, P_l and P_g were respectively 0.01, 0.05 and 0.05. All particles were initialized in the region $[0, 0.25]$ for both x and y to better show the effect of the particles converging on the maxima.

Figure 5 show the flow of particles between the two *subspaces*. The particles quickly converge on their respective maxima (not shown in the figure). For this simple problem, the population sizes of both *subspaces* can be easily calculated in the limit of the iteration using equation 2. Given that all particles will at some point have visited both *subspaces* (due to P_e), such that the global best and local best are both located in the second *subspace*. This results in a probability $P(s_1|s_2) = P_e$ and $P(s_2|s_1) = 1-(1-P_e) \cdot (1-P_l) \cdot (1-P_g)$. Given the probabilities as defined before, this results in $P(s_1|s_2) = 0.01$ and $P(s_2|s_1) \approx 0.1$. Thus the final populations would be approximately, on average as given in equation 4.

$$S1 = P(s_1|s_2) \cdot \frac{N}{P(s_1|s_2) + P(s_2|s_1)} \approx 0.1 \cdot \frac{40}{0.11} \approx 37$$
$$S2 = P(s_2|s_1) \cdot \frac{N}{P(s_1|s_2) + P(s_2|s_1)} \approx 0.01 \cdot \frac{40}{0.11} \approx 3 \qquad (4)$$

Figure 5 show the trend towards these population sizes (though the simulation would have to be prolonged further to approach these values).

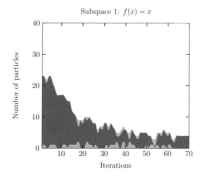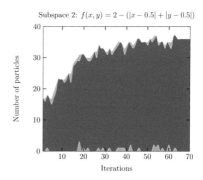

Fig. 5. Particle flow for subspace 1 (left) and subspace 2 (right). The green (upper) and orange (lower) areas show respectively the in- and out-flow of particles in each subspace.

5   Applications

Although we feel that the full details of the applications of MMPSO are out of the scope of this paper, we would like to briefly discuss one previous application and one future application of MMPSO to show how this algorithm can be applied to a specific set of robotics problems.

5.1   Automatic Gait Generation in Modular Robots

In [15] we explored the generation of locomotion gait patterns for a modular robot named Roombots [19] using MMPSO. This work did not focus on the specifics of the used optimization algorithm, but rather on the control methodology of generating locomotion for modular robots. One module of this robot has 3 degrees of freedom (DOFs) (see [19] for details about the robot structure). One particular feature which makes Roombots an interesting platform for studying gait generation is that each DOF can continously rotate, allowing a diverse array of locomotive behaviors. Two Roombots modules joint together are termed a *Metamodule*. The goal of this work was to explore locomotion modes of a Roombots *Metamodule*. The peculiar placement of the degrees of freedom of the *Metamodule* however make it hard to design locomotion controllers by hand.

If all 6 degrees of freedom of the *Metamodule* would have the same control law, then a standard PSO would have sufficed to optimize the various controller parameters. In this work however we were interested in exploring combinations of three different control modes for each of the DOFs, Oscillation (i.e. sinusoidal), continuous Rotation and Locked in which the DOF is controlled to be at a certain constant offset.

To explore combinations of these different control modes, we have successfully used MMPSO to select control modes for each of the DOFs. In MMPSO terminology, there were 6 *pools* (one for each DOF). Each *pool* consisted of three parameter *groups* (one for each control mode). The open control parameters to be optimized (for each DOF i) were the oscillation amplitude R_i, the oscillation or locked offset X_i and a phase bias ψ_{ij} controlling the phase relationship between neighboring DOFs. The MMPSO *pool* for each DOF i is given by: $[(R_i, X_i), (), (X_i)]$, with *groups* for respectively the Oscillation, Rotation and Locked modes. Note that there are no parameters for the Rotation mode and that the offset X_i is shared between the Oscillation and Rotation modes. In terms of MMPSO *subspaces*, there are a total of $3^6 = 729$ different *subspaces* to be explored. One possible MMPSO *subspace* is given in 5:

$$\{(R_1, X_1), (R_2, X_2), (X_3), (), (), (X_6)\} \tag{5}$$

Where the two DOFs are Oscillating, the third and last DOF are Locked and the fourth and fifth DOF are Rotating.

We ran MMPSO to optimize at the same time the control mode configuration and the control parameters. One of the main outcomes of that work shows that allowing optimization of so-called Hybrid control modes, selected by MMPSO, generally outperforms Pure control modes (such as only oscillatory or rotational modes for all the joints). The choice of the migration probabilities P_e, P_l and P_g gives precise control over how much iterations (on average) particles explore different *subspaces* and can be chosen properly according to the total number of *subspaces* available to the problem. For more details on this particular work, we refer to [15].

5.2   Co-design of Mechanics and Control of a Wearable Exoskeleton

MMPSO has been designed for applications where there are a certain (known) set of design choices to be made for (sub)parts of the system. This leads to a combinatory number of possible solutions to be explored. One interesting application of MMPSO for robotics is the co-design of the mechanics (or morphology) and control of a robot. We are currently exploring the use of MMPSO in the context of co-design of a wearable, non-anthropomorphic exoskeleton. The main idea here is to first assume the human body to be a given, fixed mechanical structure. This "system" is then augmented with parallel structures composed of various components (linear/revolute actuators, rigid links, passive elements such as springs and dampers), composing the exoskeleton. Mechanical parameters to be optimized are comprised of material density, mass distribution, actuator placement and rigid link lengths. At the same time, open control parameters for controlling the actuators have to be optimized, which similarly to our work described before can have different control modes. The augmented system can then be evaluated on certain tasks such as locomotion assistance. MMPSO can be used here to explore the different combinations of mechanical parts to construct the exoskeleton attached in parallel to the human body, as well as the control of this exoskeleton, simultaneously. Although we only have preliminary results of this work at this time, we hope to present results of this approach in the near future.

6   Discussion

In this contribution we have described a novel PSO based algorithm for optimizing specific optimization problems combining real-valued parameters with
certain discrete choices in the type of solution being explored. Although the
algorithm is suited only for these specific type of problems, we believe that it
provides a valuable addition to the variety of existing modifications of the base
PSO algorithm. The work explains in detail how principles of migration, inspired
by Genetic Algorithms, can be applied to PSO in a collaborative way such that
multiple, partially-overlapping parameter subsets can be explored simultaneously. The use of proper migration probabilities which separate *exploration, local
exploitation* and *global exploitation* and their semantics makes choosing values
for these probabilities well defined and understandable. The resulting behavior
can be analyzed in terms of these probabilities and makes it easier to design the
probability functions. Furthermore, the two-layer approach of the algorithm allows for any number of extensions of the base PSO algorithm to be used without
any additional modifications to the *outer* layer algorithm.

Although we have shown that the algorithm is capable of solving the problem
it was designed for, we currently lack a more extensive comparison with different algorithms capable of solving similar types of problems at this time (e.g.
Genetic Programming). Future work will investigate more thoroughly how the
performance of MMPSO compares to other algorithms capable of solving similar
problems.

References

[1] van den Bergh, F., Engelbrecht, A.: A cooperative approach to particle swarm op-
     timization. IEEE Transactions on Evolutionary Computation 8(3), 225–239 (2004)
[2] Clerc, M., Kennedy, J.: The particle swarm - explosion, stability, and convergence
     in a multidimensional complex space. IEEE Transactions on Evolutionary Com-
     putation 6(1), 58–73 (2002)
[3] Clerc, M.: The swarm and the queen: towards a deterministic and adaptive par-
     ticle swarm optimization. In: Proceedings of the 1999 Congress on Evolutionary
     Computation, CEC 1999, vol. 3. IEEE (1999)
[4] Clerc, M.: Discrete particle swarm optimization, illustrated by the traveling sales-
     man problem. In: Onwubolu, G.C., Babu, B.V. (eds.) New Optimization Tech-
     niques in Engineering. STUDFUZZ, vol. 141, pp. 219–239. Springer, Heidelberg
     (2004)
[5] Eberhart, R., Shi, Y.: Comparing inertia weights and constriction factors in par-
     ticle swarm optimization. In: Proceedings of the 2000 Congress on Evolutionary
     Computation, vol. 1, pp. 84–88 (2000)
[6] Evers, G., Ghalia, M.B.: Regrouping particle swarm optimization: A new global
     optimization algorithm with improved performance consistency across bench-
     marks. In: IEEE International Conference on Systems, Man and Cybernetics, SMC
     2009, pp. 3901–3908 (2009)
[7] Kennedy, J., Eberhart, R.: Particle swarm optimization. In: Proceedings of IEEE
     International Conference on Neural Networks, vol. 4, pp. 1942–1948 (1995)

244 J. van den Kieboom, S. Pouya, and A.J. Ijspeert

[8] Kennedy, J., Eberhart, R.: A discrete binary version of the particle swarm algorithm. In: IEEE International Conference on Systems Man and Cybernetics, vol. 5 (1997)

[9] Latiff, N.A., Tsimenidis, C., Sharif, B.: Performance comparison of optimization algorithms for clustering in wireless sensor networks. In: IEEE Internatonal Conference on Mobile Adhoc and Sensor Systems, MASS 2007, pp. 1–4. IEEE (2007)

[10] Leong, W.F., Yen, G.G.: Impact of tuning parameters on dynamic swarms in PSO-based multiobjective optimization. In: IEEE Congress on Evolutionary Computation, CEC 2008 (IEEE World Congress on Computational Intelligence), pp. 1317–1324. IEEE (2008)

[11] Monson, C.K., Seppi, K.D.: Adaptive diversity in PSO. In: Proceedings of the 8th Annual Conference on Genetic and Evolutionary Computation, pp. 59–66. ACM, Seattle (2006)

[12] Nickabadi, A., Ebadzadeh, M.M., Safabakhsh, R.: DNPSO: a dynamic niching particle swarm optimizer for multi-modal optimization. In: IEEE Congress on Evolutionary Computation, CEC 2008 (IEEE World Congress on Computational Intelligence), pp. 26–32. IEEE (2008)

[13] Ou, C., Lin, W.: Comparison between PSO and GA for parameters optimization of PID controller. In: Proceedings of the 2006 IEEE International Conference on Mechatronics and Automation, pp. 2471–2475. IEEE (2006)

[14] Parsopoulos, K.E., Vrahatis, M.N.: Particle swarm optimization method in multiobjective problems. In: Proceedings of the 2002 ACM Symposium on Applied Computing, pp. 603–607. ACM (2002)

[15] Pouya, S., van den Kieboom, J., Spröwitz, A., Ijspeert, A.J.: Automatic gait generation in modular robots: to oscillate or to rotate; that is the question. In: 2010 IEEE/RSJ International Conference on Intelligent Robots and Systems (IROS), pp. 514–520. IEEE (2010)

[16] Pugh, J., Martinoli, A.: Discrete multi-valued particle swarm optimization. In: Proceedings of IEEE Swarm Intelligence Symposium, vol. 1, pp. 103–110 (2006)

[17] Ray, T., Liew, K.M.: A swarm metaphor for multiobjective design optimization. Engineering Optimization 34(2), 141 (2002)

[18] Shi, Y., Eberhart, R.: A modified particle swarm optimizer. In: The 1998 IEEE International Conference on Evolutionary Computation, ICEC 1998, pp. 69–73 (1998)

[19] Sprowitz, A., Pouya, S., Bonardi, S., Van den Kieboom, J., Mockel, R., Billard, A., Dillenbourg, P., Ijspeert, A.J.: Roombots: reconfigurable robots for adaptive furniture. IEEE Computational Intelligence Magazine 5(3), 20–32 (2010)

[20] Worasucheep, C.: A particle swarm optimization with stagnation detection and dispersion. In: IEEE Congress on Evolutionary Computation, CEC 2008 (IEEE World Congress on Computational Intelligence), pp. 424–429. IEEE (2008)

Empirical Study of Computational Intelligence Strategies for Biochemical Systems Modelling

Zujian Wu[1], Crina Grosan[2,3], and David Gilbert[3]

[1] School of Natural and Computing Sciences
University of Aberdeen, UK
[2] Department of Computer Science
Babes-Bolyai University, Cluj-Napoca, Romania
[3] Department of Information Systems and Computing
Brunel University, London, UK
zujian.wu@abdn.ac.uk, {crina.grosan,david.gilbert}@brunel.ac.uk

Abstract. Modelling biochemical networks can be achieved by iteratively analyzing parts of the systems via top-down or bottom-up approaches. It is feasible to piece-wise model the biochemical networks from scratch by employing strategies able to assemble reusable components. In this paper, we investigate a set of strategies that can be employed in a bottom-up piece-wise modelling framework, to obtain synthetic models with similar behaviour to the target systems. A combination of evolution strategies and simulated annealing is employed to optimize the structure of the system and its kinetic rates. Simulation results of different variants of those computational methods on a standard signaling pathway show that it is feasible to obtain a tradeoff between the generation of desired behaviour and similar and alternative topologies.

1 Introduction

In theoretical chemistry and systems and synthetic biology, time-dependent chemical concentration data for large networks of biochemical reactions are important. These data are collected with the purpose to identifying the exact structure of a network of chemical reactions and their corresponding kinetic rates for which the identity of the chemical species present in the network is known but no information is available on the species interactions.

General methods for engineering biochemical networks can be divided into two main approaches: top-down or bottom-up, which allow the modelling of biochemical systems by manipulating parts of the systems. In the top-down (analytical) approach, a whole complex biochemical system is segregated into subunits that can be easily dealt with for further investigation, such as dissecting apoptotic signals [9] and tuning complex signal cascades [14]. In the bottom-up (constructionist) approach, a complex biochemical system is composed from building blocks where the relationships of involved compounds are investigated, such as building synthetic oscillators [15] and transplanting synthetic genomes [2]. The modelling of biochemical networks involves the optimisation of two main attributes: network topology and kinetic rates.

G. Terrazas et al. (eds.), *Nature Inspired Cooperative Strategies for Optimization (NICSO 2013)*, Studies in Computational Intelligence 512,
DOI: 10.1007/978-3-319-01692-4_19, © Springer International Publishing Switzerland 2014

There exist several approaches dealing with inferring biochemical systems, some of them with limitations and drawbacks [6][12][13]. They mainly include evolutionary algorithms and genetic programming (from the class of evolutionary computation models). Previous research applies a hybrid combining Evolutionary Strategies (ES) and Simulated Annealing (SA) to the optimisation of topology and the kinetic rates of a biochemical system [21]. In this paper, we investigate variants of the ES-SA heuristics for bottom-up systems modelling. Due to the flexibility of these strategies, various combinations of the evolutionary operators, evaluation criteria and design principles can be considered. These variants are presented in detail in Section 4.

2   Biochemical Systems

The modelling of biochemical systems has been investigated widely in computational biology, especially in systems biology. In biochemistry, a chemical reaction is a process of converting molecules of reactants into products within a specific time period. The reactants are usually known as substrates. Biochemical systems are composed of interacting molecules (or molecular species), whose dynamic evolution is determined by the occurrence of chemical reactions. A biochemical model is fully characterized by the initial concentration of each molecular species and the description of the reactions with their kinetic rate laws. The representation of the dynamics is given by an ordinary differential equation (ODE) as follows:

$$\frac{dX_i}{dt} = \sum_j \mu_{ij} \cdot \gamma_j \prod_k X_k^{f_{jk}} \tag{1}$$

where X_i represents one species of the model, for instance metabolite concentrations, protein concentrations or levels of gene expression; j represents the biochemical reaction affecting the dynamics of the species; μ_{ij} indicates the stoichiometric coefficient; γ_j indicates rate constants; f_{jk} stands for kinetic orders; and k denotes the number of species.

Mass action kinetics are used in chemistry and chemical engineering to describe the dynamics of chemical reactions. The mass action given in equation MA is used in this work; note that A is the substrate, B the product, E the enzyme and $A|E$ is the intermediary substrate-enzyme complex.

$$A + E \underset{k_2}{\overset{k_1}{\rightleftharpoons}} A|E \xrightarrow{k_3} B + E . \tag{MA}$$

There are different methodologies employed to describe biochemical systems in computational biology. Petri nets are one of the existing mathematical modelling structures used for the description of biochemical systems as a reaction-system behaviour descriptor, and comprise two types of nodes – places and transitions – connected via edges. The usage of Petri nets in biological systems comes as a natural solution as biochemical reactions are inherently bipartite, comprising reactions between biochemical entities [11], which are mapped onto transitions and places respectively. A continuous Petri net can be represented by a system of ODEs [8]. We focus on the automatic identification of both network structure and its corresponding kinetic rates from observed time-domain concentrations alone without assuming a given basic structure or any given reaction kinetics.

3   ES-SA Metaheuristic for Biochemical Systems

ES (as well as any of the evolutionary computation methods) are good candidates for evolving biochemical systems. A solution of the ES encodes a Petri net which is a representation of a biochemical system. SA is a powerful optimization method and it is used for optimizing the kinetic rates. The hybrid method ES-SA applied to biochemical systems is described in detail in [19][21]. We reproduce here the main characteristics. In order to understand the main constituents of an ES solution, elements such as *pattern, component, model* and *rules* are required. Any complex biochemical reactions can be described by employing instantiations from the binary patterns. The two general patterns we use describe how two species form a new species, or how one species is decomposed into two species:

- *binding pattern*: two reactants are merged into a complex with a specific kinetic rate
- *unbinding pattern*: a complex is disassociated back to reactants, or converted to a product and an enzyme with a specific kinetic rate.

A *component* for constructing biochemical models is given by $C = \langle P, T, f, v, m_0 \rangle$, which is based on the structure of Petri nets, where:

- P is a disjoint set of three continuous Places
- T is a singleton set containing one continuous Transition
- $f : ((P \times T) \cup (T \times P)) \to R_0^+$ defines a set of three directed arcs, weighted by non-negative real numbers, such that there is at least one arc of the form $p \to t$ and at least one of the form $t \to p$
- $v : T \to H$ assigns a firing rate function to the transition, whereby the set of all firing rate functions is $H := \bigcup_{t \in T} \{h_t | h_t : R^{|\bullet t|} \to R\}$, and $v(t) = h_t$ is for the transition $t \in T$
- $m_0 : P \to R_0^+$ gives the initial marking.

A *model* of a biochemical system is a generalized form of a component but with no restrictions on the number of places and transitions. The mathematical interpretation of both component and model is a system of ODEs, illustrating the nonlinear relationship among at least three involved biochemical elements.

The ES part of the ES-SA metaheuristic builds models from single components by using evolutionary mechanisms for composition operators and rules. A database has been designed and two libraries developed to store the components and models. Components are created at initial stage, according to the predefined patterns. A components library is developed as a table in the database, to preserve the generated components as atomic building blocks. The library maintains detailed information of these atomic components, such as labels of involved species, constants of associated kinetic rates and structures of created components.

The fitness function for a generated model M_G is given by:

$$f(M_G) = d_{M_T, M_G}(X_k) + \frac{1}{\eta} \sum_{k=1}^{\eta} \Phi(X_k) \tag{2}$$

where

$$d_{M_T, M_G}(X_k) = \frac{1}{\eta} \sum_{k=1}^{\eta} \sqrt{\sum_{t=1}^{P} (x_k^t - \hat{x}_k^t)} \tag{3}$$

$X_T = (X_1, X_2, ..., X_N)$ represent the behaviour of the N species, P denotes data points in each time series $X_i = (x_{1i}, x_{2i}, ..., x_{Pi})$, $i = 1, ..., N$. There are M time series $X_G = (\hat{X}_1, \hat{X}_2, ..., \hat{X}_M)$ describing the behaviour of M species in a constructed model M_G, and there are P data points for each time series $\hat{X}_j = (\hat{x}_{1j}, \hat{x}_{2j}, ..., \hat{x}_{Pj})$, $j = 1, ..., M$. The intersection between M_T and M_G of species is defined by $X_C = X_T \cap X_G = (X_1, X_2, ..., X_n)$, $1 \leq n \leq N$. $\eta = n$ if the compared species are from the intersection X_C and $\eta = n'$ if the compared substrates are from X_C', the set which contains the species for behaviour comparison specified by the user. The fitness function has to be minimized, therefore the smaller the evaluated fitness value, the better the generated model.

A set of composition operators are adapted from the evolutionary optimization to fine tune the structures of the models:

- *addition*, represented by $\oplus$: addition rules add a component to a model
- *subtraction*, represented by $\ominus$: subtraction rules remove a component from a model
- *crossover*, represented by $\otimes$: crossover rules combine two models. The crossover rules allow two models be cut and spliced by swapping parts of the models via a "cut and splice" approach.

ES builds solutions, i.e. biochemical systems represented by Petri nets, in a piece-wise manner by applying the operators above to the components library. In this way, ES optimises the topology of the biochemical system. The kinetic rates of the reactions encoded in the Petri net are optimized using simulated annealing. In order to evaluate an ES-SA solution, the fitness function includes both the topology and the kinetic rates. The topology part of the fitness function gives the number of common species and their interactions in the evolved model compared to the target one. Some of the target model species and interactions may be missing from the generated model, as well as extra species not in the target topology could be generated. For the optimization of the kinetic rates, we employed the BioNessie [10] platform to simulate the model and generate time course data as a set of target behaviour of species in the model. The measurement of behavioural distance is obtained by employing the Euclidean distance function. This part of the fitness involves solving the system of ODEs associated with the reactions. More details on the implementation of the two methods and all the parameters involved can be found in [19][21].

4   ES-SA Variants for Biochemical Systems Modelling

Due to a large variety of ways in which evolutionary methods can be designed in terms of performing genetic operators, comparing species behaviour and evaluating generated models during the construction process, we have carried out an empirical investigation of the advantages and disadvantages of some variants for the piecewise modelling, with an emphasis on the effect of genetic operators and evaluation criteria. Five sets of specific modelling variants are compared and general descriptions of these variants are given in what follows.

1. Methods of driving model composition:
   - Fixed: behaviour of a fixed set of species to be compared
   - Dynamic: behaviour of a dynamic set of species to be compared
   Time series data presenting behaviour of species in a target biochemical system is used to drive the modelling process via reducing the behaviour distance between generated and target model. Given a target biochemical system and a generated model which consist of N and M species respectively, there are two sets of time series data describing species behaviour in the target and generated model. It is easy to deduce that species to be compared can be selected via a fixed or a dynamic method.

   In the *fixed method*, the species in a fixed set are specified by users at the initial stage. They are referred to the target biochemical system. Therefore, all the information (names, concentrations and behaviour in time series data format) of these species is provided without uncertainty. Regarding the process of piecewise modelling, a model which is constructed at initial stages or evolved by mutation after many generations could only consist of less species than the target model. Thus some of the species could be absent.

   In the *dynamic method*, the species for comparison are generated and preserved in a dynamic set according to the existence of species in both generated and target models. The number of species is a dynamic variable in a range of $[0, N]$, N denoting the number of species in the target model.
2. Methods of survival selection:
   - SES: standard (1+1)-evolution strategy
   - PES: probabilistic (1+1)-evolution strategy, probabilistically accept a worse model
   A probabilistic evolution strategy (PES) is proposed, which differs from the standard evolution strategy (SES) in the sense that it can accept worse models by a probability while searching the solution space. This may be helpful in avoiding local optima.

   The *SES method* is the standard evolutionary process, selecting model candidates as offsprings for further evolution in following generations. The criteria for survival models is based on fitness value. The *PES method* introduces an acceptance probability into the stages of choosing survival models, which is integrated within the normal model selection stages of SES.

3. Methods of implementing the mutation operator (mutation consists in adding and/or subtracting a component to/from the topology):

   - Fixed: a fixed frequency of switching the addition/removal of a component to/from the model
   - Random: a random way of switching the addition/removal of a component to/from the model

   In the *fixed method*, the two mutation operators can be performed alternatively.

   In the *random method*, addition and subtraction are applied to models at every generation in a random manner.

4. Methods of performing crossover operator:

   - Best: each individual mates with the best individual in the population
   - Random: each individual mates with a randomly selected individual from the population

   The crossover operator mates two individual models under construction by a cut and splice method. New offspring are generated from the combination of parental models in terms of components (reactions and species). Parents and offspring compete and only one of them can be preserved as a model candidate in the population of the next generation. We consider two ways to performing the crossover operator: best and random methods.

   In the *best method*, each model under construction from the population is recombined with the model with best fitness. It is inspired by the elitism based individual selection in genetic algorithm.

   In the *random method*, each model in the population will be crossed over with another model chosen randomly.

5. Methods of evaluating solutions (models):

   - ED: the objective function represents the Euclidean distance function
   - ED+RP: the objective function is a combination of a reward and penalty mechanism and the Euclidean distance function

   The difference between generated and target model is calculated by employing an objective function. In the objective function, there are two methods of evaluating the composed models: Euclidean distance (ED) based method, and Euclidean distance with a reward and penalty mechanism (ED+RP) based method. ED is an ordinary distance between two points on the time series data representing the species behaviour from generated and target model. The inclusion of the reward and penalty in an objective function is intended to prioritize individuals whose components are among the ones existing in the target model. For instance, if a species is generated in a synthetic model and the species is also among the ones existing in the target model, fitness will be improved by giving a reward value.

5   Evaluation Metrics

In order to evaluate the synthetic model structures quantitatively, two measures are employed: Compression and Coverage. Both measures vary from 0 (worst) to 1 (best). If either compression or coverage is low for a particular model, it indicates the topology of the generated model is very different from the target biochemical system, even if their behaviours are similar.

Compression (adapted from [3] and [7]) measures the percentage of matched common arcs between synthetic and target model and it is given by:

$$Compression = \frac{|Intersection|}{Max(|Target|, |Generated|)}$$

where $|Intersection|$ represents the number of matched arcs between target and generated topology, $|Target|$ is the number of arcs in the target topology, $|Generated|$ denotes the number of arcs in the generated topology, and $Max(|Target|, |Generated|)$ is the maximum number of arcs in either of the target and generated model.

Coverage calculates the ratio of matched arcs in the target model and it is given by:

$$Coverage = \frac{|Intersection|}{|Target|}$$

where $|Intersection|$ represents the number of matched arcs between target and generated topology, and $|Target|$ is the number of arcs in the target topology.

6   Experiments and Comparisons

In order to quantitatively study the modelling variants, we performed statistical analysis of the performance by comparing fitness values, compression and coverage scores. One of the most important and intensively studied signaling pathways is ERK pathway (the Ras/Raf-1/MEK/ERK signaling pathway) which transfers the mitogenic signals from the cell membrane to the nucleus [17]. The ERK pathway is de-regulated in various diseases, ranging from cancer to immunological, inflammatory and degenerative syndromes and thus represents an important drug target. A brief illustration of regulations among proteins and complex based on signaling transduction in the ERK pathway is given as follows. Ras is activated by an external stimulus, via one of many growth factor receptors; it then binds to and activates Raf-1 to become Raf-1*, or activated Raf, which in turn activates MAPK/ERK Kinase (MEK) which in turn activates Extracellular signal Regulated Kinase (ERK). Cell differentiation is controlled by following cascade of protein interactions: Raf-1$\rightarrow$ Raf-1* $\rightarrow$ MEK $\rightarrow$ ERK. The effect of regulation is dependent upon the activity of ERK. The Raf-1 kinase inhibitor protein (RKIP) inhibits the activation of Raf-1 by binding to it, disrupting the interaction between Raf-1 and MEK, thus playing a part in regulating the activity of the ERK pathway [18]. A number of computational models have been developed in order to understand the role of RKIP in the pathway and ultimately to develop new therapies [4][5].

Due to the space limitation we present the analysis of a single signaling pathway but other examples could be found in [19].

Figure 1 shows a Petri net of the *RKIP* signaling pathway. Figure 2 displays the behaviour of all the species in the model of *ERK* signaling pathway regulated by *RKIP*, which is generated by simulation on a set of given ODEs and a group of original kinetic rates.

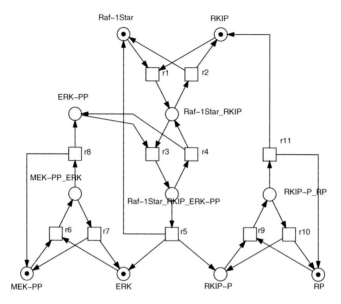

Fig. 1. A Petri net of the *RKIP* signaling pathway. Initial markings are taken from [20]

6.1   Simulation Settings

There are five pairs of ES-SA variants compared and investigated. Details of simulation settings are given in Table 1.

The hybrid ES-SA platform calls the subtraction operator at every two generations, Sub@Ge=2; SA is called to optimize kinetic rates at every 25 generations, OptRate@Ge=25; reward ε_1 and penalty ε_2 values are 0.01 and 1000 respectively.The number of generations in one run of ES is 100, GeSi=100; the number of individuals is 50, PopSi=50. Initial SA system temperature is 10, T_{ini}=10; cooling rate of SA system is 0.8, CoRate=0.8; minimum temperature for stopping simulation is 1, T_{min}=1; number of iterations at each temperature is 10, Iter=10. The mean μ and standard deviation σ of Gaussian distribution $N(\mu,\sigma)$ are 0 and 0.00001, μ=0 and σ=0.00001. Other properties of the simulation setting during the modelling process are fixed without modification except the two compared modelling variants, which allows a fair comparison between

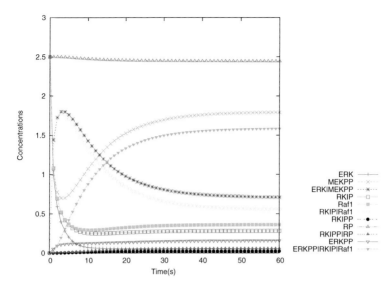

Fig. 2. Behaviour of all species in *RKIP* signaling pathway

two modelling variants in each pair in terms of performance on generation of synthetic models.

We investigate both alternative topologies and similar topologies. By analyzing the number of reactions in target pathway to the ones existing in the model generated by our method, we can quantitatively measure the difference between the alternative topology compared to the target one. A similar topology contains parts identical to the ones in the target network, but it could as well contain parts which are absent in the target network. An alternative topology has a different structure for the network, for instance could contain the same reactants as the target one, but the arcs between them are different.

Table 1. Simulation settings for running modelling variants

Modelling Variants	Hybrid Modelling	ES	SA	Gaussian $N(\mu, \sigma)$
Data Driven:	$\sharp$Runs = 10	GeSi = 100	$T_{ini} = 10$	$\mu = 0$
Fixed vs Dynamic	Sub@Ge= 2	PopSi = 50	CoRate = 0.8	$\sigma = 0.00001$
Survival Selection:	OptRate@Ge = 25		$T_{min} = 1$	
SES vs PES	$\varepsilon_1 = 0.01$		Iter = 10	
Mutation:	$\varepsilon_2 = 1000$			
Fixed vs Random				
Recombination:				
Best vs Random				
Fitness Function:				
ED vs (ED+RP)				

6.2 Statistical Analysis

Two statistical measures in the R packages [16], 'var.test(X, Y)' and 't.test(X, Y)', are employed to perform the statistical analysis.

Fitness values, compression and coverage scores are used to calculate the P-value in 'var.test(X, Y)' and 't.test(X, Y)' for further statistical analysis. The P-value is compared with a traditional significant level 'p=0.05', and the ratios of variances among generated models are also compared (see Tables 2, 3, 4). Results over 10 independent runs are summarized.

Table 2. Statistical analysis of average fitness sets

NO.	X vs Y	var.test(X, Y)		t.test(X, Y)		
		P-value	$r_{Variances}$	P-value	X	Y
1.1	Dri_{Fixed} vs Dri_{Dyn}	0.0229	0.6309	$< 2.2e\text{-}16$		3.1602
1.2	SES vs PES	0.4574	1.1616	0.837		4.2289
1.3	M_{Fixed} vs M_{Ran}	0.6821	0.9208	0.0262	4.2474	4.035
1.4	$\otimes_{Ran}$ vs $\otimes_{Best}$	1.07e-03	1.9448	0.5737		4.2019
1.5	ED vs (ED+RP)	$< 2.2e\text{-}16$	6.15e-06	$< 2.2e\text{-}16$		348.78

Table 3. Statistical analysis of average compression

NO.	X vs Y	var.test(X, Y)		t.test(X, Y)		
		P-value	$r_{Variances}$	P-value	X	Y
1.1	Dri_{Fixed} vs Dri_{Dyn}	0.0096	0.4713	$< 2.2e\text{-}16$		0.025
1.2	SES vs PES	0.0461	1.7802	6.78e-16		0.0361
1.3	M_{Fixed} vs M_{Ran}	0.75	1.0958	0.0296	0.0526	0.0567
1.4	$\otimes_{Ran}$ vs $\otimes_{Best}$	1.60e-06	0.2387	$< 2.2e\text{-}16$		0.1033
1.5	ED vs (ED+RP)	1.25e-05	3.6546	0.0004		0.0469

Table 4. Statistical analysis of average coverage

NO.	X vs Y	var.test(X, Y)		t.test(X, Y)		
		P-value	$r_{Variances}$	P-value	X	Y
1.1	Dri_{Fixed} vs Dri_{Dyn}	6.74e-12	8.4369	$< 2.2e\text{-}16$		0.0731
1.2	SES vs PES	0.4961	1.2161	0.0261		0.2065
1.3	M_{Fixed} vs M_{Ran}	0.062	1.7147	6.63e-05	0.2322	0.2765
1.4	$\otimes_{Ran}$ vs $\otimes_{Best}$	0.3373	1.3178	0.1888		0.2174
1.5	ED vs (ED+RP)	9.39e-05	0.3163	1.05e-14		0.3967

Table 5 shows a comparative example of the reactions obtained in a model generated by ES-SA strategies compared with the ones in the real (target) model. In the case presented here, four reactions marked with a star in target *RKIP* pathway are generated in the synthetic model. The synthetic model consists of 12 reactions, four of them being identical to the ones in *RKIP* pathway. The ES-SA metaheuristics can obtain alternative topologies exhibiting similar behaviour to the target ones.

Alternative topologies in synthetic models illustrate target biochemical system in a different way, providing templates to biologists in wet-lab for further experimental examination at the properties of the biochemical systems.

Table 5. Comparison of one synthetic model with *RKIP* pathway

Reactions in *RKIP* pathway	Reactions in One Generated Model
$*Raf1 + RKIP \xrightarrow{k1} RKIP|Raf1$	$ERK|RP \xrightarrow{r1} ERKP + RP$
$*RKIP|Raf1 \xrightarrow{k2} Raf1 + RKIP$	$ERKPP|MEKPP \xrightarrow{r2} ERKPP + MEKPP$
$RKIP|Raf1 + ERKPP \xrightarrow{k3} ERKPP|RKIP|Raf1$	$ERK|RP + ERKPP|RKIPP \xrightarrow{r3} ERK|ERKPP|RKIPP|RP$
$ERKPP|RKIP|Raf1 \xrightarrow{k4} RKIP|Raf1 + ERKPP$	$ERK + RKIP|Raf1 \xrightarrow{r4} ERK|RKIP|Raf1$
$ERKPP|RKIP|Raf1 \xrightarrow{k5} Raf1 + ERK + RKIPP$	$*RKIP + Raf1 \xrightarrow{r5} RKIP|Raf1$
$*ERK + MEKPP \xrightarrow{k6} ERK|MEKPP$	$*ERK + MEKPP \xrightarrow{r6} ERK|MEKPP$
$*ERK|MEKPP \xrightarrow{k7} ERK + MEKPP$	$ERKPP|MEKPP + MEKPP|RKIPP \xrightarrow{r7} ERKPP|MEKPP|RKIPP$
$ERK|MEKPP \xrightarrow{k8} MEKPP + ERKPP$	$RKIP + ERK|RP \xrightarrow{r8} ERK|RKIP|RP$
$RKIPP + RP \xrightarrow{k9} RKIPP|RP$	$*RKIP|Raf1 \xrightarrow{r9} RKIP + Raf1$
$RKIPP|RP \xrightarrow{k10} RKIP + RP$	$ERK|MEKPP \xrightarrow{r10} ERKP + MEKPP$
$RKIPP|RP \xrightarrow{k11} RKIPP + RP$	$RKIP|Raf1 + ERKP \xrightarrow{r11} ERKP|RKIP|Raf1$
	$*ERK|MEKPP \xrightarrow{r12} ERK + MEKPP$

6.3   Discussion

Details of the advantage and disadvantage of applying ES-SA variants to construct models are described below, each pair being considered separately.

Fixed vs. Dynamic – Data Driven. For generating desired behaviour and alternative topologies, dynamic variant, is better than fixed one, but for generating similar topologies, the fixed variant is better than dynamic one.

Figure 3(a) shows that the dynamic version converges more quickly in terms of fitness function than the fixed one.

SES vs. PES – Survival Selection. For generating desired behaviour, the experiments do not show any difference between SES and PES; for generating similar topologies, SES is better than PES and for generating alternative topologies, SES is better than PES.

Figure 3(b) shows that SES and PES have a similar performance regarding the convergence of fitness values.

Fixed vs. Random – Mutation Operator. For generating desired behaviour and similar topologies random variant is better than fixed one; and for alternative topologies random variant is the same as fixed one.

Figure 3(c) shows the convergence of the fitness values for the fixed and random variant.

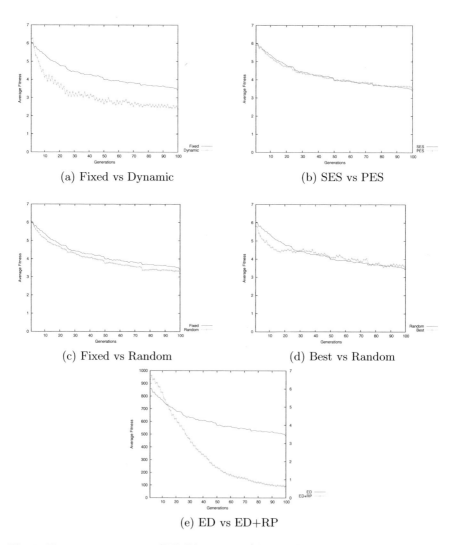

(a) Fixed vs Dynamic

(b) SES vs PES

(c) Fixed vs Random

(d) Best vs Random

(e) ED vs ED+RP

Fig. 3. Fitness convergence of ES-SA variants (the number of generations is shown on the X-axis and the average fitness values on the Y-axis): (a) variants for data driven, Fixed vs Dynamic; (b) variants for survival selection, SES vs PES; (c) variants for applying mutation operator, Fixed vs Random; (d) variants for applying crossover operator, Best vs Random; (e) variants for models estimation, ED vs ED+RP

Best vs. Random – Crossover Operator. For generating desired behaviour and similar topologies, a random selection of mate for recombination works the same as the selection of the best individual but selection of best individual for recombination is better than the random selection for generating alternative topologies. Figure 3(d) shows the convergence of the fitness values. In Table 2 (1.4) and Table 4 (1.4), the two P-values of t.test() are both larger than the significant level 0.05, indicating that the mean fitness and coverage values of the random variant are the same as the ones of the best variant. This suggests that the best and random mechanisms of selecting individual for crossover have the same performance.

ED vs. ED+RP. For generating similar topologies, ED+RP variant is better than ED but ED is better for generating alternative topologies. In Table 2 (1.5), the P-value is much smaller than 0.05, indicating a significant difference between ED and ED+RP. Figure 1(e) presents the convergence of the fitness value for ED and ED+RP. The average coverage value is larger for the models estimated by ED+RP which suggests that the ED+RP variant can be better than the ED variant in terms of generating similar topologies. However the P-value of var.test() in Table 3 (1.5) is smaller than 0.05 and the ratio of variances is larger than 1.

Note that some of the ES-SA variants are not directly comparable, because the statistical values are not in the same measurement scale. For instance, the ED and ED+RP are not comparable in terms of fitness values, since the mechanism of reward and penalty generates a different fitness scale.

We are aware that sometimes small amendments to the original methods could have an impact upon the final results; this is what we tried to prove in this paper, but with the aim of selecting those forms of operators and evaluation procedures would best fit the biochemical network design. The probabilistic ES makes no difference to the standard one (it is even worse in certain situations) which shows that accepting worse solutions will not bring additional exploration of the search space. The manner in which mutation is performed helps if additional information is known about the problem to be solved. In the case presented in this paper, the imposition of a certain number of steps for adding a component or removing a component is not helpful. This could work better than in the random case if more interaction is provided, i.e. remove a component every certain fixed number of steps only if the size of the network is too big. The step size of the application of an addition or a subtraction is also important, but that requires extra analysis. Elitism plays an important role and in our case it helped in selection the individuals for crossover.

Table 6 shows the overall pair-wise comparison of all the five variants in terms of topologies generation and behaviour.

Table 6. A summary of performance between compared modelling variants

Modelling Variants	Desired Behaviours	Similar Topologies	Alternative Topologies
Data Driven: Fixed vs Dynamic	Dynamic	Fixed	Dynamic
Survival Selection: SES vs PES	=	SES	SES
Mutation: Fixed vs Random	Random	×	=
Recombination: Best vs Random	=	×	Best
Fitness Function: ED vs (ED+RP)	×	ED+RP	ED

Notes: '×' means not comparable; '=' means the same.

7 Summary and Conclusions

The work described in this paper focuses on the empirical analysis of piece-wise modelling approaches of signalling pathways, comparing performance of different Evolutionary Strategies – Simulated Annealing variants. Alternative topologies of synthetic models obtained *in silico* can be taken as general guides for biologists to examine and understand biochemical systems by experimental techniques in wet-lab. Moreover, these can be used as templates for researchers in synthetic biology to develop specific functions of biochemical systems. The research presented here aims at guiding biomodel engineers in deciding the computational setup and selecting the right parameters. Our analysis of some of the combinations which could be considered helps in developing models that are useful for further construction with respect to specific characteristics of modelling biochemical systems.

Acknowledgements. The authors acknowledge the scientific support form Monika Heiner (Brandenburg University of Technology Cottbus). Part of this work has been supported by the Romanian National Authority for Scientific Research PN-II-PT-PCCA-2011-3.2-0917.

References

[1] Aarts, E., Korst, J., Michiels, W.: Simulated Annealing and Boltzmann Machines: a stochastic approach to combinatorial optimization and neural computing, pp. 188–202. Wiley (1989)
[2] Baker, M.: Synthetic genomes: the next step for the synthetic genome. Nature 473, 403–408 (2011)

[3] Brazma, A., Jonassen, I., Vilo, J., Ukkonen, E.: Pattern discovery in biosequences. In: Honavar, V.G., Slutzki, G. (eds.) ICGI 1998. LNCS (LNAI), vol. 1433, pp. 257–270. Springer, Heidelberg (1998)

[4] Calder, M., Gilmore, S., Hillston, J.: Modelling the influence of RKIP on the ERK signalling pathway using the stochastic process algebra PEPA. In: Priami, C., Ingólfsdóttir, A., Mishra, B., Riis Nielson, H. (eds.) Transactions on Computational Systems Biology VII. LNCS (LNBI), vol. 4230, pp. 1–23. Springer, Heidelberg (2006)

[5] Cho, K.-H., Shin, S.-Y., Kim, H.-W., Wolkenhauer, O., McFerran, B., Kolch, W.: Mathematical modeling of the influence of RKIP on the ERK signaling pathway. In: Priami, C. (ed.) CMSB 2003. LNCS, vol. 2602, pp. 127–141. Springer, Heidelberg (2003)

[6] Fogel, G., Corne, D.: Evolutionary Computation in Bioinformatics, pp. 256–276. Morgan Kaufmann (2003)

[7] Gilbert, D., Westhead, D., Viksna, J.: Techniques for comparison, pattern matching and pattern discovery: from sequences to protein topology. In: Frasconi, P., Shamir, R. (eds.) Artificial Intelligence and Heuristic Methods in Bioinformatics, pp. 128–147. IOS Press (2003)

[8] Gilbert, D., Heiner, M., Lehrack, S.: A Unifying Framework for Modelling and Analysing Biochemical Pathways Using Petri Nets. In: Calder, M., Gilmore, S. (eds.) CMSB 2007. LNCS (LNBI), vol. 4695, pp. 200–216. Springer, Heidelberg (2007)

[9] Lau, K.S., Juchheim, A.M., Cavaliere, K.R., Philips, S.R., Lauffenburger, D.A., Haigis, K.M.: In vivo systems analysis identifies spatial and temporal aspects of the modulation of TNF-alpha-induced apoptosis and proliferation by MAPKs. Sci. Signal. 4(165), 16 (2011)

[10] Liu, X., Jiang, J., Ajayi, O., Gu, X., Gilbert, D.: BioNessie(G)- A Grid Enabled Biochemical Networks Simulation Environment. Studies in Health Technology and Informatics 138, 147–157 (2008)

[11] Murata, T.: Petri nets: Properties, analysis and applications. Proceedings of the IEEE 77(4), 541–580 (1989)

[12] Rausanu, S., Grosan, C., Wu, Z., Parvu, O., Gilbert, D.: D., Evolving Biochemical Systems. In: IEEE Congress on Evolutionary Computation (CEC), Cancun, Mexico (2013)

[13] Sakamoto, E., Iba, H.: Inferring a system of differential equations for a gene regulatory network by using genetic programming. In: Proceedings of the IEEE Congress on Evolutionary Computation. IEEE Service Center, Piscataway (2000)

[14] OShaughnessy, E.C., Palani, S., Collins, J.J., Sarkar, C.A.: Tunable signal processing in synthetic MAP kinase cascades. Cell 144(1), 119–131 (2011)

[15] Elowitz, M.B., Leibler, S.: A synthetic oscillatory network of transcriptional regulators. Nature 403, 335–338 (2000)

[16] R Development Core Team. R: A language and environment for statistical computing. R Foundation for Statistical Computing, Vienna, Austria (2009)

[17] Yeung, K., Janosch, P., McFerran, B., Rose, D.W., Mischak, H., Sedivy, J.M., Kolch, W.: Mechanism of suppression of the Raf/MEK/Extracellular signal regulated kinase pathway by the Raf kinase inhibitor protein. Molecular and Cellular Biology 20(9), 3079–3085 (2000)

[18] Yeung, K., Seitz, T., Li, S., Janosch, P., McFerran, B., Kaiser, C., Fee, F., Katsanakis, K.D., Rose, D.W., Mischak, H., Sedivy, J.M., Kolch, W.: Suppression of Raf-1 kinase activity and MAP kinase signaling by RKIP. Nature 401, 173–177 (1999)

[19] Wu, Z.: A generic approach to behaviour-driven biochemical model construction, PhD Thesis, Brunel University (2013)

[20] Wu, Z., Gao, Q., Gilbert, D.: Target Driven Biochemical Network Reconstruction Based on Petri nets and Simulated Annealing. In: Proceedings CMSB 2010 (8th International Conference on Computational Methods in Systems Biology), pp. 33–42. ACM Digital Library (2010)

[21] Wu, Z., Yang, S., Gilbert, D.: A Hybrid Approach to Piecewise Modelling of Biochemical Systems. In: Coello Coello, C.A., Cutello, V., Deb, K., Forrest, S., Nicosia, G., Pavone, M. (eds.) PPSN 2012, Part I. LNCS, vol. 7491, pp. 519–528. Springer, Heidelberg (2012)

Metachronal Waves in Cellular Automata: Cilia-Like Manipulation in Actuator Arrays

Ioannis Georgilas[1,2], Andrew Adamatzky[1,2], David Barr[3],
Piotr Dudek[3], and Chris Melhuish[2]

[1] International Centre for Unconventional Computing,
University of the West of England, Frenchay Campus, BS16 1QY, Bristol, UK
ioannis.georgilas@uwe.ac.uk
http://uncomp.uwe.ac.uk/giannis/Home.html
[2] Bristol Robotics Laboratory,
University of Bristol and University of the West of England, T Block,
Frenchay Campus, BS16 1QY, Bristol, UK
http://brl.ac.uk
[3] School of Electrical & Electronic Engineering , University of Manchester,
Sackville Street Building, M13 9PL, Manchester, UK
http://personalpages.manchester.ac.uk/staff/p.dudek/default.htm

Abstract. Paramecium is covered by cilia. It uses the cilia to swim and transport food particles to its mouth. The cilia are synchronised into a collective action by propagating membrane potential and mechanical properties of their underlying membrane and the liquid phase environment. The cilia inspired us to design and manufacture a hardware prototype of a massively parallel actuator array, emulated membrane potentials via a discrete excitable medium controller and mechanical properties based on vibrating motors. The discrete excitable medium is a two-dimensional array of finite automata, where each automaton, or a cell, updates its state depending on states of its closest neighbours. A local interaction between the automata lead to emergence of propagating patterns, waves and gliders. The excitable medium is interfaced with an array of actuators. Patterns travelling on an automaton array manifest patterns of actuation travelling along the array of actuators. In computer models and laboratory experiments with hardware prototypes we imitate transportation of food towards mouth pore of the Paramecium. The hardware actuator arrays proposed could in future replace simple manipulators in demanding micro-scale application.

Keywords: cellular automata, multi-agent systems, natural collaboration.

1 Introduction

Multi-agent engineering systems are ubiquitous. Their control is challenging [9]. A specific category of multi-agent systems are the smart surfaces [4], manipulator arrays replacing traditional manipulators in industrial applications specifically at the micro-scale.

G. Terrazas et al. (eds.), *Nature Inspired Cooperative Strategies for Optimization (NICSO 2013)*, Studies in Computational Intelligence 512,
DOI: 10.1007/978-3-319-01692-4_20, © Springer International Publishing Switzerland 2014

Fig. 1. Image of the bristles fabricated to emulate natural cilia. This design is used as an alternative to the membrane design described in [13].

Typical manipulation tasks in industrial environments are being performed by classic 6-DOF manipulators. These robots are often very large, and can manipulate objects varying in size but are smaller than the manipulator itself. Typical arrangements of these manipulators are either as single units or in small groups, where collaborative tasks are achieved by properly planning the spatial trajectories of the individual units. Given the small number of agents involved the planning tasks can be solved analytical. Recently another class of manipulation tasks emerged: the micro-assembly or assembly on the molecular level. This novel field of manipulation leads to the creation of massive arrays of distributed small manipulators. The most common types are based on airjets [8], vibration based solutions [22] and micro electro-mechanical systems (MEMS) [10].

The actuators used by these systems are small, weak and inexpensive, but by using cooperative control, they are able to manipulate object much larger than their own size. Some types of cooperative control of smart surfaces has been analysed in [5, 18]. A manipulation based on cooperative behaviour in Cellular Automata (CA) has been studied in [12, 14]. Presently we explore similarities of waves propagating in CA to biological cilia, like the ones found in Paramecium, and cilia controlling signalling [21] and how the latter allows the emergence of collaborative behaviour between the actuators of the prototype system developed in [13], and the bristled version as seen in Fig. 1. The artificial cilia of the hardware behave in a coordinated manner via excitation waves and propagating self-localised excitations that are travelling in the underlying CA lattice.

2   Biological Coordination

Coordination of multiple activities is amongst most intriguing mechanisms of natural systems [16]. It is still unclear how biological systems are coordinated to

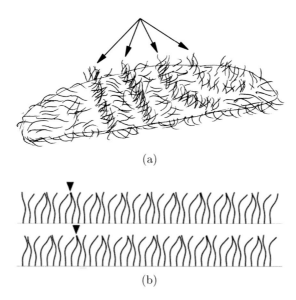

(a)

(b)

Fig. 2. (a) Image of Paramecium with cilia exhibiting metachronal waves (arrows), and (b) line of 40 cilia in the presence of a metachronal wave indicated by arrows [15]

produce specific behaviours. To answer this question the mechanics of spatial-temporal patterns have been investigated in different systems. The main idea is that chemical signalling mechanisms [7] are being used to produce these behaviours.

Of particular interest are the mechanisms that govern the operation of cilia. Cilia are organelle structures found in single-cell organisms used for propulsion, like in Paramecium Figs. 2(a), and manipulate particles of food, including other bacteria, towards the Paramecium's mouth pore. The latter is also the task by which the proposed artificial system will be evaluated. In multicellular organisms cilia are used as part of other systems, e.g. in the trachea for cleaning the incoming air from dust and small particles. The intriguing structure of a cilium has inspired building and control of artificial ones as flow generators. The fabrication technics involve, among other, magnetically actuated structures [11] and soft, electro-active polymer approaches [20].

The cilia coordination has been investigated in order to establish how the combined action of arrays of cilia produce effective fluidic flows that are propagating objects. The coordinated action of adjacent cilia and the break of symmetrical beating was analysed to establish the dynamics of macroscopic flow [15]. Also, phase-oscillator descriptions of the hydrodynamics interaction of beating cilia arrays have been formulated to allow the analytical investigation of cilia coordination [19]. The recurring subject of metachronal waves, Figs. 2(a), 2(b) and how this type of wave formation creates the maximum effective flow is analysed in [2].

Further to the signalling procedure that leads to the cooperative operation of cilia, the shape of the resulting flow, and related force vector, has influenced the current research. It has been proposed and demonstrated in modelling, that the 3D fluidic flow around an actuated cilium has the shape of a 3D vortex [6]. The 2D component of this motion generates the common shape of vortex with the vectors of acting forces rotating around a centre and converging to the generative specific spot, i.e. the cilium. The latter is the inspiration for the model that is used for describing the hardware behaviour of the system in [13].

3   Cooperative Smart Surface

3.1   Cellular Automata

It has been proposed that reaction-diffusion (RD) models can provide a framework to analyse signal propagation in biological systems [17]. The intrinsic simplicity of the RD approach facilitates simulation of the chemical and fluid dynamics of the cilia signalling methods that lead to their cooperative functionality. Based on this observation, CA as a form of modelling RD, are being used in this research to model the signals used for cilia coordination. CA are a type of distributed embedded controller that, via the state changing rules patterns emerge demonstrating collaborative performance. The investigation of the emerging patterns in CA have been extensively investigated for their information processing abilities [1].

In previous work we used the 2^+-medium ruleset as the controller of the smart surface [12, 14]. In brief, the neighbourhood of this ruleset is 3×3 cells and the rules are as follows; each cell, x of a 2D lattice $\mathbf{L}$, can take three states, resting $(\cdot)$, excited $(+)$ and refractory $(-)$, for this manuscript whenever a CA lattice is depicted in a figure resting state will be black, excited state will be grey and refractory state will be white. A resting cell is excited if the number of excited neighbours is exactly two. An excited cell takes refractory state and a refractory cell takes a resting state unconditionally:

$$x^{t+1} = \begin{cases} +, & x^t = \cdot \text{ and } \sum_{y \in u(x)} \chi(y, +) = 2 \\ -, & x^t = + \\ \cdot, & \text{otherwise} \end{cases} \quad (1)$$

where $u(x) = \{y \neq x \in \mathbf{L} : |x - y| \leq 1\}$ and $\chi(y, +) = 1$ if $y = +$ and 0, otherwise.

In this ruleset self-localisations, or wave-fragments, having an excited front, two cells being in the excited state ($+$ or grey in the pictures), and a refractory tail, two cells being in the refractory state ($-$ or white in the pictures), travel in straight lines. These excitation waves can be considered analogs of metachronal waves that are traveling in the lattice. As shown in Fig. 3, ten wave-fragments traveling in a line are behaving in a similar manner to the waves in cilia of Fig. 2(b).

Under the CA states is a side view of the 'equivalent' cilia. Each cilium can take central, left or right position. Although the beating pattern, the bending of the cilium, is simpler that in nature, the intended behaviour is emerging from the phenomenology of the wave propagation.

Fig. 3. Excitation waves travelling in a 40×2 cylindrical lattice. The 'equivalent' cilia metachonal wave is depicted below the CA states (grey is excited, white is refractory and black is resting).

3.2 Actuators and Coupling with CA

In order to achieve the vortex generation method of the cilia in 2D, we use vibrating motors interconnected using a flexible membrane. Each motor can rotate in both directions, clockwise and counter-clockwise. This directional operation modes, plus being off are mapped with the three CA states. Hence resting state (·) is mapped as *motor off*, excited (+) is mapped as *motor CW* and refractory (−) is mapped as *motor CCW*, The motors are designed to vibrate radially, but being positioned on a flexible platform allows them to vibrate slightly in the axial direction. A more detailed description of the hardware can be found in [13].

The combination of the axial and radial vibrations create the vortex shaped force field on Fig. 5(a). This is similar to the 2D components of the vortexes generated by cilia bitting in fluids [6]. The generated force field is created by the rotation of a single motor. With arrows are the vectors of the force in specific position on the top of the membrane. The length of the vector represents the magnitude of the force at the point. As it is expected, the closer a vector is to the motor the higher the exerted forces are. If the forces are negligible no vector is shown (blank space in the figures).

In order to achieve controllable collaborative behaviour the CA states and the motor commands need to be mapped. This connection is straightforward since the three states of the CA can be connected with the three states of the motor. The resting state is connected with the motor being off, while excited and refractory are arbitrarily selected as motor rotating clockwise (CW) and motor rotating counter-clockwise (CCW) respectively.

4 Actuator Cooperative Behaviour

To demonstrate similarities between manipulation methods in Paramecium and our actuator array we chosen a task of transporting 'food' object to a dedicated site of a manipulator, which imitates a 'mouth pore'.

We compare three different control actuation signals, single motor, random and metachronal waves in CA. The motor commands were generated using the software APRON and the physical behaviour was simulated in MATLAB. The location of the object is selected randomly and in the lower right corner of the lattice. Since the model is deterministic a single simulation is needed for each of the three signalling methods.

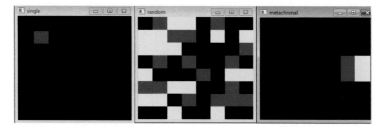

Fig. 4. Screenshot from the APRON environment where the control signals for the motors lattice are depicted in colour matrices. Single motor, random signal and metachronal wave (at frame 1).

APRON. **A***rray* **P***rocessing envi***RON***ment* [3] is used as the control signals development and simulation environment. It is a real-time, interactive and highly visual simulation platform for working with and debugging 2D arrays of data and rapidly prototyping array based algorithms. APRON scripts treat arrays as a primitive data type, where the user sequentially defines operations between them, and computation is performed in an element-wise manner. Individual elements can be disabled based upon both global and local data and conditions, yielding local autonomy and algorithmic branching with the array. The CA rules were developed in open-loop and a frame of the output from each signalling method is shown Fig. 4.

MATLAB. The information from APRON was then transferred to MATLAB where the generation of the vortex force-field for each motor was generated. To create the total force-field of a single motor a combination of coupled oscillators, (2a) and membrane wave oscillations (2b) was used. The former provided the components for the behaviour of each motor as affected by the rotation of the adjacent motors, while the latter describes the dynamics of the membrane that covers all the motors and connects them.

$$F_C(t) - kd - p\dot{d} = m\ddot{d} \tag{2a}$$

$$\frac{\partial^2 z}{\partial x^2} + \frac{\partial^2 z}{\partial y^2} = \frac{1}{c^2}\left(\frac{\partial^2 z}{\partial t^2} + p\frac{\partial z}{\partial t}\right), \; c^2 = F_W/\rho \tag{2b}$$

where, $F_C(t)$ is the actuation force (motor's torque), k and p are the spring and damping coefficients of the membrane, m is the mass of the eccentric weight and d, $\dot{d}$ and $\ddot{d}$ is the displacement, velocity and acceleration along the axis of oscillation. And c is a speed of wave travelling on the membrane, F_W is a force generated by the wave and ρ is a density of the membrane's medium. The membrane is considered homogeneous and isotropic. The term $p\dot{d} = p\frac{\partial z}{\partial t}$ is again the damping effect of the membrane as in (2a) but in the z axis. The combined force-field was then used to calculate the displacement and rotation of the object using the equations for normally accelerated motion (3).

$$x(t) = x(t - \Delta t) + u_x(t - \Delta t) \cdot \Delta t + \frac{a_x(t - \Delta t) \cdot \Delta t^2}{2}$$

$$y(t) = y(t - \Delta t) + u_y(t - \Delta t) \cdot \Delta t + \frac{a_y(t - \Delta t) \cdot \Delta t^2}{2} \tag{3}$$

$$r(t) = r(t - \Delta t) + \omega(t - \Delta t) \cdot \Delta t + \frac{\alpha(t - \Delta t) \cdot \Delta t^2}{2}$$

where $x(t)$, $y(t)$, $r(t)$ are the position and rotation of the object at time t, $u_x(t)$ and $u_y(t)$ are the linear velocity in the two directions and $\omega(t)$ is the angular velocity, and finally, $a_x(t)$, $a_y(t)$ are the linear and $\alpha(t)$ is the angular acceleration respectively, as calculated by $F_M = m \cdot a$ and $M_M = \alpha(t) \cdot I_d$, where F_M is the resultant force of F_C and F_W, M_M the rotational momentum of this force, and I_d the inertia and m the mass of the object. All calculations are performed in specified simulation time intervals, Δt.

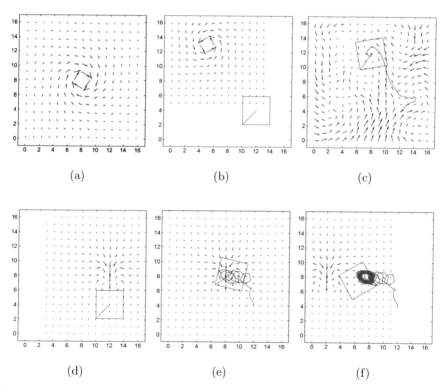

Fig. 5. Simulation frames from MATLAB with APRON generated control signals. The vectors of the force field, the rectangular object with a rotation indication line and the trajectory are depicted. (a) Is the vortex force-field created by a single motor, (b) is the single motor force field 'pulling' the object towards the centre of the lattice (frame 60), (c) is the trajectory of the object under a randomly generated force field (frame 400), (d)-(f) are frames 0, 60 and 400 of the trajectory under the metachronal wave signal. Object is placed at coordinates (12,4). Axis are simulation based units.

The three different control signals are representing three methods for achieving the task. The first is to enable a single motor (cilium) symmetrically from the location of the object and try to 'pull' the 'food' towards the 'mouth pore', Fig. 5(b). The second is a randomly generated control signal, Fig. 5(c), where the 'cilia' are flapping randomly to try to move the 'food' towards the 'mouth pore'. Finally, the third control signal is an excitation wave traveling horizontally on the lattice. The lattice is assumed being a torus and as such the wave is 'wrapping' around from left to right, Fig. 5(d)-5(f).

4.1   Discussion

Analysing the trajectories from the different control signals we can deduce some interesting results regarding the cooperative attributes of the various control signalling.

The single motor approach, with the motor trying to 'pull' the object towards the centre fails even to start moving the object. Investigating the exhibited forces we find that the single motor is not able to exert the friction between the object and the membrane. This is a demonstration that a single 'cilium' fails to achieve the task and some form of cooperative action needs to take place.

The random motor approach overcome the lack of power, since it does move the object from its initial position. Nonetheless, the object is moving in a random path. Furthermore, it might be locked in local attractors that will not coincide with the intended target. Hence, the use of multiple 'cilia' is necessary to produce manipulation, but random 'beating', hence signalling, does not create controllable behaviour.

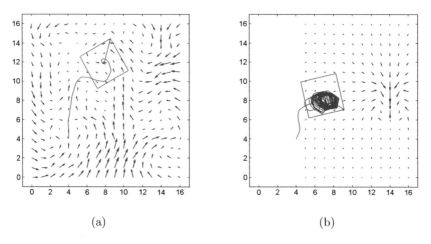

(a)   (b)

Fig. 6. Object trajectory for random and CA metachronal wave control signal. Initially the object is placed in coordinates (4,4) in this experiment.

Finally, the CA excitable/metachronal wave both moves the object and reaches the intended target. The synchronised movement of the motors create the collaborative behaviour that natural cilia create. Hence, this type of control signalling can create controllable collaborative behaviour.

In order to demonstrate the generalisation of the proposed method, in Fig. 6 the trajectory paths of the object from a different location are depicted. As can be easily noted, the random force-field, Fig. 6(a), exhibits the same random/local attractor behaviour, while the CA metachronal wave, Fig. 6(b), moves the object towards the target.

5 Conclusions

Coordination is crucial for the function of biological systems. There is a great deal of research being done in order to unravel the process with which systems of multiple agents are communicating and collaborating, like the case of cilia in the Paramecium. One of the most prominent explanations is the use of metachronal waves to co-ordinate the ciliate movements.

With this work we demonstrated a method to replicate the metachronal waves of cilia utilising reaction-diffusion models of CA. We demonstrated how excitation waves travelling in the CA lattice can phenomenologically model waves in cilia arrays and how this waves can be used as signals to control a prototype actuators array, a smart surface.

Using these metachronal waves in the CA medium the actuators exhibited a controlled collaborative behaviour. The ability to succeed in a simple manipulation task, has been demonstrated through a set of experiments. The metachronal wave control signals were compared with single actuator activation, lacking sufficient force, and random control signalling, lacking controllable manipulation.

Future work could involve developing machine learning and evolutionary computing approaches to investigate the possible collaborative behaviour of the system. Hence, the solutions to the optimal cooperation method can emerge by the analysis of the potential solution space. Furthermore, a consistent understanding of the dynamics of the cooperative behaviour must be investigated. This will be achieved by developing further the vortex modelling for the actuators. Also, a connection of the modelled behaviour with the hardware described in [13] will prove the suitability of the approach under real-world conditions. Furthermore, the potential of the proposed method to be used as a generic controller for other system should be investigated. Especially as a solution for scalability issues in multi-agent systems, since the metachronal CA-generated wave can inherently scale without increasing the complexity of the underlying hardware.

References

[1] Adamatzky, A.: Computing in non-linear media and automata collectives. Institute of Physics Publishing (2001)

[2] Aiello, E., Sleigh, M.A.: The metachronal wave of lateral cilia of mytilus edulis. The Journal of Cell Biology 54(3), 493–506 (1972)

[3] Barr, D.R., Dudek, P.: Apron: A cellular processor array simulation and hardware design tool. EURASIP Journal on Advances in Signal Processing, 9 (2009)

[4] Bohringer, K.F., Donald, B.R., Mihailovich, R., MacDonald, N.C.: Sensorless manipulation using massively parallel microfabricated actuator arrays. In: Proceedings of the 1994 IEEE International Conference on Robotics and Automation, pp. 826–833. IEEE (1994)

[5] Bohringer, K.F., Bhatt, V., Donald, B., Goldberg, K.: Algorithms for sensorless manipulation using a vibrating surface. Algorithmica 26(3-4), 389–429 (2000)

[6] Chen, D., Norris, D., Ventikos, Y.: The active and passive ciliary motion in the embryo node: A computational fluid dynamics model. Journal of Biomechanics 42(3), 210–216 (2009), http://www.sciencedirect.com/science/article/pii/S0021929008005472, doi:10.1016/j.jbiomech.2008.10.040

[7] Christensen, S.T., Pedersen, L.B., Schneider, L., Satir, P.: Sensory cilia and integration of signal transduction in human health and disease. Traffic 8(2), 97–109 (2007)

[8] Delettre, A., Laurent, G., Le Fort-Piat, N.: 2-dof contactless distributed manipulation using superposition of induced air flows. In: 2011 IEEE/RSJ Int. Conf. Intelligent Robots and Systems, IROS 2011, San Francisco, CA, pp. 5121–5126 (2011), doi:10.1109/IROS.2011.6048251

[9] Ferber, J.: Multi-agent systems: an introduction to distributed artificial intelligence, vol. 1. Addison-Wesley, Reading (1999)

[10] Fujita, H., Ataka, M.: System configuration and fabrication technology for distributed mems. In: 1st Worksh. Hardw. Softw. Impl. Contr. Distr. MEMS, dMEMS 2010, Besancon, pp. 1–5 (2010)

[11] Gauger, E.M., Downton, M.T., Stark, H.: Fluid transport at low reynolds number with magnetically actuated artificial cilia. The European Physical Journal E 28(2), 231–242 (2009)

[12] Georgilas, I., Adamatzky, A., Melhuish, C.: Manipulating objects with gliders in cellular automata. In: 2012 IEEE International Conference on Automation Science and Engineering (CASE), pp. 936–941. IEEE (2012)

[13] Georgilas, I., Adamatzky, A., Melhuish, C.: Towards an intelligent distributed conveyor. In: Herrmann, G., Studley, M., Pearson, M., Conn, A., Melhuish, C., Witkowski, M., Kim, J.-H., Vadakkepat, P. (eds.) TAROS-FIRA 2012. LNCS, vol. 7429, pp. 457–458. Springer, Heidelberg (2012)

[14] Georgilas, I., Adamatzky, A., Melhuish, C.: Manipulating with excitations: Waves or gliders? In: Workshop Notes of the ICRA Workshop in Unconventional Approaches to Robotics, Automation and Control Inspired by Nature, Karlruhe, International Conference in Robotics and Automation, ICRA (2013)

[15] Guirao, B., Joanny, J.F.: Spontaneous creation of macroscopic flow and metachronal waves in an array of cilia. Biophysical Journal 92(6), 1900–1917 (2007)

[16] Kelso, J., Schöner, G.: Toward a physical (synergetic) theory of biological coordination. In: Lasers and Synergetics, pp. 224–237. Springer (1987)

[17] Kondo, S., Miura, T.: Reaction-diffusion model as a framework for understanding biological pattern formation. Science 329(5999), 1616–1620 (2010)

[18] Konishi, S., Fujita, H.: System design for cooperative control of arrayed microactuators. In: Proceedings of the Micro Electro Mechanical Systems, MEMS 1995, p. 322. IEEE (1995)

[19] Lenz, P., Ryskin, A.: Collective effects in ciliar arrays. Physical Biology 3(4), 285 (2006)

[20] Sareh, S., Rossiter, J., Conn, A., Drescher, K., Goldstein, R.E.: Swimming like algae: biomimetic soft artificial cilia. Journal of the Royal Society Interface 10(78) (2012)

[21] Satir, P., Christensen, S.T.: Structure and function of mammalian cilia. Histochemistry and Cell Biology 129(6), 687–693 (2008)

[22] Vose, T., Umbanhowar, P., Lynch, K.: Toward the set of frictional velocity fields generable by 6-degree-of-freedom oscillatory motion of a rigid plate. In: 2010 IEEE Int. Conf. Robot, ICRA 2010, Anchorage, AK, pp. 540–547 (2010)

Team of A-Teams Approach for Vehicle Routing Problem with Time Windows

Dariusz Barbucha

Department of Information Systems
Gdynia Maritime University
Morska 83, 81-225 Gdynia, Poland
d.barbucha@wpit.am.gdynia.pl

Abstract. The paper proposes a new approach for Vehicle Routing Problem with Time Windows, which integrates the asynchronous team paradigm with the island-based evolutionary algorithm concept. The process of solving the problem is carried-out by the set of agents, each representing a heuristic algorithm, operating on population of individuals (solutions) stored in the common sharable memory. Agents are grouped in teams working on islands. Each team of agents periodically communicates with other teams by sharing a promising results. Computational experiment confirmed effectiveness of the proposed approach.

Keywords: asynchronous teams, multi-agent systems, island-based evolutionary algorithm, cooperative problem solving, vehicle routing problem with time windows.

1 Introduction

Vehicle Routing Problems (VRPs) class consists of the family of problems where a set of customer requests has to be served by the set of available vehicles in order to minimize (or maximize) a given goal function (typicaly reflecting the cost), and satisfying several customers and vehicles constraints. Because of their practical importance, VRPs have attracted a lot of attention during recent years [13] and the field is still an active field of research.

Among the approaches proposed for solving difficult optimization problems including VRP, special interest of researchers and practitioners is focused on hybridization of various methods which are able to produce a synergetic effect while solving instances of the problem. Different kinds of methods, forms of combining them into the effective problem-solving strategies, and technological advances, where jointly used, may offer effective tools for solving such problems.

Last years, one of the promising and intensively expanding directions of research, is the field of agent and multiple-agent systems [26]. A number of multiple-agent approaches integrated with some nature-inspired methods, proposed to solve different types of optimization problems grows systematicaly. One of them, where paradigms of the population-based methods, multiple agent systems and cooperative problem solving have been integrated, is the concept of an asynchronous team (A-Team), originally introduced in [25].

G. Terrazas et al. (eds.), *Nature Inspired Cooperative Strategies for Optimization* (NICSO 2013), Studies in Computational Intelligence 512,
DOI: 10.1007/978-3-319-01692-4_21, © Springer International Publishing Switzerland 2014

The goal of the paper is to propose (and validate) a new cooperative Team of A-Teams approach for solving instances of the Vehicle Routing Problem with Time Windows (VRPTW). The idea of Team of A-Teams, introduced in [2] is to integrate the team of asynchronous agent paradigm [25] with the island-based genetic algorithm concept [17]. Technicaly the proposed approach is implemented in a multi-agent environment presented in [1] and extended in [2]. Its main functionality focuses on organizing and conducting the process of search for the best solution using a set of search procedures (implemented as software agents) executed in parallel, where each search program is an implementation of a single-solution method. During such execution, agents representing search procedures communicate asynchronously with each other but this communication is performed indirectly via the common, sharable memory (also called warehouse or pool of solutions). The novelty of the approach presented here is that agents are grouped in teams working on islands. Each team of agents periodically communicates with other teams by sharing a promising results with them.

Team of A-Teams architecture can be also viewed as a cooperative search approach, where cooperation takes place between software agents within a single A-Team and/or between A-Teams. In the former case, agents work on results obtained by another agents, but information is not exchanged directly between agents but through the common memory containing the population of individuals. The later case involves a periodical exchange of some solutions between A-Teams belonging to the Team of A-Teams.

The rest of the paper includes the following sections. Section 2 contains the VRPTW problem formulation. Section 3 provides the background of the asynchronous teams concept. In Section 4 details of the proposed dedicated Teams of A-Teams designed for solving the VRPTW instances are presented. Section 5 describes the computational experiment and reports on the results. Finally, Section 6 concludes the paper and suggests directions for future research.

2   Problem Formulation

The Vehicle Routing Problem with Time Windows can be formulated as the problem of determining optimal routes passing through a given set of locations (customers) and defined on the undirected graph $G = (V, A)$, where $V = \{0, 1, \ldots, n, n+1\}$ is the set of nodes and $A = \{(i, j)|i, j \in V\}$ is the set of edges. Nodes 0 and $n+1$ represent a central depot with a set $K = \{1, 2, \ldots, NV\}$ of identical vehicles of capacity C. Each node $i \in V \setminus \{0, n+1\}$ denotes a customer characterized by a non-negative demand d_i, and a service time s_i at the customer i. It is assumed that $d_0 = d_{n+1} = 0$, and $s_0 = s_{n+1} = 0$.

Moreover, with each customer $i \in V$, a time window $[a_i, b_i]$ wherein the customer has to be supplied, is associated. Here a_i is the earliest possible departure (ready time), and b_i - the latest time the customer request has to be started to be served. The time window at the depot $([a_0, b_0]$ and $[a_{n+1}, b_{n+1}])$ is called the scheduling horizon, and denoted as $[E, L]$.

Each link $(i,j) \in E$ denotes the shortest path from customer i to j and is described by the cost c_{ij} and time t_{ij} of travel from i to j by shortest path $(i,j \in V)$.

The goal is to minimize the cost of travel needed to supply all customers, such that each route starts and ends at the depot (0 and $n+1$, respectively), each customer $i \in V \setminus \{0, n+1\}$ is serviced exactly once by a single vehicle, the total load on any vehicle associated with a given route does not exceed vehicle capacity, and each customer $i \in V$ has to be supplied whithin the time window $[a_i, b_i]$ associated with it (the vehicle arriving before the lower limit of the time window causes additional waiting time on the route).

Following [9], the VRPTW can be formally described as the following multi-commodity network flow model with time window and capacity constraints, where two types of variables are involved: binary flow variables x_{ijk} $((i,j) \in A, k \in K)$, equal to 1, if arc (i,j) is used by vehicle k, and 0 otherwise; and time variables $w_{ik}(i \in V, k \in K)$, specifying the start of service at node i when serviced by vehicle k:

$$\min z = \sum_{k \in K} \sum_{(i,j) \in A} c_{ij} x_{ijk} \tag{1}$$

subject to

$$\sum_{k \in K} \sum_{j \in \Delta^+(i)} x_{ijk} = 1, \quad i \in V \setminus \{0, n+1\} \tag{2}$$

$$\sum_{j \in \Delta^+(0)} x_{0jk} = 1, \quad k \in K \tag{3}$$

$$\sum_{i \in \Delta^-(j)} x_{ijk} - \sum_{i \in \Delta^+(j)} x_{jik} = 0, \quad k \in K, j \in V \setminus \{0, n+1\} \tag{4}$$

$$\sum_{i \in \Delta^-(n+1)} x_{i,n+1,k} = 1, \quad k \in K \tag{5}$$

$$x_{ijk}(w_{ik} + s_i + t_{ij} - w_{jk}) \leq 0, \quad k \in K, (i,j) \in A \tag{6}$$

$$a_i \left(\sum_{j \in \Delta^+(i)} x_{ijk} \right) \leq w_{ik} \leq b_i \left(\sum_{j \in \Delta^+(i)} x_{ijk} \right), \quad k \in K, i \in V \setminus \{0, n+1\} \tag{7}$$

$$E \leq w_{ik} \leq L, \quad k \in K, i \in \{0, n+1\} \tag{8}$$

$$\sum_{i \in N} d_i \sum_{j \in \Delta^+(i)} x_{ijk} \leq C, \quad k \in K \tag{9}$$

$$w_{ik} \geq 0, \quad k \in K, i \in V \tag{10}$$

$$x_{ijk} \in \{0,1\}, \quad k \in K, (i,j) \in A \tag{11}$$

The objective function (1) expresses the total cost of servicing all customers. Constraints (2) guarantee the assignment of each customer to exactly one vehicle route. Constraints (3)-(5) characterize the flow on the path to be followed by vehicle k. Additionally, constraints (6)-(8) and (9) assure schedule feasibility with respect to time considerations and capacity aspects, respectively. Finally, (10)-(11) impose conditions on the variables.

There have been important advances in the development of exact and approximate algorithms for solving VRPTW. Because of NP-hardness of the problem, most of approaches are of heuristic nature. The general classification of existing heuristics distinguishes a group of *classical approaches* (route construction and local search algorithms) and *metaheuristics* [7, 8]).

Classical heuristics, like Solomon's *I1* algorithm [22], or various *local search* algorithms of Russell [20], Shaw [21], and Bräysy [6] are often relatively simple and fast but the quality of solutions obtained by them is as a rule not satisfactory. On the other hand, although metaheuristics require much more computational resources and have to be fine-tuned in order to fit a particular problem, they provide much better solutions, especially in case of large-scale instances. Implementations of Rochat and Taillard [19], and Taillard et al. [24] (tabu search), Homberger and Gehring [14], and Berger et al. [4] (evolutionary algorithms), and Gambardella et al. [11] (ant colony optimization approach) are worth mentioning in this respect.

As mentioned, although there exists a lot of heuristics solving effectively instances of VRPTW (a review of different methods can be found in two papers of Bräysy and Gendreau [7, 8]), a relatively small number of papers aim at using methods based on agent paradigm for solving VRPTW (see, for example [5, 16]. This paper attempts to fill this gap by combining asynchronous teams of agents [25] concept with the island evolutionary algorithm idea [17].

3   Asynchronous Teams

The concept of asynchronous teams (A-Teams) was originally introduced by Talukdar [25] as a result of integration of paradigms of the population-based methods and multiple agent systems. According to this concept, an asynchronous team can be seen as a collection of *autonomous agents* that cooperate to solve a problem by dynamically evolving a population of solutions stored in the *common memories*. Within an A-Team, each agent encapsulates a particular problem-solving method (exact or heuristic) along with the methods to decide when to work, what to work on and how often to work. The role of memories is to accumulate results or trial solutions forming populations, processed by working agents. During an A-Team activity, memories are time varying: new members (solutions improved by agents) are continually added by construction agents, while older members (worse solutions) are being erased by destruction agents, so the quality of the solutions gradually evolves over the time.

An important advantage of applications of the A-Team concept for solving particular optimization problem, stems from a problem-solving method encapsulated in each optimization agent. Although the whole approach belongs to a group of the population-based methods, each optimizing agent is, in fact, an implementation of a single-solution method. The general assumptions about A-Team do not indicate a need of using methods with specific features. Local search methods, dedicated improvement heuristics, or nature-inspired metaheurisics belong to the most frequently used ones.

Besides the set of effective methods (represented by agents), the ground principle of asynchronous teams rests on combining these methods, which alone could be inept for the task, into effective problem-solving organizations, possibly creating a synergy effect. The observed combined effect of agents teamwork is often greater than the sum of their separate effects.

The existence, within A-Team, of shared memory, a mechanism of management of population of solutions and a set of autonomous agents, provide a basis for cooperation between agents. As Rachlin et al. [18] observed agents working within A-Team cooperate by sharing access to populations of candidate solutions. Solutions obtained by one agent are shared, through the central memory mechanism, with other agents, which can exploit these solutions in order to guide the search through new promising region of the search space, thus increasing chances for reaching the global optimum. It is expected that such a collective of agents can produce better solutions than individual members of such collective, thus, achieving a synergetic effect.

4   Team of A-Teams Approach for Solving the VRPTW

The architecture of the presented approach includes: a *set of single A-Teams* (Team of A-Teams) working within a network, and the *communication protocol* assuring effective communication between the A-Teams while solving instances of the problem.

4.1   Single A-Team

Main part of the architecture of the proposed approach is a single A-Team which consists of (see Figure 1):

- A common, sharable *memory*, which store a population of individuals (solutions),
- A set of *agents* (called *OptiAgents*), representing a single-solution method, which operate on individuals during the process of search, and an agent (*SolutionManager*), which act as an intermediary between common memory and *OptiAgents*. It maintains the common memory and is responsible for managing the population of solutions,
- A *population management strategy* combining these agents into a single effective problem-solving strategy.

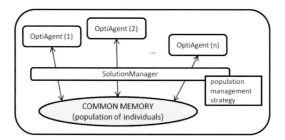

Fig. 1. Architecture of the single A-Team

The main functionality of such A-Team is organizing and conducting the search process, which is organized as a sequence of steps, including initialization and improvement phases. At first the initial population of solutions is generated and stored in the memory. Next, at the following computation stages, individuals forming the initial population are successively improved by autonomously working *OptiAgents*, each executing an improvement algorithm. The main steps of the proposed approach, repeated in loop until a stopping criterion is met, include:

1. Selecting a particular individual (solution) from the common memory by *SolutionManager* and sending it to autonomous, independently acting *OptiAgents*, which have already announced their readiness to act,
2. Improvement of solutions by these agents and sending them to back to the *SolutionManager*, and
3. Storing back (by *SolutionManager*) the potentially improved solution returned by *OptiAgents* in the common memory.

When the stopping criterion is met, the best solution in the population is taken as the final solution of the given problem instance.

Common Memory. Common sharable memory stores a population of *popSize* individuals (solutions). Each individual stored in the memory is represented in a form that reflects the characteristics of the problem being solved, as well as which is convenient to handle the calculations performed on it by search procedures. In Teams of A-Teams approach for VRPTW, each solution is represented as a list of routes $R = (R_1, R_2, ..., R_m)$, (m - the number of routes), where each selected route includes customer requests in the order in which customers are visited by a single vehicle.

Set of Agents. Each optimizing agent is, in fact, an implementation of a single-solution method. Using different improvement algorithms executed by different agents supposedly increases chances for reaching the global optimum. Four kinds of optimizing agents have been used in a single A-Team (see Table 1). Two of them are based on crossover evolutionary operator, remaining two are implementations of tabu search [12], and simulated annealing [10] metaheuristics.

Table 1. Agents and their characteristics

Agent	Description
Cross1	An agent which is implementation of the one-point crossover operator, where initially one point is randomly selected on each route R_i and R_j, dividing these routes on two subroutes. Next, the first subroute of R_i is connected with the second subroute of R_j, and the first subroute of R_j is connected with the second subroute of R_i. If the resulting routes are feasible and the solution has been improved, it is accepted.
Cross2	An implementation of the two-point crossover operator, where initially two points are selected randomly on each route R_i and R_j, dividing these routes on three subroutes. Next, the middle parts (between crossing points) of each route are exchanged between considered routes. Only feasble and improving moves are accepted.
TSSwap	An implementation of tabu search metaheuristic, where *swap* operator has been used as the move. Up to 2 randomly selected customers from routes R_i and R_j are swapped. The feasible move giving the greatest decrease of the goal function value is accepted.
SAMove	An implementation of simulated annealing algorithm with *move* operator included in it. Here, up to 2 randomly selected customers from R_i are moved to R_j. The criterion of acceptance of the new solution is the same as in TSSwap.

All methods operate on two randomly selected routes R_i and $R_j \in R$ ($i \neq j$, and $i, j = 1, \ldots, m$).

Population Management Strategy. It defines the set of rules determining: how the inital population is created and how many solutions does it include, how to choose solutions which are to be sent to the search programs for improvement, how to merge the improved solutions returned by the search procedures with the whole population and when to stop the process of searching [1].

In the presented Teams of A-Teams approach for VRPTW, all individuals including in the initial population of solutions are created using modified version of Solomons *I1* constructive heuristic for VRPTW [22]. Opposite to the Solomons approach, where two initialization criteria based on the farthest unrouted customer and the uncounted customer with the earliest deadline, were tested, here, creation of each route starts from a randomly selected unrouted customer. The process of creating the whole initial population is repeated until *popSize* individuals have been generated.

The *OptiAgent*, which anouncess its readiness to act, receives from *Solution-Manager* the individuals randomly selected from the population. However once selected individual can not be selected again until all other individuals have been tried.

After improvement, returning individual are incorporated into the common memory by replacing the first found worse individual. If a worse individual can not be found within a certain number of reviews (where review is understood as a search for the worse individual after an improved solution is returned) then the worst individual in the common memory is replaced by the new, randomly generated one, representing a feasible solution. In the approach the number of reviews after which a random solution is generated is set to 5.

The process of solving the instance by a single A-Team stops after a predefined amount of time (after the preliminary investigation, in the experiment which has been carried out and described in section 5, it has been decided to stop the process after 3 minutes).

4.2   Team of A-Teams

The Team of A-Teams allows for a number of single A-Teams to solve the same task in parallel by exploring different regions of the search space with the added process of communication between A-Teams (see Figure 2). The process of communication between A-Teams is supervised by a specialized agent called *MigrationManager* and defined by a *migration strategy* including a number of parameters [15]: an architecture in which an A-Team receives communication from another A-Team and sends communication to some other A-Team (*migration topology*), number of individuals sent between common memories of A-Teams in a single cycle (*migration size*), length of time between migrations (*migration frequency*), and a rule determining how the received solution is incorporated into a common memory of the receiving A-Team (*migration policy*).

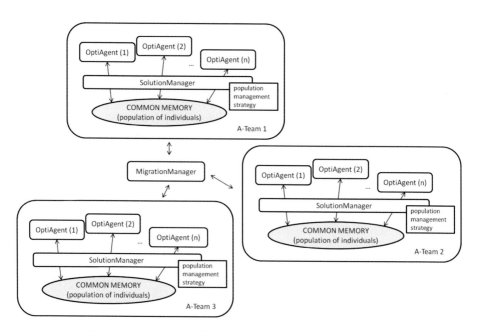

Fig. 2. Architecture of the Team of A-Teams for VRPTW

The migration strategy in the proposed implementation for VRPTW is based on the randomized topology [15] in which the one source A-Team asks for a new solution when the current best solution in its common memory has not been changed by a fixed part of no improvement time gap. The source A-Team sends appropriate message to the *MigrationManager*. It chooses randomly one other target A-Team and asks it for sending its best solution to the source A-Team. The best solution taken from the source A-Team replaces the worst solution in the common memory of the target A-Team.

When one of the single A-Teams stops due to its population management strategy, the whole Team of A-Teams stops its computation, regardless of recent improvements in best solutions of the others A-Teams. The overall best result from common memories of all A-Teams in Teams of A-Teams is taken as the final solution found for the task.

5   Computational Experiment

Computational experiment has been carried out in order to evaluate the effectiveness of the proposed approach. The experiment aimed at answering the following question: To what extent (if any) different configurations of teams of A-Teams, including number of islands, number of optimizing agents, and number of individuals forming the population, influence computation results? The quality of the results obtained by the proposed approach has been evaluated using two measures: the number of vehicles needed to serve all requests, and the total distance needed to pass by vehicles in order to supply all customers. The lexicographic preference ordering on the number of vehicles, and the total distance has been used. This means that a solution with fewer routes is preferred over one with more routes and that in the case of a tie in the number of routes, the solution with the shortest distance is chosen. The results were compared with the best known solutions obtained by heuristics and reported in [23].

The experiment involved 56 instances of Solomon [22] (available at [23]) with 100 customers each. The whole set of instances is divided into six groups (R1, R2, C1, C2, RC1, RC2) including customers with randomly generated coordinates (R1, R2), clustered coordinates (C1, C2) or both (RC1, RC2). Additionally, instances belonging to R1, C1, and RC1 have a short scheduling horizon, whereas the instances from R2, C2 and RC2 have a long scheduling horizon.

All computations have been carried out on PC computer with Intel Core i5-2540M CPU 2.60 GHz and 8 GB RAM running under MS Windows 7 operating system.

In the reported experiment four different configurations have been tested, each defined by the number of A-Teams and the population size of the single A-Team. It has been assumed that the total number of individuals within Teams of A-Teams is constant. In the first configuration, the computation has been performed using a single A-Team with population of 120 individuals (solutions). In the remaining configurations, the number of A-Teams has been set to 2, 4, and 8, and the population size of the single A-Team has been set to 60, 30, and

15, respectively. A single copy of each optimizing agent has been used in each A-Team.

Results of the experiment are presented in Tab. 2, 3, and 4, separately for each group of dataset. Each table includes six columns: name of the instance, best known results (number of vehicles/distance) obtained by heuristics [23], and results produced by the proposed approach for all tested configurations. The best results obtained for each instance are emphasized using bold font.

Table 2. Results (min. number of vehicles/distance) obtained by the proposed approach for all tested configurations (groups R1, R2)

		Number of A-Teams (islands)			
		1	2	4	8
		Population size of the single A-Team			
Instance	Best known	120	60	30	15
R101	19/1645.79	19/1732.86	**19/1707.70**	19/1735.26	19/1720.13
R102	17/1486.12	17/1606.36	17/1593.42	17/1570.23	**17/1564.35**
R103	13/1292.68	14/1310.90	**14/1281.85**	14/1300.49	14/1301.75
R104	9/1007.24	10/1067.12	**10/1027.46**	10/1044.45	11/1040.83
R105	14/1377.11	15/1447.42	15/1459.15	15/1432.47	**14/1500.71**
R106	12/1251.98	13/1343.87	**13/1289.64**	13/1314.73	13/1308.63
R107	10/1104.66	11/1170.33	11/1151.97	**11/1122.77**	11/1142.04
R108	9/960.88	10/1003.13	**10/991.45**	10/1000.56	10/1000.92
R109	11/1194.73	12/1262.69	12/1228.03	12/1255.31	**12/1220.74**
R110	10/1118.59	12/1154.45	12/1148.20	12/1157.47	**12/1141.44**
R111	10/1096.72	11/1174.21	**11/1147.95**	11/1152.74	12/1113.61
R112	9/982.14	10/1058.03	**10/1011.92**	10/1040.85	10/1025.53
R201	4/1252.37	**4/1311.02**	4/1322.18	4/1324.11	4/1322.14
R202	3/1191.70	4/1130.89	4/1128.37	**4/1115.51**	4/1122.65
R203	3/939.54	**3/984.84**	3/994.11	3/988.19	3/990.67
R204	2/825.52	3/796.43	**3/778.64**	3/790.01	3/792.84
R205	3/994.42	3/1087.50	3/1093.76	**3/1064.72**	3/1089.78
R206	3/906.14	**3/958.78**	3/964.83	3/985.53	3/961.19
R207	2/893.33	**3/845.57**	3/848.80	3/855.77	3/854.15
R208	2/726.75	**2/753.70**	2/767.52	2/796.53	2/815.64
R209	3/909.16	**3/958.00**	3/970.02	3/981.70	3/977.19
R210	3/939.34	3/1007.85	**3/986.45**	3/1036.23	3/1012.19
R211	2/892.71	3/831.57	**3/805.04**	3/818.85	3/820.94

Analysis of the results allows for several observations. The first one is that the results produced by the proposed approach are competitive with the best known approximate solutions produced by heuristics. For most cases, the number of vehicles needed to serve all requests is the same as in the best known results. Also, taking into account the distance needed to be covered by all vehicles, the relative errors from the best solutions do not exceed a few percent. The overall best results are obtained for instances belonging to group C, where majority of results is equal to the best known ones.

Table 3. Results (min. number of vehicles/distance) obtained by the proposed approach for all tested configurations (groups C1, C2)

		Number of A-Teams (islands)			
		1	2	4	8
		Population size of the single A-Team			
Instance	Best known	120	60	30	15
C101	10/828.94	**10/828.94**	**10/828.94**	**10/828.94**	**10/828.94**
C102	10/828.94	**10/828.94**	10/838.37	**10/828.94**	**10/828.94**
C103	10/828.06	10/848.31	10/853.48	10/860.31	**10/845.32**
C104	10/824.78	10/877.28	10/872.09	**10/861.71**	10/868.00
C105	10/828.94	**10/828.94**	**10/828.94**	**10/828.94**	**10/828.94**
C106	10/828.94	**10/828.94**	**10/828.94**	**10/828.94**	**10/828.94**
C107	10/828.94	**10/828.94**	**10/828.94**	**10/828.94**	**10/828.94**
C108	10/828.94	10/839.07	10/830.60	10/832.67	**10/828.94**
C109	10/828.94	**10/833.83**	10/837.68	10/841.96	10/835.06
C201	3/591.56	**3/591.56**	**3/591.56**	**3/591.56**	**3/591.56**
C202	3/591.56	**3/591.56**	**3/591.56**	**3/591.56**	**3/591.56**
C203	3/591.17	**3/600.21**	3/614.71	3/608.39	3/610.99
C204	3/590.60	**3/613.07**	3/619.60	3/658.00	3/620.49
C205	3/588.88	**3/588.88**	3/589.89	3/589.89	3/603.64
C206	3/588.49	**3/588.49**	3/589.51	**3/588.49**	3/602.18
C207	3/588.29	3/601.19	3/595.34	**3/588.29**	3/595.26
C208	3/588.32	3/612.87	3/596.06	3/608.30	**3/595.85**

Table 4. Results (min. number of vehicles/distance) obtained by the proposed approach for all tested configurations (groups RC1, RC2)

		Number of A-Teams (islands)			
		1	2	4	8
		Population size of the single A-Team			
Instance	Best known	120	60	30	15
RC101	14/1696.94	15/1728.05	**15/1703.56**	15/1722.02	15/1729.49
RC102	12/1554.75	14/1535.75	13/1594.40	14/1528.88	**13/1549.09**
RC103	11/1261.67	12/1357.84	**12/1323.42**	12/1338.77	12/1326.16
RC104	10/1135.48	11/1199.85	10/1208.22	10/1240.12	**10/1199.68**
RC105	13/1629.44	15/1640.31	15/1608.74	15/1628.93	**15/1598.11**
RC106	11/1424.73	13/1454.21	**13/1429.15**	13/1437.53	13/1455.49
RC107	11/1230.48	12/1282.55	12/1289.02	12/1276.98	**12/1262.66**
RC108	10/1139.82	11/1180.96	11/1193.75	11/1184.25	**11/1178.24**
RC201	4/1406.91	**4/1503.65**	4/1509.85	4/1505.59	4/1530.84
RC202	3/1367.09	**4/1228.29**	4/1257.35	4/1249.07	4/1249.74
RC203	3/1049.62	3/1182.38	3/1172.67	**3/1166.21**	3/1180.79
RC204	3/798.41	**3/831.37**	3/873.16	3/851.26	3/853.82
RC205	4/1297.19	**4/1399.91**	4/1418.68	4/1402.22	4/1431.41
RC206	3/1146.32	3/1303.50	4/1135.37	**3/1195.36**	3/1281.89
RC207	3/1061.14	4/1053.00	4/1045.70	**3/1177.31**	4/1063.92
RC208	3/828.14	3/902.15	**3/874.22**	3/910.45	3/900.86

The second observation refers to comparison of results produced by a single A-Team with results obtained by Teams of A-Teams. Fig. 3 presents the percentage of the instances for which the best results have been obtained by a single A-Team and by the Teams of A-Teams, grouped by the form of customers' coordinates distribution, and the length of the scheduling horizon.

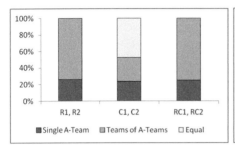 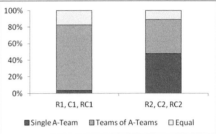

Fig. 3. Percentage of the instances for which the best results have been obtained by the single A-Team and by Teams of A-Teams, grouped by the form of customers' coordinates distribution (left), and the length of the scheduling horizon (right)

Through analysis of the results, one can conclude that their quality strongly depends on the group, the instance belongs to. Taking into account the customers' coordinates distribution (Fig. 3, left), it is easy to see, that only 25% of all tested instances were solved better by a single A-Team (regardless of the group of instances), whilst Teams of A-Teams obtained better results for 61% of instances. The same results have been obtained for remaining 14% of instances.

By focusing observation on scheduling horizon (Fig. 3, right), one can conclude, that whereas instances with the short scheduling horizon (R1, C1, and RC1) have been solved more effectively using Teams of A-Teams (79% of instances), the instances with long scheduling horizon (R2, C2 and RC2) have been solved with approx. the same effectiveness by both, a single A-Team and Teams of A-Teams (48% - single A-Team, 41% - Teams of A-Teams, 11% - equal). Unfortunately, in case of outperformance of Teams of A-Teams, the experiment does not provide a decisive arguments for answering the question: how many A-Teams should be used in order to obtain the best results. It can suggests, that using 2 or 4 A-Teams seems to guarantee better results than using 8 A-Teams.

6   Conclusions

The paper proposes a new approach for Vehicle Routing Problem with Time Windows, integrating the asynchronous team paradigm with the island-based evolutionary algorithm concept. The process of solving the problem is carried-out by a set of agents, each representing a heuristic algorithm, operating on population of individuals (solutions) stored in the common sharable memory.

Agents are grouped in teams, which periodically communicates with other by sharing a promising results.

Computational experiment confirmed effectiveness of the proposed approach in terms of relative errors from the best known solutions obtained by heuristics. For most tested instances, Teams of A-Teams outperform a single A-Team. To what extent this takes place depends however on features of the instance in question (customers' distribution and the length of the scheduling horizon).

Future research will focus on considering different migration topologies (ring, torus, etc.) and investigating their influence on results produced by the proposed approach. The investigation will be extended to other vehicle routing problems, like Pickup and Delivery Vehicle Routing Problem, etc. Also, the planned extended experiment will focus on comparison the proposed approach with other techniques known from the literature.

Acknowledgments. The research has been supported by the Polish National Science Centre grant no. 2011/01/B/ST6/06986 (2011-2013).

References

[1] Barbucha, D., Czarnowski, I., Jędrzejowicz, P., Ratajczak-Ropel, E., Wierzbowska, I.: Web accessible A-Team middleware. In: Bubak, M., van Albada, G.D., Dongarra, J., Sloot, P.M.A. (eds.) ICCS 2008, Part III. LNCS, vol. 5103, pp. 624–633. Springer, Heidelberg (2008)

[2] Barbucha, D., Czarnowski, I., Jędrzejowicz, P., Ratajczak-Ropel, E., Wierzbowska, I.: Parallel Cooperating A-Teams. In: Jędrzejowicz, P., Nguyen, N.T., Hoang, K. (eds.) ICCCI 2011, Part II. LNCS, vol. 6923, pp. 322–331. Springer, Heidelberg (2011)

[3] Bellifemine, F., Caire, G., Poggi, A., Rimassa, G.: JADE. A White Paper. Exp. 3(3), 6–20 (2003)

[4] Berger, J., Barkaoui, M.: A route-directed hybrid genetic approach for the vehicle routing problem with time windows. Information Systems and Operations Research 41, 179–194 (2003)

[5] Boudali, I., Fki, W., Ghedira, K.: How to Deal with the VRPTW by using Multi-Agent Coalitions. In: Proceedings of the Fourth International Conference on Hybrid Intelligent Systems (HAIS 2004), pp. 416–421. IEEE Press, Washington (2004)

[6] Bräysy, O.: Fast Local Searches for the Vehicle Routing Problem with Time Windows. Information Systems and Operations Research 40(4), 319–330 (2002)

[7] Bräysy, O., Gendreau, M.: Vehicle Routing Problem with Time Windows, Part I: Route Construction and Local Search Algorithms. Transportation Science 39, 104–118 (2005)

[8] Bräysy, O., Gendreau, M.: Vehicle Routing Problem with Time Windows, Part II: Metaheuristics. Transportation Science 39, 119–139 (2005)

[9] Cordeau, J.-F., Desaulniers, G., Desrosiers, J., Solomon, M.M., Soumis, F.: The VRP with Time Windows. In: Toth, P., Vigo, D. (eds.) The Vehicle Routing Problem. SIAM Monographs on Discrete Mathematics and Applications, pp. 157–193. SIAM, Philadelphia (2002)

[10] Eglese, R.W.: Simulated annealing: A tool for operational research. European Journal of Operational Research 46, 271–281 (1990)

[11] Gambardella, L.M., Taillard, E., Agazzi, G.: MACS-VRPTW: A Multiple Ant Colony System for Vehicle Routing Problems with Time Windows. In: Corne, D., Dorigo, M., Glover, F. (eds.) New Ideas in Optimization, pp. 63–76. McGraw-Hill, Maidenhead (1999)

[12] Glover, F., Laguna, M.: Tabu Search. Kluwer, Boston (1997)

[13] Golden, B.L., Raghavan, S., Wasil, E.A. (eds.): The Vehicle Routing Problem: Latest Advances and New Challenges. Operations Research Computer Science Interfaces Series, vol. 43. Springer, Heidelberg (2008)

[14] Homberger, J., Gehring, H.: Two Evolutionary Metaheuristics for the Vehicle Routing Problem with Time Windows. Information Systems and Operations Research 37, 297–318 (1999)

[15] Jędrzejowicz, P., Wierzbowska, I.: Impact of Migration Topologies on Performance of Teams of A-Teams. In: Grana, M., et al. (eds.) Advances in Knowledge-Based and Intelligent Information and Engineering Systems, pp. 1161–1170. IOS Press, Amsterdam (2012)

[16] Leong, H.W., Liu, M.: A multi-agent algorithm for vehicle routing problem with time window. In: Proceedings of the ACM Symposium on Applied Computing, SAC 2006, pp. 106–111. ACM, New York (2006)

[17] Martin, W.N., Lienig, J., Cohoon, J.P.: Island (migration) models: evolutionary algorithms based on punctuated equilibria. In: Bäck, T., et al. (eds.) Handbook of Evolutionary Computation, pp. C6.3:1–C6.3:16. Oxford University Press, New York (1997)

[18] Rachlin, J., Goodwin, R., Murthy, S., Akkiraju, R., Wu, F., Kumaran, S., Das, R.: A-Teams: An Agent Architecture for Optimization and Decision-Support. In: Papadimitriou, C., Singh, M.P., Müller, J.P. (eds.) ATAL 1998. LNCS (LNAI), vol. 1555, pp. 261–276. Springer, Heidelberg (1999)

[19] Rochat, Y., Taillard, E.D.: Probabilistic Diversification and Intensification in Local Search for Vehicle Routing. Journal of Heuristics 1, 147–167 (1995)

[20] Russell, R.A.: Hybrid Heuristics for the Vehicle Routing Problem with Time Windows. Transportation Science 29, 156–166 (1995)

[21] Shaw, P.: Using Constraint Programming and Local Search Methods to Solve Vehicle Routing Problems. In: Maher, M.J., Puget, J.-F. (eds.) CP 1998. LNCS, vol. 1520, pp. 417–431. Springer, Heidelberg (1998)

[22] Solomon, M.: Algorithms for the Vehicle Routing and Scheduling Problems with Time Window Constraints. Operations Research 35, 254–265 (1987)

[23] Solomon, M.: VRPTW Benchmark problems, http://w.cba.neu.edu/~msolomon/problems.htm

[24] Taillard, E., Badeau, P., Gendreau, M., Geurtin, F., Potvin, J.Y.: A Tabu Search Heuristic for the Vehicle Routing Problem with Time Windows. Transportation Science 31, 170–186 (1997)

[25] Talukdar, S., Baerentzen, L., Gove, A., de Souza, P.: Asynchronous Teams: Co-operation Schemes for Autonomous Agents. Journal of Heuristics 4(4), 295–321 (1998)

[26] Wooldridge, M.: An Introduction to MultiAgent Systems. John Wiley & Sons, Chichester (2009)

Self-adaptable Group Formation of Reconfigurable Agents in Dynamic Environments

Ruby L.V. Moritz and Martin Middendorf

Parallel Computing and Complex Systems Group Institute of Computer Science,
Universität Leipzig, Germany
{ruby.moritz,middendorf}@informatik.uni-leipzig.de

Abstract. The formation of groups in heterogeneous swarms is important whenever there is benefit from cooperation between the members of a group. Here we investigate decentralized group formation strategies for a set of moving agents that have to collect resources. The agents are reconfigurable and can adapt themselves to the needs of their group. It is assumed that the agents are simple and communication between groups is limited to sensing each other. Members of one group are able to share information about their capabilities and movement decision. Also the decision strategies for integrating a new agent into a group and for the moving direction of a group are simple. Several versions of the system are compared experimentally for a dynamic situation where the number of available resource types changes. It is shown that the costs for reconfiguration influences the optimal strategy for the integration of new agents into a group. The system can adapt the average group size to the number of available resource types.

Keywords: agent simulation, group formation, reconfigurable agents, dynamic environments.

1 Introduction

The problem of decentralized group formation for a heterogeneous set of agents is studied in this paper. The particular version of the problem studied here is to dynamically partition a set of moving agents that have different capabilities with respect to a resource collection task such that a utility function is maximized. Each set of the partition is called a group. It is known that the general problem of group formation between agents in order to maximize a utility function is NP hard and several heuristics have been developed for solving it (for an overview see [1] or [2]). Important aspects of group formation problems are how much communication is possible between the agents, how much information is shared between them and if the problem can be solved by some central agent or has to be solved decentralized (e.g., [3, 4]). Communication is particularly helpful for group formation when the agents have different skills and an ideal group contains different types of agents. But communication is also useful between the

G. Terrazas et al. (eds.), *Nature Inspired Cooperative Strategies for Optimization (NICSO 2013)*, Studies in Computational Intelligence 512,
DOI: 10.1007/978-3-319-01692-4_22, © Springer International Publishing Switzerland 2014

agents within a group in order to make agreements or compromises, e.g., about where the group should move to or which task to execute next. Typically, there exists a trade-off between the benefit of cooperation and its potential costs due the necessary communication for making compromises.

In this paper decentralized and distributed multi agent systems are investigated where communication is restricted. The task of the agents is to collect resources of different types, e.g. food and construction material. It is assumed that collecting resources requires different skills of the agents, e.g. cutting, carrying, or digging out material. The resources that the movable agents have to collect are located within an arena and the agents can form groups that collect the resources together. In a previous work [5] we have studied the group formation problem for a scenario where the number of available resource types is constant. Here we study a dynamic situation were the number of available resource types varies over time. Moreover, different from [5] the agents are reconfigurable such that they can adapt their skills to the need of their group. A simple mutation strategy that is motivated by evolutionary concepts is studied for the reconfiguration process. We are particularly interested in the problem whether the agents are able to form groups that can adapt their common skills and size to the available resource types.

Distributed strategies for the coalition formation problem for a set of heterogeneous agents have already been investigated in the literature. The main aim of [6] was to investigate the computational effort of the distributed group formation algorithm in relation to the quality of the resulting groups. Similar to our model an agent was characterized by a vector $(a_1, \ldots, a_k)$ where a_r, $r \in [1 : k]$ describes its capability or talent to perform a specific action of type r and the capability of the group is the joint capability of the agents. Improved algorithms have been presented in [7]. In [8] an auction process was used for coalition formation with the aim to make local decisions within the coalition to schedule the execution of tasks in order to reduce or avoid global communication between agents.

Coalition formation for moving agents has been studied in [9] where the agents could use global information about all other existing agents and their capabilities. The agents have to execute a hierarchy of differently located tasks and a task requires several agents with different capabilities for its execution. Therefore, a subset of the agents that are located closely to a task and altogether have the required capabilities can form a coalition and then move to the tasks and execute it. A self-organized coalition formation scheme for agents that are located within an arena has been studied in [10]. In this work an agent can form a coalition with neighbored agents when this is profitable with respect to a payoff function. It was assumed that a coalition has to pay coordination costs. The focus of [10] was to investigate how the coalition sizes change with different pay off functions and different coordination costs. In [11] the agents move in a 2-dimensional arena in order to form clusters of cooperating agents. Trial and error is used by the agents to find out which agents to cooperate with and to solve the conflict between achieving a social optimum in the long term or an individual optimum

$\rightarrow$ $(a_{i1}, a_{i2}, a_{i3}) = (2, 5, 1)$

Fig. 1. The slice model used for the agents configuration in an example state of agent a_i. The number of slices s is set to eight, there are three different types of resources and therefore three states a slice can obtain.

in the short term. A special application where the agents form coalitions to traverse unsafe areas after a disaster was studied in [12].

2   Design of the Model

Our model consists of a set of agents $A = \{a_1, \ldots, a_n\}$ and an arena F that is a two dimensional torus of $d \times d$ fields (i.e. the last field in every row and column is adjacent to the corresponding first field in its row respectively column). Each agent is located on a field of the arena. An agent can move within the arena by changing its location from its current field to a neighbored field. The task of the agents is to collect resources. There exist k different types of resources and $R = \{r_1, \ldots, r_k\}$ is the set of resource types. For a field $f \in F$ the amount of resources of type r_i that is available on f is denoted by f_i. The agents have different capabilities for collecting the different types of resources. An agent that is very skilled in collecting a resource of type r_i, might be unable in collecting a resource of type r_j, $j \neq i$. However, an agent can recruit other agents to form a group of agents. For ease of description we assume that a single agent forms a group of size 1. Agents within a group help each other in collecting resources by combining their skills. The number of resource types that are available in the arena and the amount of the different resources on the fields locations of the resources can change over time and provide a challenging environment for the agents. Details are described in the following subsections.

2.1   Agents

To describe the capabilities of an agent and their reconfiguration we apply a simple slice model (similar models are used in the field of reconfigurable hardware). Each agent has s slices and each slice has one of k possible states. A slice with state $i \in [1, k]$ helps the agent to collect resources of type r_i. Thus the number of slices with state i determines the agents skill to collect resources of type r_i. The initial state of each slice is chosen randomly with uniform distribution from the k possible states. An agent can change the states of its slices, i.e. the agent can reconfigure itself. How reconfiguration is done is described in Section 2.3.

If an agent a_i is alone, i.e. it is not member of a group of size ≥ 2, then the amount of resources that it can collect per simulation turn is determined as described in the following. If a_{ij} of the s slices of agent a_i are in state j, i.e. configured for collecting resource type r_j, then the agent is able to collect the fraction a_{ij}/s of the available amount f_j of resource type r_j on the field f

where a_i is located. Thus per simulation turn the agent can collect the amount $(a_{ij}/s) \cdot f_j$ of resource type r_j. For agents that are member of a group with more than one agent the amount of resources that they can collect is described in Subsection 2.2.

2.2   Groups

An agent is member of exactly one group and all agents of a group are located on the same field and share a common moving direction. The moving direction points to one of the eight neighbored fields. A group can increase its size by recruiting new members. Newly recruited members change their moving direction into that of the group.

The advantage of group formation lies in the diversity of its members. Here we assume that the amount of resources that an agent of the group can collect for a resource type r_i per simulation turn is determined by the maximum number of slices that some member of the group has in state i. Thus the resource capabilities of each agent in a group equal the capabilities of the best agent in the group for the corresponding resource type. Clearly, other models for the synergetic effect of a group would be possible. Yet this specific model that is used here has the interesting (and for applications realistic) consequence that the positive effect of group formation fades away when the group becomes too large. Ideally, a group has a highly qualified member for each resource type that is available. The performance $P(G)$ of a group G is defined by the total amount of resources that each member of the group can collect per simulation turn when the group is on a field f. Formally,

$$P(G) = \frac{1}{s \cdot k} \sum_{j=1}^{k} \max_{a_i \in G} a_{ij} \cdot f_j \qquad (1)$$

Observe, that the possible values of $P(G)$ range from 0 to 1, because they are normalized by the number of slices s and the number of resource types k. The groups performance also gives the performance of all its member agents.

The movement of a group is inspired by bacterial movement along a light (phototaxis) or nutrition gradient (chemotaxis). Hence, the movement of a group depends on its performance gradient that is derived from the difference of the groups current performance $P(G)$ and the performance that the group would have when it would move from the current location f to a field $f' \in F$ that is neighbored in direction of the movement direction of the group. Formally,

$$gradient(G, f, f') = \frac{1}{s \cdot k} \sum_{j=0}^{k} \max_{a_i \in G} a_{ij}(f_j - f'_j) \in [-1, 1] \qquad (2)$$

Observe that, $-1 \leq gradient(G, f, f') \leq 1$. If the gradient is positive or zero and no other group is located on f' the group moves to f'. Otherwise the group changes its movement direction by turning 45°. The decision to turn clockwise or counterclockwise is done randomly, unless the group did already make a turn in the previous simulation turn. In the latter case the group continues to turn in the same direction as in the simulation turn before.

2.3   Reconfiguration

An agents capability to change the state of its slices, i.e. to reconfigure itself, is an essential feature of our model. A state change seems particularly promising when the old state of a slice was not really useful. Therefore, an agent checks if the resource type for which a slice is configured, i.e. the state of the slice, is available on the current field of the agent and if there is no other agent in its group that has a higher capability for collecting the respective resource type. If neither of these two cases hold the state of the slice does not increase the groups performance $P(G)$. Therefore, the slice is called idle during that simulation turn. In the special case that more than one agent has a maximum number of slices in state i for a resource type r_i then for one of these agents that is chosen randomly the slices with state i are not idle whereas for the other agents the slices with state i are idle. A state change of an idle slices can at worst keep the groups performance at the same level. Clearly, a reconfiguration, i.e. the change of the state of a slice, cannot come for free. Therefore, it is assumed here that a reconfiguring agent is able to collect only a certain percentage of the resources as it would otherwise collect during the simulation turn.

Inspired by evolutionary concepts, an agent reconfigures a slice with a certain probability to a randomly chosen new state. Thus, a reconfiguration can be considered as a mutation of the agent. The reconfiguration probability for a slice is defined as follows. Initially the reconfiguration probability for a slice is zero. In every simulation turn where a slice is idle its reconfiguration probability is increased by adding a value $\mu \in [0, 1]$. Parameter μ called the mutation strength. After a reconfiguration of a slice or when an idle slice becomes not idle any more its reconfiguration probability is reset to zero.

Whenever the size of a group is larger than the number of available resource types the group contains agents that do not increase the performance of the group, i.e. removing the agent from the group would not change the performance of the group. As the group contains at most one agent per resource type that is the best in collecting that resource type (ties are resolved randomly) there are at least $|G|$ minus the number of available resource types agents with constantly idle slices. These agents reconfigure their slices often and lose performance. Then, the average performance of the group members and hence the overall performance of the system, i.e. the sum of the performance of all agents, decreases.

2.4   Simulation Phases

Each agent has two phases of activity in every simulation turn: the recruitment phase and the working phase (see Figure 2). During each turn of a simulation all agents go first through their recruitment phase. During this phase each group

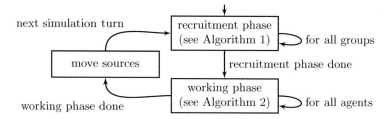

Fig. 2. Flow diagram of the two phases making up a simulation turn

---

Algorithm 1. Recruitment Phase

---

1 **for** *each group G* **do**
2   f' is the neighboured field in the current movement direction
3   **if** *there is a single agent a_i on f', i.e. a_i forms a group of size one* **then**
4   draw a random number ϕ
5   **if** $\phi \leq$ *recruitment rate* **then**
6   recruit agent from f'

---

recruits a single agent when it is on the neighbor field in the movement direction with a certain probability called *recruitment rate* (see Algorithm 1).

Once all agents as part of their group have finished the recruitment phase, they go through their working phase. The course of action for the working phase is shown in Algorithm 2 (part of the algorithm concerning the group behavior is explained later in the following subsection). During the working phase an agent is able to move within the arena and either collect resources or reconfigure itself as described before. For its actions during the working phase an agent needs energy that it takes from its battery. If the battery has not enough energy, the agent becomes inactive in order to reload its battery. An inactive agent reloads its battery by a certain rate, until the battery is fully loaded. During this time an agent is not part of any group. When an agent reactivates it forms a group of its own, unless its former group did not move throughout the time the agent took to reload. In the latter case the agent rejoins its old group as they are located on the same field.

3 Experimental Settings

In order to investigate the general characteristics of the system and the benefit of the reconfiguration we performed several tests.

Each experimental run lasts 3000 simulation turns and includes $n = 100$ agents on an arena F of size 50×50. The agents are placed initially randomly onto a free field of the arena, with random moving direction and a random battery loading state.

Algorithm 2. Working Phase

---

1 **if** *inactive* **then**
2 | reload battery **if** *full* **then**
3 | | turn active

4 **else**
5 | lose energy **if** *battery is empty* **then**
6 | | turn inactive as off next turn

7 | determine next field f';
8 | **if** *group did not move yet* **then**
9 | | **if** *gradient* ≥ 0 **and** f' *is empty* **then**
10 | | | move group on f'
11 | | **else**
12 | | | turn group in previous or random turning direction

13 | collect resources and reconfigure idle slices

Each type of resource $r_i \in R$ has one field in the arena where its availability is maximal ($f_i = 1$). This field is called the source of r_i. The availability of a resource of a certain type on the fields of the arena decreases with growing distance from the source. Here the distance between two fields f and f' is the Euclidean distance between the center points of the two fields and is denoted $\Delta(f, f')$. The function giving the radial decrease of the availability of a resource type r_i, $r_i \in R$, on a field f is given by $f_i = \max(0, (radius - \Delta(f, r_i))/radius)$ where $radius = 1/\sqrt{2\pi}$. Thus the size of the area where r_i is available is half the size of the arena. Fields that are farther away than this radius do not contain any resource of type r_i.

The locations of the source fields for the different resource types are initially chosen randomly. Let $r_i(t) \in [0 : 50]^2$ denote the coordinates of the source of resource type $r_i \in R$ at time t. In each turn of the simulation the source of a resource type slightly changes its position. This is achieved by the following formula using the random numbers $\rho_x, \rho_y \in [0 : 49]$ and the velocity $v = 0.25$

$$r_i(t+1) = \left((1 - v) \cdot r_i(t) + v \cdot \begin{bmatrix} \rho_x \\ \rho_y \end{bmatrix} \right) \bmod 50 \qquad (3)$$

The number of available resource types influences the optimal group size for the agents. To infer their adaptability the number of available resource types is changed after a specific number of simulation turns. This is achieved by reducing the availability of such a resource type r_i on all fields f to 0.

To investigate the influence of the number k of resource types we conducted two types of experiments. In the first type of experiments a fixed number $k = 4$ or $k = 5$ was used. In the second type of experiments $k = 10$ resource types are available during the simulation turn interval $[1001 : 2000]$, whereas during the intervals $[1 : 1000]$ and $[2001 : 3000]$ only two of the ten resource types are available. The different reconfiguration costs that have been tested are $\{1, 0.9, 0.8, 0.5\}$.

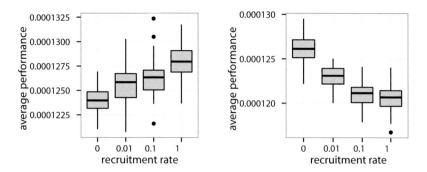

Fig. 3. Average performance of the agents for different numbers of resource types and different recruitment rates; left: system with four resource types; right: system with five resource types

Recall that the reconfiguration costs denote the percentage of resources that the agent can collect less during the simulation turn (compared to what it would have collected without the reconfiguration). The mutation strength $\mu = 1/20$ is used in the tests. It is not hard to show that this value results in an expected number of approximately 5.29 simulation turns, before an idle slice changes its state.

Of special interest is also the recruitment rate of the system. For the tests recruitment rate values in $\{0, 0.01, 0.1, 1\}$ were used. Each simulation run was repeated 50 times.

4   Results

In the following we analyze the performance of the agents (defined in Equation 1) of the tested systems. Different numbers of resource types favor different systems with different recruitment rates when all resource types are available (see Figure 3). If there are four or less resources in the system the agents are fast to adapt themselves to these resources and the price for reconfiguration is cheap, as an agent needs less turns to switch all its slices to the same state. The more states a slice can switch into, the lower the probability for the slice to switch into a specifically desired state $(1/k)$. With five or more resource types the benefit from the groups fades, because the agents need too much time to specialize towards one specific resource type still missing in the group. An agent does have a better performance when it runs by itself, i.e. it forms a group of size 1, and thus reconfigures only if a resource type is not available on the field of its current location. In general the deviation of the systems performances between different runs is high. The outliers of the plots show how large the impact of the random part of the model is.

We can decrease the reconfiguration costs by letting the agent collect a larger fraction of the resources during the simulation turn where he reconfigures its

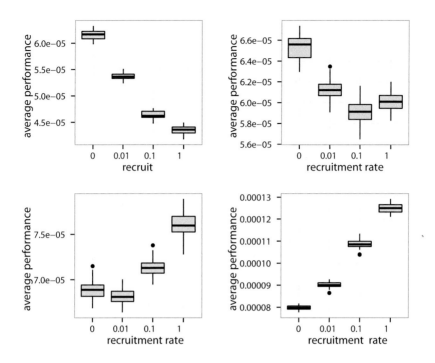

Fig. 4. Average performance of the agents in a systems with a dynamically changing number of available resource types; the system has ten resource types, but in the intervals [1 : 1000] and [2001 : 3000] only two are available; reconfiguration costs are 1 (top left), 0.9 (top right), 0.8 (bottom left) or 0.5 (bottom right)

slices instead of nothing. In the following a reconfiguring agent can still collect 0.5, 0.2, 0.1 or 0.0 of the resources it can usually collect in a simulation turn. Thus the corresponding reconfiguration costs are a fraction of 0.5, 0.8, 0.9, respectively 1 of the resources the agent collects in the respective turn. Throughout these tests there are ten different types of resources and therefore ten different possible states for the agents slices. But only in the interval [1001 : 2000] all ten resources are available. Before and after this time interval only two of the ten resources are actually available to collect. Thus an agent should reconfigure its slices to one of the two available resource types during the first and final thousand simulation turns. Yet this takes a considerable time, as the slices often randomly change their state into one of the eight states corresponding to the eight unavailable resources.

In Figure 4 the effects of different reconfiguration costs on the performance of the agents are shown. With high reconfiguration costs (1 or 0.9) it pays off to avoid the recruitment of other agents whereas lower reconfiguration costs

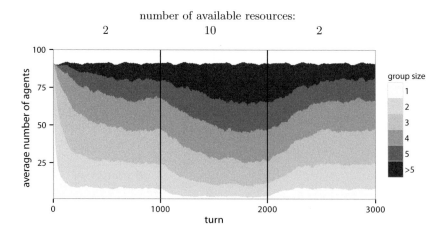

number of available resources:

Fig. 5. Stacked histogram of the number of agents in groups of a specific size per simulation turn; The system has ten resources, but in the intervals [1 : 1000] and [2001 : 3000] only two are available for collection. The reconfiguration costs are set to one simulation turn. The recruitment rate is 1.

(0.8 or 0.5) favor high recruitment rates. In the two intermediate systems with reconfiguration costs 0.8 or 0.9 the worst performing system is neither the one where groups recruit all agents they meet, nor the system where the groups do not recruit at all.

The cause for this phenomenon is that with reconfiguration costs of 0.9 the membership in a group with more than one agent decreases the performance. Indeed with recruitment rates smaller than one, it seems that the larger the group is, the worse is the performance. But if the group does recruit every single agent it meets and increases its size as far as possible the performance increases again. For a single agent it only pays off to be member of a group if the group is very large. The group should contain highly specialized agents for most of the resource types present. That way it collects enough resources in a more or less parasitic fashion to make up for the reconfiguration costs. If the groups were allowed to increase their size indefinitely by saturating the system with a large number of single, recruit-able, agents, such a system will eventually show the highest performance.

With reconfiguration costs of 0.8 we observe the inverse case. Agents are more effective by themselves, than when they are member of a small group of size two or three. In such groups the agents are often forced to reconfigure some of their slices and lose precious time for collecting resources. Yet the few other members of the group do not cover a lot of other resource types, or do so with mediocre capabilities. Thus, the profit of shared capabilities does not pay for the reconfiguration costs, unless the group has enough specialists.

In Figure 5 the average number of different group sizes over the test runs is shown for a system with reconfiguration costs 1 and recruitment rate 1. In

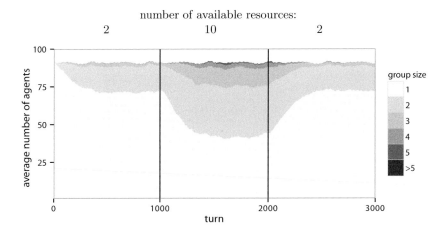

Fig. 6. Stacked histogram of the number of agents in groups of a specific size per simulation turn; The system has ten resource types, but in the intervals [1 : 1000] and [2001 : 3000] only two are available for collection. The reconfiguration costs are set to one simulation turn. The recruitment rate is 0.01.

the beginning half of the agents are in groups with less than five members. When more resource types become available at simulation turn 1001 the groups grow considerably until the number of single active agents diminishes. Unable to recruit more agents, the group sizes stabilize after approximately 500 simulation turns. Note that inactive agents are not counted to be in groups in this histogram. Thus only about 90 agents are considered to be in groups at any time.

A smaller recruitment rate yield smaller groups as can be seen in Figure 6. An agent is expected to recruit about every hundredth agent with a recruitment rate of 0.01. If there are only two resource types available for collection, the agents stay mostly by themselves and only about a fifth of the agents share their group with other agents. The number of single agents decreases with an increasing number of available resource types.

It is desirable for a group to have more members in environments with many resource types than in barren regions. Larger groups have more specialists and can cover the greater spectrum of resource types more effectively. If there are only two resource types available, groups do not need more than two agents. In such an environment a group of three agents contains at least one agent with constantly idle slices. The group sizes in the analyzed systems adapt to the available number of resources. This phenomenon is a result of the gradient walk of the agents. The following deeper analysis of the moving behavior gives more insight on this adaptive behavior.

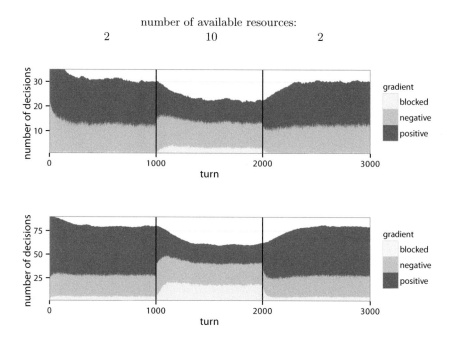

Fig. 7. Stacked histogram of the number of movement decisions per simulation turn by type; A positive gradient results in the groups movement, a negative gradient or a blocked field leads the group to turn on the same field. A field is blocked if another group of agents stands on it. The system has ten resource types, but in the intervals [1 : 1000] and [2001 : 3000] only two are available for collection. The reconfiguration costs are set to one simulation turn. The recruitment rate is 1 (top) and 0.01 (bottom). The different scales are caused by the different number of groups.

Figure 7 shows how the different number of resource types influences the gradient perceived by the agents and results in different movement behavior. With 10 resource types groups move less as a result of their movement decision, i.e. the perceived gradient is negative more frequently. With ten resource types and the position of their sources chosen randomly it is possible but improbable that there are fields without any resources of any type at any time. The frequency of such fields is much higher when eight of the ten sources stop providing their respective resource type to the environment. The gradient between two such fields is zero resulting in the groups movement. This allows groups to leave empty regions of the arena as fast as possible. With ten active sources of ten different resource types the agents configurations are less specialized, especially in small groups of one or two agents. The slices do not turn idle as often as

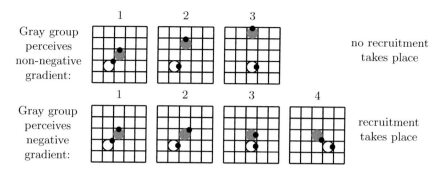

Fig. 8. Exemplary situation of a collision between a single agent (white) and group of more than one agent (gray); If the perceived gradient of the larger group is negative, it detects the single agent and is able to recruit it. Otherwise the groups go their separate ways.

with less active sources and consequently less available resource types. Groups are inclined to follow the source of the resource type they can collect the best. If the source moves into the opposite direction of the groups walking direction, it needs several turns to get back on the sources tracks all the while facing a negative gradient. But the gradient is rarely zero.

The agents average velocity, i.e. average number of fields they traverse in one simulation turn, decreases due to this behavior. The agent then spends many simulation turns turning instead of moving on to the next field. For a single agent it is now harder to avoid the recruitment by the groups, if that were its desire. This phenomenon is based on the actions taking place in case of a collision of two groups of agents. We only consider the case were a single agents path is blocked by a group. The two other cases, were two groups of two or more agents collide, or a groups path is blocked by a single agent result in the same actions of the agents, whether they remain on the same field or not.

Whenever a single agent can not follow its moving direction because another group of at least two agents is blocking the way the larger groups action is crucial. The single agent is unable to recruit, because the other group is too large and the large group is unable to recruit, because the single agent faces the wrong direction. If the larger group faces a non-negative gradient it moves on and the situation passes. But if the group faces a negative gradient it turns around. The single agent needs at least three simulation turns to leave the neighborhood of the larger group. This gives the larger groups three simulation turns to detect the single agent in its neighborhood while it is turning and checking all neighboring fields for a non-negative gradient. An example of such a situation is given in Figure 8. The resulting increase of collisions between groups and them turning away from each other can be seen in Figure 7 as well.

With ten available resource types much fewer movement decisions are made compared to the case with only two available resource types. As there are larger groups during the time with ten available resource types, consequently there are less groups in total. Each group makes one decision during each simulation turn that is valid for all group members.

5 Conclusion and Outlook

We have investigated simple decentralized group formation strategies for a heterogeneous set of moving and reconfigurable agents that have to collect resources. Different recruitment rates and reconfiguration costs have been compared experimentally for a dynamic situation where the number of available resource types changes. We obtained a system that can adapt the group sizes to the number of available resource types and has a good resource collection performance, if the reconfiguration costs are chosen appropriately.

Although the group sizes adapt themselves to the available number of resource types, group formation is not advisable when there are high reconfiguration costs. The development of a dynamic approach, where the agents can measure their current performance and react accordingly when they have the possibility to recruit new agents is our aim for future work.

Acknowledgment. This work has been supported by the European Social Fund (ESF) and the Free State of Saxony within Nachwuchsforschergruppe "Schwarm-inspirierte Verfahren zur Optimierung, Selbstorganisation und Ressourceneffizienz".

References

[1] Gerkey, B.P., Mataric, M.J.: A formal analysis and taxonomy of task allocation in multi-robot systems. The International Journal of Robotics Research 23, 939–954 (2004)
[2] Vig, L., Adams, J.A.: Multi-robot coalition formation. IEEE Transaction on Robotics 22, 637–649 (2006)
[3] Li, C., Sycara, K.: A stable and efficient scheme for task allocation via agent coalition formation. In: Algorithms for Cooperative Systems, p. 20 (2004)
[4] Procaccia, A.D., Rosenschein, J.S.: The communication complexity of coalition formation among autonomous agents. In: Proceedings of the 5th International Joint Conference on Autonomous Agents and Multiagent Systems, pp. 505–512 (2006)
[5] Moritz, R.L., Middendorf, M.: Self-organized cooperation between agents that have to solve resource collection tasks. In: Proc. IEEE Swarm Intelligence Symposium, p. 12 (2013)
[6] Shehory, O., Kraus, S.: Methods for task allocation via agent coalition formation. Artificial Intelligence 101, 165–200 (1996)
[7] Rahwan, T.: Algorithms for Coalition Formation in Multi Agent Systems. University of Southhampton, PhD Thesis (2007)

[8] Kutanoglu, E., Wu, S.D.: Coalitions in coordinated multi-agent production scheduling: A computational study. J. Manuf. Syst. 26, 12–21 (2007)

[9] Khalouzadeh, L., Nematbakesh, N., Zamanifar, K.: A decentralized coalition formation algorithm among homogeneous agents. Journal of Theoretical and Applied Information Technology 22, 35–42 (2010)

[10] Singh, V.K., Husaini, S., Singh, A.: Self-organizing agent coalitions in distributed multi-agent systems. In: 2010 International Conference on Computational Intelligence and Communication Networks, pp. 650–655 (2010)

[11] Abramson, M.: Coalition formation of cognitive agents. In: Proc. of the Second International Conference on Computational Cultural Dynamics, p. 7 (2008)

[12] Bölöni, L., Khan, M.A., Turgut, D.: Agent-based coalition formation in disaster response applications. International Journal of Intelligent Control and Systems 12, 107–117 (2007)

A *Choice Function* Hyper-heuristic for the Winner Determination Problem

Mourad Lassouaoui and Dalila Boughaci

USTHB- FEI- Department of Computer Science
Laboratory of Artificial Intelligence LRIA
BP 32 EL-Alia, Beb-Ezzouar, Alger, 16111
lassouaoui.mourad@gmail.com,
dboughaci@usthb.dz

Abstract. A hyper-heuristic is a high-level method that incorporates a set of low-level heuristics to handle classes of problems rather than solving one problem. In this paper, we propose *a choice function* hyper-heuristic (CFH) for the winner determination problem in combinatorial auctions (WDP). The proposed method is evaluated on various benchmark problems, and compared with the well-known Stochastic Local Search (SLS) for WDP. The experimental study shows that the CFH algorithm is able to find good solution for the winner allocation compared to the SLS.

Keywords: Winner Determination Problem, Combinatorial Auctions, Optimization Problems, Hyper-heuristic, Meta-Heuristic, Heuristic, Stochastic Local Search.

1 Introduction

The combinatorial auctions (CA) is a mechanism that permits to bid on bundles of items. By using such mechanisms, the bidders can express both complementarities and substitutabilities of their preferences within bids [15, 22].

Combinatorial auctions have been used in several domains such as economics, game theory, task allocation in multi-agent systems and in real-world applications such as the sale of spectrum licenses in America's Federal Communications Commissions (FCC)[1] auctions.

The optimal winner determination problem (WDP) in combinatorial auctions is the problem of finding winning bids that maximize the auctioneer's revenue under the constraint that each item can be allocated to at most one bidder. The WDP is known to be an *NP-Complete* [9, 17].

In this work, we propose a hyper-heuristic approach for WDP.

A hyper-heuristic is a high-level method that incorporates a set of low-level heuristics to handle classes of problems rather than solving one problem. The hyper-heuristic method permits to select automatically and during the search

---

[1] http://wireless.fcc.gov/auctions

G. Terrazas et al. (eds.), *Nature Inspired Cooperative Strategies for Optimization* 303
(NICSO 2013), Studies in Computational Intelligence 512,
DOI: 10.1007/978-3-319-01692-4_23, © Springer International Publishing Switzerland 2014

process the heuristic that should be applied for finding good quality solutions and avoiding search stagnation. The low-level heuristics can be either constructive or perturbative heuristics.

The constructive hyper-heuristic that uses a set of constructive heuristics starts with an empty solution and tries to complete it at each step while the perturbative hyper-heuristic starts with a complete initial solution and tries to find better ones from it. In general, a hyper-heuristic functions as follow: Given an instance of problem, the high level method used a certain selection or choice function strategies to choose the adequate low-level heuristic at any given time.

Several works have been done on hyper-heuristics. Some researchers were interested in methods permitting to generate the low-level heuristics for the hyper-heuristic. These methods are based generally on genetic programming [6–8].

On the other hand, several works focused on the methods of selection of the low-level heuristics of a hyper-heuristic. We cite *choice function* and *random* strategies. For more detail on hyper-heuristics and the different strategies proposed in the literature, the reader can refer to [8].

In this paper, we develop a choice function hyper-heuristic for the WDP.

The rest of the paper is organized as follows. The second section gives a background on the WDP problem. The third section presents our contribution for the WDP. The implementation and some numerical results are given in the fourth section. Finally, the fifth section concludes the work.

2   Background

In this work, we are interested in the optimal winner determination problem (WDP) in combinatorial auctions. The problem can be stated as follows:

Let's consider a set of m items, $M = 1, 2 \ldots m$ to be auctioned and a set of n bids, $B = B_1, B_2 \ldots B_n$ submitted by n buyers where each bid can cover a set of items, $S_k \sqsubseteq M$, and result in a profit to the supplier, $P_k \succeq 0$, $k \in 1, 2 \ldots n$, if the bid B_k is accepted. Further, let $X_k = 1$ if bid B_k is accepted (a winning bid) and $X_k = 0$ otherwise (a losing bid). The WDP is the problem of finding winning bids that maximize the auctioneer's revenue under the constraint that each item can be allocated to at most one bidder.

The WDP can be given as the following integer program [21]:

$$\textbf{Maximize} \quad \sum_{k=1}^{n} P_k X_k \qquad (1)$$

$$\textbf{Subject to}: \quad \sum_{k|i \in S_k, X_k \in 0,1} X_k \leq 1, \quad i = 1 \ldots m \qquad (2)$$

The objective function (1) maximizes the auctioneer's revenue which is computed as the sum of prices of the winning bids. The constraints (2) mean that the item can be allocated to at most one bidder. The inequality (≤ 1) allows that some item could be left uncovered. This is due to the free disposal assumption.

The winner determination problem is a known to be an *NP-Complete* [9]. Several methods have been proposed to solve it with primary contribution from Tuomas Sandholm [21].

The main methods proposed for the WDP can be summarized as follows: the Branch-on-Items (BoI) [18], the Branch on Bids (BoB) [20], and the Combinatorial Auctions BoB (CABoB) [19]. These methods can find reasonable optimal allocation with hundreds of items. Other methods are also investigated for the WDP such as: the CASS (Combinatorial Auction Structural Search) which is a Branch-and-Bound algorithm proposed by [9]. The authors in [14] proposed CA-MUS (Combinatorial Auctions Multi-Unit Search) which is a new version of the CASS for determining the optimal set of bids in general multi-unit combinatorial auctions. In [16], authors proposed a linear programming method. Anderson *et al.* proposed another exact algorithm based on integer programming [1].

Besides these exact algorithms, inexact methods are studied for the WDP. Among them, we cite Hybrid Simulated Annealing SAGII [10, 11], Casanova [12], Stochastic local search [5], Memetic algorithm [3] and Deferential Evolution [4].

3   The Proposed *Choice Function* Hyper-heuristic for the WDP

In the following we start with the main components of the proposed approach: the solution representation, the random key encoding strategy used to generate initial solutions and the objective function used to measure the quality of solutions. Then, we give the five low-level heuristics used by the Choice Function Hyper-heuristic.

3.1   The Solution Representation

A solution for the WDP is an allocation A which can be represented by a Vector with a variable length. Each of whose components A_i receives the winning bid number.

3.2   The Random Key Encoding

The initial solution is generated at random by using the strategy of the random key encoding (RK) introduced by Bean [2] and used mainly for ordering and scheduling problems. The RK encoding mechanism permits to generate and manipulate only feasible solutions. The Random Key Encoding operates as follows: we generate n real numbers sequenced by an r order, where n is the number of bids and the r order is a permutation of keys values. Initially we start with an empty allocation. Then we select the bid having the highest order value to include in the allocation. Secondly, the bid having the second-highest order value is accepted if its acceptance with accepted bid currently in the allocation verifies the constraint that means that for any good i, the sum of units of i over all the winning bids in the current allocation does not exceed mu_i, otherwise it is discarded. The process continues until having examined the n bids. We obtain a subset of bids that can be a feasible solution to the WDP.

3.3 The Objective Function

The objective function permits to measure the quality of an allocation of winning bids. The objective function is equal to the overall price of the winning bids of the allocation $A = \{B_1, B_2, \ldots, B_L\}$.

$$F(A) = \sum_{i=1}^{L} Price(B_i) = \sum_{i=1}^{L} P_i \tag{3}$$

where L is the number of element of the allocation A.

3.4 The Conflict Graph

To ensure feasibility of allocations during the heuristic search process, we have created a conflict graph where bids are vertices and edges connect bids that cannot be accepted together. This graph permits to detect directly the conflict bids that share a good. According to this graph we can ensure and maintain the feasibility of our allocations.

3.5 The Five Low-Level Heuristics for WDP

The Heuristic h_1. This heuristic h_1 is a local search method which starts with a solution X. Then it selects randomly from the current solution X a bid B_1. We choose after that a new bid not in X that is not in conflict with B_1 to be added in the current solution X. All conflicting bids that can be occurred in the current allocation are removed to maintain the feasible allocation. The process is iterated for a certain number of iterations in the hope to improve the quality of the solution X.

The Heuristic h_2. The mechanism used in the heuristic h_2 is the combination of the currently best solution with a current solution created according to the conflict graph. The resulting solution is mutated and then enhanced by using a local search method.

The Heuristic h_3. The heuristic h_3 is a one iteration of the well-known stochastic local search method (SLS). The heuristic h_3 is an iterative approach that starts with the best current allocation A. Then, it performs a certain number of local steps that consists in selecting a *bid* to be added in the current allocation A and in removing all conflicting bids that can be occurred in the current allocation. The bid to be accepted is selected according to one of the two following criteria:

1. The first criterion consists in choosing the bid in a random way with a fixed probability $wp > 0$.
2. The second criterion consists in choosing the best bid (the one maximizing the auctioneer's revenue when it is selected) to be accepted.

The Heuristic h_4. In the heuristic h_4, the mutation operator is applied on the current solution. The mutation is done with a certain probability called mutation rate. The resulting solution is improved by using a local search method.

The Heuristic h_5. In the heuristic h_5, we combine the best solution with a new solution generated randomly according to the key encoding strategy. As done in heuristic h_4, the resulting solution is improved by using a local search method.

3.6   The *Choice Function* Hyper-heuristic for WDP

The *Choice Function* hyper-heuristic consists of a selection method called *Choice Function* as well as a method of acceptance of solutions. The acceptance method validates only the new solutions that improve the current ones.

We note that *Choice function* is a score-based technique which assigns a weight to each low-level heuristic. Indeed, this technique allows us to measure the effectiveness of a low-level heuristic to decide which one should be selected for the next execution. This technique is based on three parameters which are: the CPU time consumed by an heuristic during the search process, the quality of the solution, and the time elapsed since the low level heuristic had been called.

In this work, we have used the same *Choice Function* defined in [8] and given as follows:

$$\forall i, g_1(h_i) = \sum_n \alpha^{n-1} \frac{I_n(h_i)}{T_n(h_i)}$$
$$\forall i, g_2(h_{ID}, h_i) = \sum_n \beta^{n-1} \frac{I_n((h_{ID}, h_i))}{T_n(h_{ID}, h_i)}$$
$$\forall i, g_3(h_i) = elapsedTime(h_i)$$
$$\forall i, score(h_i) = \alpha g_1(h_i) + \beta g_2(h_{ID}, h_i) + \delta g_3(h_i)$$
$$\alpha, \beta \in [0, 1], \delta \in R.$$

where h_i is a low-level heuristic and h_{ID} is the last low-level heuristic recently launched. α, β and δ values are fixed empirically.

The CFH method is sketched in Algorithm 1.

Algorithm 1. The CFH method.

Require: a WDP instance, a set of low-level heuristics, the choice function :
    HBN, α, β, δ, *maxiter*
Ensure: an allocation S
  1: Generate an initial solution S according to RK.
  2: Evaluate , the quality of the solution S
  3: $S = S*$; $F* = F$; // F* is the quality of the best solution S* found
  4: **for** $I = 1$ to *maxiter* **do**
  5:     For each heuristic, calculate its score by using the *choice function HBN*
  6:     Select the heuristic h_i having the highest score.
  7:     Apply the heuristic h_i on S to obtain new solution S with a quality F //
        solution acceptance method.
  8:     **if** (f(S)> f(S*)) **then**
  9:         S* = S; F*=F
 10:     **end if**
 11: **end for**
 12: **return** the best allocation found.

---

4   Experiments

To show the effectiveness of the proposed approach we compared it with the
stochastic local search [5] which we have implemented in order to do a fair
comparison.

The C programming language is used to implement the proposed approaches
for the WDP. The source codes are run on Intel COREI7, 8GB of RAM.

The adjustment of the different parameters is fixed by an experimental study.

The CFH parameters are fixed as follows: the maximum number of iterations
= 500 000, $\alpha = 0.9$, $\beta = 0.1$ and $\delta = 1.5$.

The SLS parameters are: the maximum number of iterations = 1000000 and
$wp = 0.3$.

4.1   Benchmarks

In this study, we used realistic data of various sizes consisting of up to 1500 items
and 1500 bids provided by Lau and Goh [13]. These data sets allow for several
factors such as a pricing factor, bidder preference factor and a fairness factors
in distributing items among bids. We noted that CPLEX was unable to solve
these problems within reasonable time[10, 11]. These instances can be divided
into five different groups of problems where each group contains 100 instances,
m is the number of items and n is the number of bids as following.

- G1: 100 instances from in101 to in200: $m=500$, $n=1000$.
- G2: 100 instances from in201 to in300: $m=1000$, $n=1000$.
- G3: 100 instances from in401 to in500: $m=1000$, $n=500$.

- G4: 100 instances from in501 to in600: $m=1000$, $n=1500$.
- G5: 100 instances from in601 to in700: $m=1500$, $n=1500$.

The input format is:
m n following n lines. The first number of each line i is the profit of bid i, following a number of integers, which are the items covered by the bid i.

4.2   A Comparison between CFH and SLS

Due to the non-deterministic nature of CFH and SLS algorithms, 10 runs have been considered for each instance and for each algorithm. The average and best results found by CFH are reported and compared with the best results found by SLS.

We give the comparative histograms in Fig. 1 to 5 to show the effectiveness of CFH in reaching good quality solutions on several instances of the 5 different groups of problems compared to SLS for the WDP.

We compared both the average and the best results found by CFH with the best results found by SLS. The CFH method succeeds in finding good results for all the checked instances. The average solutions found by CFH are better than the best ones found by SLS for almost the instances. In some cases, the best results found by SLS are better than the average solutions found by CFH.

When we compared the best results found by SLS and CFH, we remarked that CFH surpassed SLS on all the tested instances.

Fig. 1 to 5 show clearly the effectiveness of the proposed hyper-heuristic in finding good solutions to the WDP for all the checked instances.

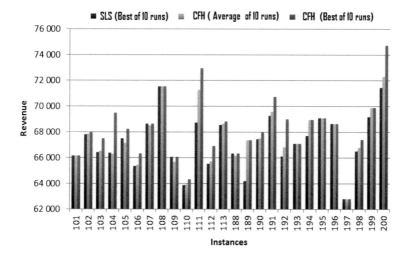

Fig. 1. CFH.vs. SLS on some instances from Group 1

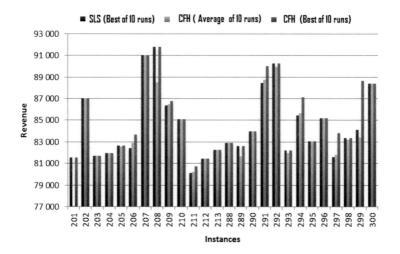

Fig. 2. CFH.vs. SLS on some instances from Group 2

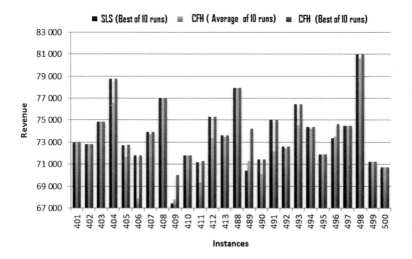

Fig. 3. CFH.vs. SLS on some instances from Group 3

We can conclude that CFH is much better than SLS on almost the tested instances. The effectiveness of the CFH method is due to the good *choice function* which permits to select adequate heuristics for the search process.

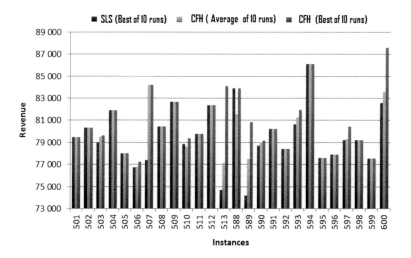

Fig. 4. CFH.vs. SLS on some instances from Group 4

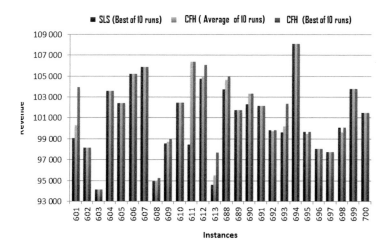

Fig. 5. CFH.vs. SLS on some instances from Group 5

4.3   Further Comparison with SLS

In this section, we compared CFH and SLS for the 500 test instances with the 5 different problem sizes. Each method computed the arithmetic average solution of the 100 instances in each group denoted μ. For each group of 100 instances, 10 runs have been considered.

Table 1 shows the numerical results found by CFH and SLS methods for the five groups of problems where $\mu_{average}$ corresponds to the average value of μ of the 10 runs for each group obtained by the CFH method and $time_{average}$ is the average time of the method in second.

The column μ_{best} of Table 1 gives the best value of μ of the 10 runs for each group, the column $time_{best}$ is the needed time in second to obtain the best solution and Imp is $(\mu_{CFH}- \mu_{SLS})/ \mu_{CFH}$ which represents the improvement in results of CFH in comparison to SLS.

Table 1. CFH.vs. SLS

	CFH				SLS		
	Average of 10 runs		Best of 10 runs		Best of 10 runs		$Imp\%$
Test Set	$time_{average}$	$\mu_{average}$	$time_{best}$	μ_{best}	$time_{best}$	μ_{best}	
REL-500-1000	32.06	66733.65	32.98	67407.44	30.49	66460.99	1.40
REL-1000-500	9.69	73137.22	9.00	73854.63	9.07	73541.06	0.42
REL-1000-1000	23.19	84147.87	25.23	84979.35	20.28	84108.96	1.02
REL-1000-1500	22.88	80414.73	22.41	80940.16	22.61	79975.69	1.19
REL-1500-1500	25.46	101071.41	24.99	101530.07	25.20	100828.27	0.69

When we compared the average results found by CFH with the best results of SLS, we remarked that CFH is better than SLS on the REL-500- 1000, REL-1000-1000, REL-1000-1500 and REL-1500-1500 groups of problems. For REL-1000-500 there is a slight difference between the two methods.

When, we examined the best results found by both CFH and SLS, we see clearly the superiority of CFH in finding good results for the five groups of problems.

As shown in Table 1, the CFH performs better than SLS on all the five groups of instances. It finds better solutions in reasonable CPU time on all the five checked benchmarks. Table 1 shows that CFH always gives a 0.42 to 1.40 percent improvement in results in comparison to SLS.

5   Conclusion

In this paper, we proposed a choice function hyper-heuristic for the winner determination problem in combinatorial auctions. The proposed method is implemented and tested on five groups of well-known benchmarks. According to the experimental study, the choice function hyper-heuristic was shown great performance and effectiveness in solving WDP. The results are very competitive. We plan to validate the proposed approach on other complex problems such Maximum satifiability (MAX-SAT), travelling tournament problem (TTP) and frequency assignment problem (FAP)in GSM networks.

References

[1] Anderson, A., Tenhunen, M., Ygge, F.: Integer programming for combinatorial auction winner determination. In: Proceedings of 4th International Conference on Multi-Agent Systems, pp. 39–46. IEEE Computer Society Press (July 2000)

[2] Bean, J.C.: Genetics and random keys for sequencing and optimization. ORSA Journal of Computing 6(2), 154–160 (1994)

[3] Boughaci, D., Benhamou, B., Drias, H.: A Memetic Algorithm for the Optimal Winner Determination Problem. Soft Computing - A Fusion of Foundations, Methodologies and Applications 13(8-9), 905–917 (2009)

[4] Boughaci, D.: A Differential Evolution Algorithm for the Winner Determination Problem in Combinatorial Auctions. In: Electronic Notes in Discrete Mathematics, vol. 36, pp. 535–542 (2010)

[5] Boughaci, D., Benhamou, B., Drias, H.: Local Search Methods for the Optimal Winner Determination Problem in Combinatorial Auctions. Journal of Mathematical. Modelling and Algorithms 9(2), 165–180 (2010)

[6] Burke, E.K., Hyde, M., Kendall, G., Ochoa, G., Ozcan, E., Woodward, J.R.: A classification of hyper-heuristic Approaches. In: International Series in Operations Research and Management Science (2010)

[7] Burke, E.K., Hyde, M.R., Kendall, G., Ochoa, G., zcan, E., Woodward, J.R.: Exploring Hyper-heuristic Methodologies with Genetic Programming. In: Collaborative Computational Intelligence (2009)

[8] Burke, E.K., Hyde, M.R., Kendall, G., Ochoa, G., zcan, E., Qu, R.: Hyperheuristics: A Survey of the State of the Art. Technical Report, School of Computer Science and Information Technology, University of Nottingham (2010)

[9] Fujishima, Y., Leyton-Brown, K., Shoham, Y.: Taming the computational complexity of combinatorial auctions: optimal and approximate approaches. In: Sixteenth International Joint Conference on Artificial Intelligence, pp. 48–53 (1999)

[10] Guo, Y., Lim, A., Rodrigues, B., Zhu, Y.: Heuristics for a brokering set packing problem. In: Proceedings of Eighth International Symposium on Artificial Intelligence and Mathematics, pp. 10–14 (2004)

[11] Guo, Y., Lim, A., Rodrigues, B., Zhu, Y.: Heuristics for a bidding problem. Computers and Operations Research 33(8), 2179–2188 (2006)

[12] Hoos, H.H., Boutilier, C.: Solving combinatorial auctions using stochastic local search. In: Proceedings of the 17th National Conference on Artificial Intelligence, pp. 22–29 (2000)

[13] Lau, H.C., Goh, Y.G.: An intelligent brokering system to support multi-agent webbased 4th-party logistics. In: Proceedings of the 14th International Conference on Tools with Artificial Intelligence, pp. 54–61 (2002)

[14] Leyton-Brown, K., Tennenholtz, M., Shoham, Y.: An Algorithm for Multi-Unit Combinatorial Auctions. In: Proceedings of the 17th National Conference on Artificial Intelligence, Games 2000, Bilbao, and ISMP 2000, Austin, Atlanta (2000)

[15] McAfee, R., McMillan, P.J.: Auctions and bidding. Journal of Economic Literature 25, 699–738 (1987)

[16] Nisan, N.: Bidding and allocation in combinatorial auctions. In: Proceedings of ACM Conference on Electronic Commerce (EC 2000), pp. 1–12. ACM SIGecom, ACM Press (October 2000)

[17] Rothkopf, M.H., Pekee, A., Ronald, M.: Computationally manageable combinatorial auctions. Management Science 44(8), 1131–1147 (1998)

[18] Sandholm, T.: Algorithms for Optimal Winner Determination in Combinatorial Auctions. Artificial Intelligence 135(1-2), 1–54 (1999)

[19] Sandholm, T., Suri, S., Gilpin, A., Levine, D.: CABoB: a fast optimal algorithm for combinatorial auctions. In: Proceedings of the International Joint Conferences on Artificial Intelligence, pp. 1102–1108 (2001)

[20] Sandholm, T., Suri, S.: Improved Optimal Algorithm for Combinatorial Auctions and Generalizations. In: Proceedings of the 17th National Conference on Artificial Intelligence, pp. 90–97 (2000)

[21] Sandholm, T.: Optimal Winner Determination Algorithms. In: Cramton, P., et al. (eds.) Combinatorial Auctions. MIT Press (2006)

[22] de Vries, S., Vohra, R.: Combinatorial auctions a survey. INFORMS Journal of Computing 15, 284–309 (2003)

Automatic Generation of Heuristics for Constraint Satisfaction Problems

José Carlos Ortiz-Bayliss[1], Jorge Humberto Moreno-Scott[2],
and Hugo Terashima-Marín[2]

[1] Automated Scheduling, Optimisation and Planning (ASAP)
School of Computer Science, University of Nottingham, UK
Jose.Ortiz_Bayliss@nottingham.ac.uk
[2] Tecnológico de Monterrey, Campus Monterrey, Mexico
Jorge.Moreno@arrisi.com, terashima@itesm.mx

Abstract. The constraint satisfaction problem (CSP) is a generic problem with many applications in different areas of artificial intelligence and operational research. When solving a CSP, the order in which the variables are selected to be instantiated has a tremendous impact in the cost of finding a solution. In this paper we explore a novel type of heuristic that combines different features that describe the current state of the instance to decide which variable to instantiate next. A generational genetic algorithm is used to automatically tune the parameters used by these new heuristics. This paper contributes to the development of new heuristics that can be either very specialized to one class of instances, or general enough to deal with different classes of instances with an acceptable performance.

Keywords: Constraint Satisfaction, Heuristics, Genetic Algorithms.

1 Introduction

A CSP is defined by a set of variables X, where each variable is associated a domain D of values subject to a set of constraints C [39]. The goal is to find a consistent assignment of values to variables in such a way that all constraints are satisfied, or to show that a consistent assignment does not exist. There is a wide range of theoretical and practical applications of CSPs that include scheduling, frequency assignment, micro-controller selection and pin assignment, among others (see for example [13] and [3]).

In CSPs, the selection of the next variable to instantiate determines the way the solution space is explored. Different orderings for instantiation of the variables produce different exploration patterns, and different patterns have different costs[1]. Then, if we decide correctly, we can find a solution which cost is

---

[1] In this research, we refer to the cost of finding a solution, not the cost of the solution itself. All the solutions to one instance are equally valid because the problem is treated as a combinatorial one. The cost of the search can be measured in terms of time, expanded nodes or consistency checks, to mention some.

G. Terrazas et al. (eds.), *Nature Inspired Cooperative Strategies for Optimization*
(NICSO 2013), Studies in Computational Intelligence 512,
DOI: 10.1007/978-3-319-01692-4_24, © Springer International Publishing Switzerland 2014

smaller than the others. The general idea in this investigation is to describe a methodology to automatically produce variable ordering heuristics based on the combination of the criteria of existing ones. The approach produces a linear combination of the features that describe the variables within a CSP instance and uses that combination to rank each variable and decide the next one to be instantiated.

This paper is organized as follows. Section 2 presents a brief survey of works relevant to this research. In Sec. 3 some important heuristics and the features they use to order the variables are discussed. The methodology proposed is described in detail in Sec. 4. The experimental set up and the main results are presented in Sec. 5. Finally, the conclusion and future work of this investigation are discussed in Sec. 4.

2   Background

In the last years we have witnessed a rapid growth of developments oriented to improve how heuristics work. Two trends are clearly identified: methods that optimize the use of existing heuristics and methods that construct new heuristics. Regarding the methods that optimize the use of existing heuristics they produce a mapping between the states of the problem and a feasible heuristic. These methodologies maintain a set of heuristics and then, as the problem changes, decide which heuristic to apply. Examples of these methodologies include dynamic algorithm portfolios like CP-Hydra [28, 31] and ACE [10] and selective hyper-heuristics [11, 29, 37]. On the other hand, methodologies that produce new heuristics identify critical parts of existing heuristics to produce new ones [7, 7]. In this paper we will explore the second trend, the one that produces new heuristics based on some components of existing ones.

Our approach is related to the automated parameter tuning problem, which consists in adjusting the parameters of an algorithm without the intervention of the user [22]. As we will see later, the approach proposed represents heuristics as functions. These functions are very similar to the ones used by linear regression (in linear regression these functions are known as hypotheses). Although the representation of the functions is similar, the mechanisms to adjust the parameters are completely different. In the case of linear regression, as a supervised learning mechanism, it requires training examples. In our approach, we do not need any training examples, because our model is unsupervised. To perform the tuning of the parameters, we use a generational genetic algorithm [11, 13].

In the domain of CSPs, one of the first ideas about automatic heuristic generation was proposed by Minton et al. [16]. Their system, MULTI-TAC, produced programs that represented heuristics designed for systematic algorithms. More recently, Bain et al. [2, 3] proposed the use of genetic programming to generate heuristics for CSPs. The authors proposed a representation that allows the generation of heuristics by combining individual functions and terminals that required some existing heuristics to be broken down into their component parts. The main difference between their work and the one presented in this document is the set of features used to construct the heuristics. Bain et al. [2, 3] use features that obtain information from the whole instance, while we use information of each individual variable. Although both approaches rely on evolutionary algorithms to produce heuristics, the representation of the heuristics produced is completely different for both approaches.

3   Descriptors and Variable Ordering Heuristics

Although heuristics that decide the next variable to instantiate are referred to as "variable ordering heuristics" in the literature, we do not think this is the correct term to describe what these heuristics currently do. The term comes from the first works on CSPs where the variables were ordered before the search and the order was kept until the search was over. Nowadays, this 'ordering' is performed via a dynamic fashion, where the heuristics decide, at each stage of the search, which variable will be instantiated next. Then, if a heuristic orders all the variables at a certain point of the search, it will only use the first one of the list, because it has no guarantee that once that variable is instantiated, the ordering at the next stage of the search will remain as it was in the previous one. The properties of the instances change as the search progresses and so the decisions made by the heuristics. For this reason, and to help explaining some concepts in this investigation, we will assume that all the variable ordering heuristics described in this document return only one variable at the time, the next one to be instantiated. This assumption simplifies our analysis without loss of generality.

Require: $X = \{x_0, x_1 \ldots, x_n\} , f(x)$
    $[\text{index, value}] \leftarrow \min(f(x_0), f(x_1), \ldots, f(x_n)))$
    return x_{index}

Fig. 1. Generic heuristic model

In a general way, we can see any variable ordering heuristic as a procedure that receives a set of uninstantiated variables X and a heuristic function $f(x)$; and returns the variable to instantiate. Thus, each specific heuristic ranks the variables in X according to the values returned by $f(x)$, $\forall x \in X$. Depending on how the heuristic is designed, the heuristic will prefer variables with large values of $f(x)$ over smaller ones, or vice versa. By changing the sign of the values returned by $f(x)$ we can automatically invert the preference of the heuristic. With this idea on mind, we propose the generic heuristic model shown in Fig. 1. Given the proper function $f(x)$, this generic heuristic interpreter is able to represent any specialized heuristic. For example, the min-domain heuristic [4] prefers the variable with the minimum domain size. In our generic heuristic model, the heuristic function for min-domain is $f(x) = dom(x)$, where $dom(x)$ returns the domain size of variable x. Thus, min-domain will instantiate first the variable that minimizes $f(x)$ among all the uninstantiated variables. If we decide to select the variable that maximizes $f(x)$, we obtain a different heuristic known as max-domain but, based also on the domain size. By using the generic heuristic interpreter described in Fig. 1, max-domain should be implemented by the function $f(x) = -dom(x)$. Similar functions can be defined for other variable ordering heuristics by using different features. This simple example shows how we can represent two simple heuristics by using our generic heuristic model and the proper function $f(x)$. In the case of min-domain and max-domain, $f(x)$ only considers one feature of the variables, the domain size. To produce more complex heuristics, more information needs to be obtained from the variables. Each one of these pieces of information is gathered by what we call 'descriptors' that extract information from the variables at a certain point of the search. In the next lines we will present the descriptors proposed for this investigation.

Constraint Density, $p_1(x)$. The constraint density of a variable is defined as the proportion of constraints with other variables over the maximum number of bidirectional constraints the variable can participate in. Given a CSP instance with n variables, the maximum number of bidirectional constraints for a variable is $n-1$. No unary constraints are considered for this calculation. Then, $p_1(x)$ is calculated as the degree of the variable, $deg(x)$ (the number of constraints with other uninstantiated variables) over the maximum number of possible constraints: $p_1(x) = deg(x)/(n-1)$. If we select the variable with the largest constraint density we obtain the max-density heuristic (also known as deg [38]), which prefers the variable involved in the largest number of constraints.

Constraint Tightness, $p_2(x)$. The constraint tightness indicates the proportion of conflicts within the constraints in which the variable is involved. A conflict is a pair of values $\langle a, b \rangle$ that is not allowed for two variables at the same time. Should we prefer the variable with the largest constraint tightness we would obtain the max-tightness heuristic.

Domain size, $\hat{dom}(x)$. As mentioned before, selecting the variable that minimizes $dom(x)$ gives place to the min-domain heuristic [4]. To restrict the range to the interval $[0, 1]$, we use $\hat{dom}(x)$ instead of $dom(x)$, where $\hat{dom}(x)$ is defined as the domain size of variable x divided by the maximum domain size among all the currently uninstantiated variables. This normalization does not modify the behaviour of the heuristic but improves the automatic learning process.

Kappa, $\hat{\kappa}(x)$. Inspired in the κ factor that estimates how restricted a problem is [20], we propose a similar measurement to be used as a descriptor for each variable. $\kappa(x)$ is calculated as:

$$\frac{-\sum_{c_j \in C_x} \log_2(1 - p_{c_j})}{\log_2(dom(x))} \tag{1}$$

where c_j is a constraint where x is involved and prohibits a fraction p_{c_j} of tuples in the constraint. If we prefer the variable that maximizes the value of $\kappa(x)$, we obtain the max-kappa heuristic [20]. To normalize the values of $\kappa(x)$

we proposed that, for values of $\kappa(x) \leq 5$, the normalized value $\hat{\kappa}(x) = \kappa(x)/5$. Larger values of $\kappa(x)$ will produce $\hat{\kappa}(x) = 1$. This normalization was inspired in results obtained from preliminary studies with the descriptors.

Additionally to these heuristics, we have also included the standard heuristic min-domain/max-density (referred to as dom/deg in [4]). This heuristic became popular because of its simplicity and because it combines two features by using a quotient, $f(x) = \hat{dom}(x)/p_2(x)$. We have decided also to include this heuristic for the experimental phase because of the fact that it is indeed, a heuristic that exploits the use of more than one descriptor to discriminate among variables. Thus, a total of five variable ordering heuristics have been considered for this investigation: max-density, max-tightness, min-domain, max-kappa and min-domain/max-density. In addition to these variable ordering heuristics, the values are ordered according to the min-conflicts heuristic [26]. Once a variable is selected for instantiation, the first value to be tried is the one more likely to success, the one that participates in the fewer conflicts (forbidden pairs of values between two variables). In all cases, lexical ordering is used to break ties.

4   Automatic Generation of Heuristics

We have observed that different descriptors and their combinations allow us to produce different heuristics. Nevertheless, it is not clear which descriptor is more suitable to describe the current problem state and then, make a good decision about the next variable to instantiate. In this paper we propose a new heuristic representation that uses the linear combination of the descriptors presented in Sec. 3 to decide which variable to try next. All the descriptors are used to obtain information about the variables, but a vector of weights determines the importance of the descriptors to make the decision. Let $s(x)$ be the vector that contains all the values of the descriptors for variable x at a certain point of the search, and w a vector of weights (the tuned parameters of the heuristic). Thus, we define the heuristic function as: $f(w, s(x)) = w \cdot s(x)$. As we can observe, the values of the vector of weights w are the same for all the variables, regardless of the instance. What changes at each step of the search are the values of the vector of descriptors $s(x)$. Each heuristic makes decisions based on its internal heuristic function and the heuristic generic interpreter that decides how to deal with the values of the heuristic function. Depending on how the heuristic is designed, it may prefer large or small values of $f(x)$. The representation proposed for the heuristic function allows both preference schemes. Let us assume that our generic heuristic model prefers the variables with small values of $f(w, s(x))$. Then, min-domain (which prefers the variable with the smallest domain size) will be defined by $w = (1)$ and $s(x) = (\hat{dom}(x))$. When $f(w, s(x))$ is calculated, variables with small domain sizes will return small values, and because the heuristic prefers small values of $f(w, s)$, it will behave exactly as min-domain. On the contrary, if $w = (-1)$, variables with large domain sizes will obtain large negative values from the heuristic function $f(w, s(x))$ and the heuristic will behave as max-domain.

All the weights in w lie in the range $[-1, 1]$. Also, all the components of the vector of descriptors $s(x)$ lie in the range, $[0, 1)$. Because of this, given k descriptors to define the state of the variables, the values $f(w, s(x))$ can produce are always in the range $(-k, k)$.

4.1 The Genetic Algorithm

To generate new heuristics, we propose the use of a genetic algorithm to adjust the values in w according to the set of descriptors provided. In our implementation, a generational genetic algorithm with memory was used, but other implementations may be considered in the future (for example a steady state genetic algorithm or a messy one). The memory is implemented as a mechanism to always keep the best configuration so far. In this way, if the evolutionary process removes a good individual from the population, we can always use the memory to recall that the best configuration was found before but it is no longer part of the population.

In our genetic algorithm, each individual encodes the values of the vector of weights w. Then, each individual determines the way in which the descriptors are to be considered by the heuristic function $f(x)$ and how the heuristic will behave. The individuals are coded by using binary strings, as in standard genetic algorithms. This representation allows us to use standard crossover and mutation operators. Each weight in w is given 10 bits of the individual. As we can see, the length of the individuals is fixed to $10k$ given the number of descriptors, k (where $k = 4$ in this investigation). Because there are 10 bits to represent each weight of the descriptor, there are 1024 possible values that can be represented. We divided the range $[-1, 1)$ in 1024 uniform steps and (each one of 0.001953125) and according to the value coded in the individual for that weight, the individual is interpreted. For example, if the string corresponding to one component in w is 1100110110 (822 in base 10), the decimal value of that weight is calculated as $-1 + (2/1024 \times 822)$, which results in -0.6055. When the initial population is created, all the individuals are randomly initialized ('0' and '1' have the same probability of occurring in the string). The fitness of the individuals is calculated as the inverse of the cost of using the vector of weights w coded in that individual to solve all the instances in a training set. The cost is given in terms of consistency checks (the number of revisions of the constraints). Thus, the best individual should minimise the number of consistency checks to solve the whole set. Three genetic operators are used in this investigation. Tournament selection of size two is used to select to parents for crossover. Once the parents have been selected, there is a probability of 0.9 that they are combined. If crossover takes place, the parents are combined by using a standard one-point crossover operator and the new individuals are included in the new population. If crossover does not occur, the parents are incorporated to the new population without any changes. For mutation, all the bits in the strings have the same probability of being affected, 0.0001. When this is the case, the value of the bit is changed to its complement.

5   Experiments and Results

In this section we present the instances used and a detailed description of the experiments conducted. In all cases, the CSP solver used is one implemented in Java by the authors. For all the experiments, the constraint propagation method used was AC3 [25], while backjumping [11] was always used as the strategy for backtracking. All the experiments were conducted on an Intel 8 Core Windows machine with 16 GB of memory.

5.1   Description of the Instances Used

For this research we have only considered those CSPs in which the domains are discrete, finite sets and the constraints involve only one or two variables (binary constraints). Rossi et al. [21] proved that for every general CSP there is an equivalent binary CSP. Thus, all general CSPs can be reduced into a binary CSP. All the random instances used in this investigation were produced with model F. In model F [14], we select uniformly, independently and with repetitions, a proportion $p_1 \times p_2$ conflicts out of the $m^2 n(n-1)/2$ possible. We then generate a constraint between each pair of connected variables in the graph until we have exactly $p_1 \times n \times (n-1)/2$ edges and throw out any conflicts that are not between connected variables in this graph. Model F has proved to be one of the most robust random CSPs generators because it is a generalization of the well studied model E [1]. The values of p_1 and p_2 used for the generation of the instances should not be confused with the descriptors $p_1(x)$ and $p_2(x)$. In model F, p_1 and p_2 determine the constraint density and tightness of the instance generated while our descriptors provide information regarding each variable.

With model F we produced three simple classes of random instances based on the constraint density (sparse or dense) and tightness (high or low) of the whole instance. All the instances have 25 variables and 10 values in their domains. The values of p_1 and p_2 were randomly selected according to one of the following: class A (sparse constraints, low tightness) $p_1 = [0.2, 0.3]$, $p_2 = [0.2, 0.3]$; class B (dense constraints, low tightness) $p_1 = [0.7, 0.8]$, $p_2 = [0.2, 0.3]$ and class C (dense constraints, high tightness) $p_1 = [0.7, 0.8]$, $p_2 = [0.7, 0.8]$. Sets A and B both contain instances with low tightness while classes B and C contain instances with dense constraints. For each class we produced two sets, one for training the new heuristics and the other used exclusively for testing purposes. Each set is named according to the class of instances it contains. Then, we produced six instance sets: training sets A, B and C; and test sets A, B and C. Each training set contains 25 instances while the test sets contain 500 instances each. Thus, a total of 1575 instances were generated and analysed in this research.

5.2   Generating New Heuristics

For the first experiment we produced three heuristics for each set of instances. Each run of the genetic algorithm produces a heuristic. Thus, nine runs of the genetic algorithm were conducted to obtain the nine heuristics analysed in the first experiment. For each run, the genetic algorithm ran for 50 generations with a population size of 30 individuals. The heuristics produced were trained with instances from one specific class. Table 1 shows the results obtained from this experiment. Each cell in the table indicates the percentage of consistency checks saved or added by using each heuristic produced with respect to the best standard heuristic for each set. For example, H_{A1}, which is the first heuristic produced by our approach when set A was for training, reduced the number of consistency checks required by the best heuristic for class A in 5.08% (the max-density heuristic obtained the best results on class A). The same heuristic, H_{A1}, when applied to test set A achieved a reduction of 4.52% with respect to the best heuristic. Negative numbers indicate a reduction (the heuristic produced was better than the best standard heuristic) and positive numbers indicate additional consistency checks with respect to the best standard heuristic. The best results from the heuristics produced by our approach are marked in bold. The best heuristic for class A was max-density, both on the training and test set. Training set B was best solved by using min-domain/max-density, but max-kappa showed the best performance on test set B. Max-kappa obtained the best results for class C among all the standard heuristics.

Table 1. Performance of each heuristic produced against the best standard heuristic on each set (positive numbers indicate the percentage of consistency checks saved with respect to the best heuristic and negative numbers indicate the percentage of additional consistency checks with respect to the best heuristic)

	Training			Test		
Heuristic	Set A	Set B	Set C	Set A	Set B	Set C
H_{A1}	**-5.08%**	1056.17%	-2.04%	**-4.52%**	381.61%	-1.03%
H_{A2}	**-5.09%**	283.01%	-2.04%	**-4.25%**	166.83%	-2.18%
H_{A3}	**-5.20%**	309.27%	-2.04%	**-4.28%**	173.00%	-2.14%
H_{B1}	7.34%	**-8.79%**	5.46%	6.36%	**-20.16%**	2.06%
H_{B2}	7.83%	**-10.95%**	3.98%	6.51%	**-19.72%**	2.03%
H_{B3}	7.24%	**-10.28%**	0.58%	6.50%	**-24.33%**	1.63%
H_{C1}	19.08%	22.41%	**-7.40%**	15.86%	11.41%	**-4.70%**
H_{C2}	20.76%	21.65%	**-7.34%**	15.75%	16.12%	**-5.23%**
H_{C3}	15.72%	18.09%	**-7.65%**	12.89%	-1.52%	**-4.86%**

It is not surprising that heuristics trained for one specific class obtain the best results on that class in most of the cases, both on training and test sets. The approach proposed is able to produce very competent heuristics for specific classes of instances, but fails (in general) to produce heuristics that can be applied to different classes of instances. These heuristics seem to deal correctly with unseen instances of the same class that was used for training, but tend to perform poorly when tested on instances from other classes. In general, the approach produces very specialized heuristics, suitable for instances of the same class used for training. This is useful if we have, in advance, some idea about the classes of instances that are required to be solved. Thus, we have found evidence that supports the idea that the approach can be used to train on a small subset of instances from a specific class and then, use the specialized heuristic to solve the rest of the instances.

5.3   Generalizing Heuristics

In the previous experiment we found evidence that supports the idea that the approach produces competent heuristics for specific classes of instances. Now, the question is how to make them more general and reusable for different classes of instances.

Heuristics produced by using only one class during the training phase suffer from being over-specialized for such class. They have problems to generalize and being reusable for other classes of instances. In this experiment, we deal with this problem by using combination of classes during the training. In this way, six composed sets of instances were constructed with the instances defined in Sec. 5.1. Training sets AB, BC, and ABC; and test sets AB, BC and ABC were constructed to be used in this new experiment. The sets are formed by the instances used in the previous experiment. For example, test set BC contains the instances from training sets B and C. As we mentioned before, some classes share some properties (for example, classes A and B contain instances with low tightness constraints). We want to observe if the heuristics trained with these new sets are better adapted for dealing with instances of different classes. Also, training set ABC and test set ABC were generated for this experiment. Sets ABC, contain all the instances used so far. We have the belief that heuristics trained on this set would be less specialized (less reduction with respect to the best heuristic, but very consistent among the three classes of instances). As in the previous experiment, three heuristics were produced for each composed set.

Table 2. Performance of each heuristic produced against the best standard heuristic on the composed sets (positive numbers indicate the percentage of consistency checks saved with respect to the best heuristic and negative numbers indicate the percentage of additional consistency checks with respect to the best heuristic)

Heuristic	Training			Test		
	Set AB	Set BC	Set ABC	Set AB	Set BC	Set ABC
H_{AB1}	**-6.49%**	-4.75%	-4.86%	**-3.15%**	-2.82%	-2.77%
H_{AB2}	**-6.68%**	-5.01%	-4.96%	**-9.85%**	-7.23%	-6.73%
H_{AB3}	**-9.96%**	-6.32%	-6.66%	9.81%	6.15%	5.25%
H_{BC1}	**-7.12%**	-5.24%	-5.19%	**-10.05%**	-7.38%	-6.94%
H_{BC2}	**-8.23%**	-5.31%	-5.76%	0.90%	0.33%	**-0.09%**
H_{BC3}	**-9.71%**	-5.62%	-5.28%	**-20.86%**	-13.65%	-12.58%
H_{ABC1}	**-8.85%**	-5.73%	-6.08%	**-4.39%**	-3.24%	-3.36%
H_{ABC2}	**-8.87%**	-5.82%	-6.09%	**-4.97%**	-3.66%	-3.72%
H_{ABC3}	**-9.63%**	-5.57%	-5.23%	**-17.87%**	-11.57%	-10.63%

The results suggest that sets AB (both training and test) are very easy to solve by most of the heuristics produced when the genetic algorithm uses the composed training sets as inputs. Only H_{AB3} and H_{BC2} fail to generalize on test set AB, where they do not achieve reductions in the cost when compared against the best heuristic for the same set.

With the change in the training sets, the approach produces heuristics that, in general, reduce the cost of the search on all the instances used for training, regardless of the set used. Nevertheless, when analysed on the test sets, these heuristics are not always as competent as they were on the training sets (see for example H_{AB1}, H_{AB2} and H_{AB3}) The problem with these heuristics is that they are very specialized for some hard instances in the training sets. For example, if one heuristic gets specialized for the hardest instances in one of the sets, it is likely that that heuristic also performs well on a composed set that contains those instances. But, once we present the heuristic new instances, the heuristic may be so specialized for the hard instances it has already solved that fails to solve the new instances.

In order to understand how these heuristics perform, we present a brief analysis of H_{BC3} and H_{ABC3}. These heuristics obtained the most relevant results on the composed sets.

$$
\begin{aligned}
H_{BC3} &= -(0.877)p_1(x) + (0.410)p_2(x) + (0.283)\hat{dom}(x) - (0.193)\hat{\kappa}(x) \\
H_{ABC3} &= -(0.994)p_1(x) + (0.547)p_2(x) + (0.326)\hat{dom}(x) - (0.232)\hat{\kappa}(x)
\end{aligned}
\tag{2}
$$

These heuristics have very similar vectors of weights and then, we expect their decisions to be similar. H_{BC3} and H_{ABC3} should be interpreted in the following way. First, recall that the general heuristic model used in this investigation is using a minimisation approach. Because it has the largest coefficient, $p_1(x)$ is the most relevant descriptor, followed by $p_2(x)$. Because of the differences in the values of the two most important coefficients (a proportion of 2 to 1), most of the decisions will be mainly based on the values of $p_1(x)$ of the remaining variables. Nevertheless, the combination of the values of the remaining descriptors makes these heuristics to behave in a different way than the max-density heuristic. In a very abstract way (and based only on the signs of the coefficients), these heuristics will instantiate first the variable involved in the largest number of constraints, with the fewer conflicts among the constraints where it is involved, with a small domain and with a large value of $\hat{\kappa}(x)$. Because these heuristics are combining the criteria of the single heuristics we expect the combination to be somehow similar to the decisions made by the single heuristics (see Sec. 3 for more details). It is interesting to notice that the way to consider the constraint tightness is inverse to the recommendation of the single and accepted heuristic (max-tightness prefers the variables with the largest values of this descriptor). Also, we are presenting a very simple and straight forward interpretation of the heuristic. The exact interactions between the descriptors –defined by the values of the weights and not only their signs– is what gives the new heuristics their strength.

6   Conclusion and Future Work

This paper describes a methodology to produce heuristics for variable ordering within CSPs by using components of standard heuristics taken from the literature. The components of these standard heuristics are the descriptors of the variables. The approach is able to generate specialized heuristics that reduce the cost of the search when compared against good quality standard heuristics on the same instances, but more research is still needed to prove the real advantages and limits of the approach. The genetic algorithm produces heuristics that perform well on unseen instances of the same classes used for training. We observed that including more diverse instances in the training sets is, in general, a way to produce more flexible heuristics that are capable of performing well on different classes of instances.

We have represented the heuristics as general procedures that take two inputs: the set of uninstantiated variables and a heuristic function. In this investigation, we used a generic heuristic interpreter for a minimisation problem. One important question derives from the use of this interpreter for the heuristic function. How would the use of a more complex interpreter affect the performance of the heuristic? Let us assume that the generic heuristic model prefers large values of $f(x)$ at the beginning of the search and small ones for the last stages. This slightly modified version of the generic heuristic interpreter can produce an algorithm that switches from one heuristic to the opposite one without any other change to the model. We consider this flexibility of the approach another important contribution of this investigation.

By analysing the results, we have identified a potential drawback in the current implementation of the genetic algorithm to update the vector of weights. The fitness function, as it is currently designed, aims to reduce the cost of a heuristic on the instances in the training set. This seemed to be the 'natural' way to measure the quality of the heuristics. But, this approach tends to produce over-fitted heuristics for the instances in the training set. The reason is simple: attempting to reduce the overall cost, the heuristic that best solves the hardest instances in the training set receives the highest values from the fitness function. Thus, if there is one hard instance in the training set, the fitness function will reward competent heuristics for such instance, regardless of their performance on other instances from the training set that may be easier to solve. The problem with the fitness function is even more important when we deal with composed training sets. In general, the use of composed sets facilitates the generalization of heuristics. But, if there is the case that a small subset of really hard instances are contained in two or more of the sets that form the composed one, and the evolutionary process specializes the heuristics for such instances, then we can expect those heuristics to perform well on the training sets but to fail to generalize to unseen instances. In this case, the heuristics only learn how to solve well a very small subset of specific hard instances. We consider to change the design of the fitness function as part of the future work.

Finally, we are interested in extending our results to include real instances to prove the effectiveness of the model to solve structured instances.

Acknowledgments. This research was supported in part by ITESM Strategic Project PRY075 "Intelligent Learning for Pattern Recognition and its Application in Medicine and Logistics" and the CONACyT Project under grant 99695.

References

[1] Achlioptas, D., Molloy, M.S.O., Kirousis, L.M., Stamatiou, Y.C., Kranakis, E., Krizanc, D.: Random constraint satisfaction: A more accurate picture. Constraints 6(4), 329–344 (2001)

[2] Bain, S., Thornton, J., Sattar, A.: Evolving algorithms for constraint satisfaction. In: Congress on Evolutionary Computation 2004 (CEC 2004), vol. 1, pp. 265–272 (2004a)

[3] Bain, S., Thornton, J., Sattar, A.: Methods of automatic algorithm generation. In: Zhang, C., Guesgen, H.W., Yeap, W.K. (eds.) PRICAI 2004. LNCS (LNAI), vol. 3157, pp. 144–153. Springer, Heidelberg (2004)

[4] Berlier, J., McCollum, J.: A constraint satisfaction algorithm for microcontroller selection and pin assignment. In: Proceedings of the IEEE SoutheastCon 2010 (SoutheastCon), pp. 348–351 (2010)

[5] Bessière, C., Régin, J.C.: Mac and combined heuristics: Two reasons to forsake FC (and CBJ) on hard problems. In: Freuder, E.C. (ed.) CP 1996. LNCS, vol. 1118, Springer, Heidelberg (1996)

[6] Burke, E.K., Hyde, M.R., Kendall, G., Ochoa, G., Ozcan, E., Woodward, J.R.: Exploring hyper-heuristic methodologies with genetic programming. In: Mumford, C.L., Jain, L.C. (eds.) Computational Intelligence. ISRL, vol. 1, pp. 177–201. Springer, Heidelberg (2009)

[7] Burke, E.K., Hyde, M.R., Kendall, G., Woodward, J.: Automatic heuristic generation with genetic programming: evolving a jack-of-all-trades or a master of one. In: Proceedings of the 9th Annual Conference on Genetic and Evolutionary Computation, GECCO 2007, pp. 1559–1565. ACM, New York (2007)

[8] Crawford, B., Soto, R., Castro, C., Monfroy, E.: A hyperheuristic approach for dynamic enumeration strategy selection in constraint satisfaction. In: Ferrández, J.M., Álvarez Sánchez, J.R., de la Paz, F., Toledo, F.J. (eds.) IWINAC 2011, Part II. LNCS, vol. 6687, pp. 295–304. Springer, Heidelberg (2011)

[9] Dunkin, N., Allen, S.: Frequency assignment problems: Representations and solutions. Tech. Rep. CSD-TR-97-14, University of London (1997)

[10] Epstein, S.L., Freuder, E.C., Wallace, R.J., Morozov, A., Samuels, B.: The adaptive constraint engine. In: Van Hentenryck, P. (ed.) CP 2002. LNCS, vol. 2470, pp. 525–542. Springer, Heidelberg (2002)

[11] Gaschnig, J.: Experimental case studies of backtrack vs. waltz-type vs. new algorithms for satisficing assignment problems. In: Proceedings of the Canadian Artificial Intelligence Conference, pp. 268–277 (1978)

[12] Gent, I., MacIntyre, E., Prosser, P., Smith, B., Walsh, T.: An empirical study of dynamic variable ordering heuristics for the constraint satisfaction problem. In: Freuder, E.C. (ed.) CP 1996. LNCS, vol. 1118, pp. 179–193. Springer, Heidelberg (1996)

[13] Holland, J.: Adaptation in Natural and Artificial Systems. The University of Michigan Press (1975)

[14] MacIntyre, E., Prosser, P., Smith, B.M., Walsh, T.: Random constraint satisfaction: Theory meets practice. In: Maher, M.J., Puget, J.-F. (eds.) CP 1998. LNCS, vol. 1520, pp. 325–339. Springer, Heidelberg (1998)

[15] Mackworth, A.K.: Consistency in networks of relations. Artificial Intelligence 8(1), 99–118 (1977)
[16] Minton, S.: An analytic learning system for specializing heuristics. In: Proceedings of the 13th International Joint Conference on Artificial Intelligence (IJCAI 1993), pp. 922–929. Morgan Kaufmann (1993)
[17] Minton, S., Johnston, M.D., Phillips, A., Laird, P.: Minimizing conflicts: A heuristic repair method for CSP and scheduling problems. Artificial Intelligence 58, 161–205 (1992)
[18] O'Mahony, E., Hebrard, E., Holland, A., Nugent, C., O'Sullivan, B.: Using case-based reasoning in an algorithm portfolio for constraint solving. In: Proceedings of the 19th Irish Conference on Artificial Intelligence and Cognitive Science (2008)
[19] Ortiz-Bayliss, J.C., Terashima-Marín, H., Conant-Pablos, S.E.: Learning vector quantization for variable ordering in constraint satisfaction problems. Pattern Recognition Letters 34(4), 423–432 (2013)
[20] Petrovic, S., Qu, R.: Case-based reasoning as a heuristic selector in a hyper-heuristic for course timetabling problems. In: Proceedings of the 6th International Conference on Knowledge-Based Intelligent Information Engineering Systems and Applied Technologies (KES 2002), vol. 82, pp. 336–340 (2002)
[21] Rossi, F., Petrie, C., Dhar, V.: On the equivalence of constraint satisfaction problems. In: Proceedings of the 9th European Conference on Artificial Intelligence, pp. 550–556 (1990)
[22] Schwartz, S., Wah, B.: Automated parameter tuning in stereo vision under time constraints. In: Proceedings., Fourth International Conference on Tools with Artificial Intelligence, TAI 1992, pp. 162–169 (1992)
[23] Soto, R., Crawford, B., Monfroy, E., Bustos, V.: Using autonomous search for generating good enumeration strategy blends in constraint programming. In: Murgante, B., Gervasi, O., Misra, S., Nedjah, N., Rocha, A.M.A.C., Taniar, D., Apduhan, B.O. (eds.) ICCSA 2012, Part III. LNCS, vol. 7335, pp. 607–617. Springer, Heidelberg (2012)
[24] Wallace, R.J.: Analysis of heuristic synergies. In: Hnich, B., Carlsson, M., Fages, F., Rossi, F. (eds.) CSCLP 2005. LNCS (LNAI), vol. 3978, pp. 73–87. Springer, Heidelberg (2006)
[25] Williams, C.P., Hogg, T.: Using deep structure to locate hard problems. In: Proceedings of AAAI 1992, pp. 472–477 (1992)

Branching Schemes and Variable Ordering Heuristics for Constraint Satisfaction Problems: Is There Something to Learn?

José Carlos Ortiz-Bayliss[1], Hugo Terashima-Marín[2],
and Santiago Enrique Conant-Pablos[2]

[1] Automated Scheduling, Optimisation and Planning (ASAP)
School of Computer Science, University of Nottingham, UK
Jose.Ortiz_Bayliss@nottingham.ac.uk
[2] Tecnológico de Monterrey
Campus Monterrey, Mexico
{terashima,sconant}@itesm.mx

Abstract. When solving a constraint satisfaction problem by using systematic algorithms it is needed to expand and explore a search tree to find a solution. In this work we study both binary and k-way branching schemes while they interact with various variable ordering heuristics, and how those interactions affect the cost of finding a solution to different instances. Both branching schemes have been used in previous investigations and it is not straight forward to determine the conditions that make one branching scheme better than the other. But we provide evidence that, in order to decide, variable ordering heuristics play a major role in the performance of these branching schemes. This study is intended to work as a preliminary study to develop hyper-heuristics for branching schemes in combination with variable ordering heuristics. The final part of the analysis presents a very simple naive hyper-heuristic that randomly applies binary and k-way branching as the search progresses in combination with some well known variable ordering heuristics. The scope of this paper is to explore the interactions between different variable ordering heuristics and these two branching schemes, in order to produce some relations between their performance. We expect these relations to be used in further studies as the basis for more robust hyper-heuristics that take into consideration the information gathered in this investigation.

Keywords: Constraint Satisfaction, Hyper-heuristics, Branching Schemes, Variable Ordering Heuristics.

1 Introduction

A constraint satisfaction problem (CSP) is defined by a set of variables X, where each variable is associated a domain D_x of values subject to a set of constraints C [23, 39]. The goal is to find a consistent assignment of values to variables in such a way that all constraints are satisfied, or to show that such

G. Terrazas et al. (eds.), *Nature Inspired Cooperative Strategies for Optimization*   329
(NICSO 2013), Studies in Computational Intelligence 512,
DOI: 10.1007/978-3-319-01692-4_25, © Springer International Publishing Switzerland 2014

consistent assignment does not exist. CSP belong to the NP-Complete class [18] and there is a wide range of theoretical and practical applications like scheduling, timetabling, cutting stock, planning, machine vision, temporal reasoning, among others (see for example [17], [27], [13] and [3]).

Several modern methods to solve CSP exist [14, 37], and solutions are found by searching systematically through the possible assignments to variables or by slightly modifying an initial complete and unfeasible solution. In all cases the solution process is guided by heuristics. It is a common practice to use depth first search (DFS) to solve CSPs. When using DFS to solve a CSP, every variable represents a node in the tree and the deeper we go in that tree, the larger the number of variables that have already been assigned a feasible value. Every time a variable is instantiated, a consistency check occurs to verify that the current assignment does not conflict with any of the previous assignments given the constraints within the instance. When an assignment produces a conflict with one or more constraints, the instantiation must be undone, and a new value must be assigned to that variable. When all the feasible values for a variable have been tried and failed, the value of a previously instantiated variable must be changed (this is known as backtracking [5]). Backtracking always goes up one single level in the search tree when a backward move is needed. One of the many ways to reduce the search space is using constraint propagation, where the idea is to propagate the effect of one instantiation to the rest of the variables due to the constraints among them. Thus, every time a variable is instantiated, the values of the other variables that are not allowed due to the current instantiation are removed from the respective domains.

The two most used approaches used to expand the search tree are binary and k-way branching. Binary branching is in fact, a particular case of domain splitting branching [30] when the pivot is selected to be the first value in the domain of the selected variable. The performance of domain splitting for different pivots has not properly been studied and is a topic beyond the scope of this investigation. Here we will use binary branching instead of domain splitting because binary branching is easier to explain, more used in practice, and requires fewer parameters to be tuned.

With k-way branching an instance P is solved as follows. Select a variable x with domain $D_x = \{v_1, v_2, \ldots, v_m\}$. For each $v \in D_x$, we restrict P by setting $x = v$, and recursively try to solve the remaining instance. P has no solution if and only if none of the m possible values for variable x produces a feasible solution given the current instantiated variables. In binary branching, the first choice point creates two alternatives, $x = v_1$ and $x \neq v_1$. The left branch is explored; if the branch fails, or if all solutions are required, the search backtracks to the choice point, and the right branch is followed instead. Crucially, the constraint $x \neq v_1$ is propagated, before a second point is created between $x = v_2$ and $x \neq v_2$, and so on.

With binary branching, the subtrees resulting from successive assignments to a variable are not explored independently; propagating the removal of a value from the current domain of the variable on the right branch can lead to further domain reductions. This propagation affects the search when future values of the variable are considered. Hence, the order in which values are assigned has more effect in the search compared to k-way branching [35]. Nevertheless the apparent difference in the way they work, most of the research performed on CSPs has used k-way branching, and only a few extra studies on the comparison between some branching strategies have been reported [2, 22, 30, 36].

In some cases, because of the ordering heuristics used, binary branching ends up simulating k-way branching [30]. To see why, consider a pure binary backtracking search with variable ordering based on the MRV heuristic [21] (a heuristic that selects the variable with the smallest domain size). If we select variable x because it has the smallest domain size m, the right branch will produce an instance where x has also the minimum domain size $(m-1)$. Although, this may be the case, the cost of the search may still be different because of the way the values are to be removed from the domains. Thus, we need a way to analyse if these branching schemes produce different search trees or not, and under which circumstances the differences are statistical significant.

The effect of the value ordering heuristics in binary branching, comparing the performance to k-way branching was already studied in [34, 36]. According to those results, binary branching is not, even with the worst value ordering heuristic, worse than k-way branching [34, 36]. Thus, it seems that we should always prefer binary branching over k-way branching; but this may not be correct. In this investigation we have explored the other part of the ordering problem in CSPs, the one related to variable ordering. We provide evidence that there are cases where the use of one branching scheme is preferable to the other when used with a specific variable ordering heuristic, but we have found no evidence that supports that one branching scheme is always better than the other for all variable ordering heuristics.

The research described in this paper is closely related to the work done by Balafoutis and Stergiou [1] and the one conducted by Lagoudakis and Littman [24]. Balafoutis and Stergiou [1] proposed two adaptive branching heuristics to modify the behaviour of binary branching once a variable ordering heuristic has been invoked. Their branching heuristics accept or reject the advise of the variable ordering heuristics once the right branch is to be evaluated. They considered some variable ordering heuristics and compared their branching heuristics on them proving that it is possible to obtain reductions in the cost by modifying the branching strategy as the search progresses. Lagoudakis and Littman [24] described an approach to select branching rules for the Davis-Putnam-Logemann-Loveland procedure for SAT [12]. The results obtained in [24] suggest that it is possible to improve traditional search methods by introducing some decision making and reasoning on top of them, to produce more robust branching rules.

This paper is organized as follows. Section 2 provides information about the variable ordering heuristics used in this investigation, which along with the branching scheme, determine the way the tree is expanded and the cost associated to the search. Later, in Sec. 3 we present the experiments that support the idea that binary and k-way branching are good for different regions of the search space and we explore the idea of producing a simple hyper-heuristic that uses both branching schemes during the search to analyse its performance. Finally, in Sec. 4 we present the conclusion and the future work of this investigation.

2 Background

Before moving onto the variable ordering heuristics used in this investigation, we need to explain how the CSP instances will be characterized in this document. There are many features that can be used to describe the instances (see for example [20]) but the most important properties used for this purpose are the constraint density p_1 and the constraint tightness, p_2.

The constraint density is a measure of the proportion of constraints within the instance; the closer the value of p_1 to 1, the larger the number of constraints in the instance. The constraint tightness (p_2) represents a proportion of the conflicts within the constraints. A conflict is a pair of values $\langle x, y \rangle$ that is not allowed for two variables at the same time[1]. The higher the number of conflicts, the more unlikely an instance has a solution.

2.1 Variable and Value Ordering Heuristics

This investigation includes a small set of ordering heuristics, four for variable ordering and one for value ordering. For variable ordering we include Minimum Remaining Values (MRV) [21], Kappa (K) [20], degree (DEG) [4] and Max-Conflicts (MXC). For value ordering we use Min-Conflicts (MNC) [26]. In all cases, the tie breaking strategy used is the lexical ordering of the variables and values. In the next lines we briefly describe each one of these heuristics.

Minimum Remaining Values (MRV) [21, 32]. MRV selects the variable with the smaller number of available values in its domain.

Kappa (K) [20]. K selects the variable that minimizes the value of κ of the remaining subproblem. κ is a measure of constrainedness which serves as an indicator of the hardness of the instances with respect to their sizes. For example, instances with $\kappa \ll 1$ have many solutions while instances with $\kappa \gg 1$ are likely to be unsatisfiable. κ is calculated as follows:

$$\kappa = \frac{- \sum_{c \in C_x} \log_2(1 - p_c)}{\log_2(D_x)} \qquad (1)$$

where D_x represents the domain size of variable x, C_x all the constraints where variable x participates and p_c the fraction of unfeasible tuples of values in constraint c.

Degree (DEG) [16, 38]. DEG prefers the variables connected to the maximum number of uninstantiated variables (forward degree of the variable).

---

[1] In this investigation we have used conflicts that involve only pairs of values because we are using binary CSPs. For constraints of arity a, an a-tuple has to be used. The approach is able to work with constraints of any arity without additional modifications.

Max-Conflicts (MXC) is a very simple and fast heuristic, and the main idea is to select the variable that is involved in the largest number of conflicts among the constraints in the instance. This instantiation will produce a subproblem that minimises the number of conflicts among the variables left to instantiate.

The value ordering heuristic Min-Conflicts (MNC) [26], is one simple heuristic that prefers the value involved in the minimum number of conflicts. This heuristic is trying to leave the maximum flexibility for subsequent variable assignments. If we select the value that is involved in the minimum number of conflicts, we can suppose that the resulting subproblem will have more solutions than the other subproblems. This heuristic is the direct implementation of the 'most promising' principle for value ordering [19].

2.2   Hyper-heuristics

Hyper-heuristics are motivated with the goal of automating the design of heuristic methods to solve hard computational search problems [6]. Although 'hyper-heuristic' is a relatively new term [10], the idea of automating the design/selection of heuristics can be traced back to the early 1960's, when Fisher and Thompson [15] suggested that combining priority dispatching rules would produce a superior performance than using any of the rules in isolation. According to Burke et al. [8], hyper-heuristics can be divided into two main categories: methodologies that select from a fixed set of heuristics and generate new heuristics. Regarding hyper-heuristics that select among existing heuristics, they produce a mapping between the states of the problem and a feasible heuristic. These methodologies maintain a set of heuristics and then, as the problem changes, decide which heuristic to apply. Examples of these methodologies (although not all of them use the term 'hyper-heuristic' to refer to their approaches) include [31],[28],[11] and [29]. On the other hand, hyper-heuristics that produce heuristics identify critical parts of existing heuristics to create new ones [7]. This study intends to obtain information about the interactions between branching schemes and variable ordering heuristics that could be used in the future to produce hyper-heuristics for branching and variable ordering within CSPs.

3   Experiments and Results

In this section we present the set of experiments conducted during this investigation. These experiments were designed to observe the behaviour of binary and k-way branching on different instances and working together with distinct variable ordering heuristics. The first experiment is designed to identify whether there are significant differences in the performance of the two branching schemes in different regions of the search space, when using distinct variable ordering heuristics. The second experiment explores the impact of the variable ordering strategy in the performance of each branching scheme when tested on more specific and larger instances. In the final experiment we study the effect of combining the two branching schemes as the search progresses by using a very simple hyper-heuristic approach.

3.1   Is One Branching Scheme Better Than the Other?

Our first experiment is designed to confirm that there are differences in the performance of the two branching schemes according to the initial values of p_1 and p_2 of the instance to solve. For this reason, we produced a grid of random instances where each instance contains 20 variables, and each variable has 10 values in its domain. Each cell in the grid covers a region of 0.05×0.05 in the space $p_1 \times p_2$ (each grid contains 20×20 cells). The purpose of such division is to cover all the space $p_1 \times p_2$ in such a way that we can map the performance of some regions of the space to one suitable branching scheme given one variable ordering heuristic. The resolution of the grid is enough to identify these regions and it can also be explored in an acceptable time.

For each cell, 30 instances were generated with random values of p_1 and p_2 within the range of the cell. In this way, 12000 instances were generated and solved for each grid in this experiment. Because four variable ordering heuristics are analysed in this investigation, four grids were produced and solved. In all cases, the cost of the search is measured by the number of consistency checks required to find the first solution or to prove than none exists.

Random model B [33] was used to generate the instances. The random instances are generated in two stages. In the first stage, a constraint graph G with n nodes is randomly constructed and then, in the second stage, the incompatibility graph C is formed by randomly selecting a set of edges (incompatible pairs of values) for each edge (constraint) in G. The parameter p_1 determines how many constraints exist in the CSP instance and corresponds to the constraint density, whereas p_2 determines how restrictive the constraints are, and corresponds to constraint tightness of the resulting instance. In model B, there should be exactly $p_1 n(n-1)/2$ constraints (rounded to the nearest integer), and for each pair of constrained variables, the number of inconsistent pairs of values should be exactly $m^2 p_2$ (where m is the uniform domain size of the variables). This generation model was selected because it provided the flexibility to produce instances in the exact region of the space $p_1 \times p_2$ where we needed them.

Each grid was solved with a distinct variable ordering heuristic. We solved the 30 instances in each cell in each grid by using binary branching and k-way branching. MNC was used as value ordering heuristic in all cases. The constraint propagation method used in all the experiments was AC3 [25]. Then, we statistically compared the cost of the two branching schemes on each cell in the grid to observe whether the differences in the performance were statistically significant or not (the cost is given by the number of consistency checks required by the search; this is, the number of times the constraints are revised). To compare the means at each point in the grid we used a bilateral hypothesis test based on a normal distribution. In each test, we use the null hypothesis $H_0 : \mu_{2B} = \mu_{kB}$

and the alternative hypothesis $H_1 : \mu_{2B} \neq \mu_{kB}$, where μ_{2B} and μ_{kB} stand for the real means of binary branching and k-way branching, respectively.

Figure 1 presents the results of the the statistical test for each variable ordering heuristic by using the two branching schemes under study. The values of each cell correspond to the p-values resulting from the test. Thus, the smaller the value, the stronger the statistical evidence that indicates that both approaches differ in performance. To be consistent with the statistical notation, the cells with values below 0.05 confirm the idea that one branching scheme is better than the other with 5% of significance.

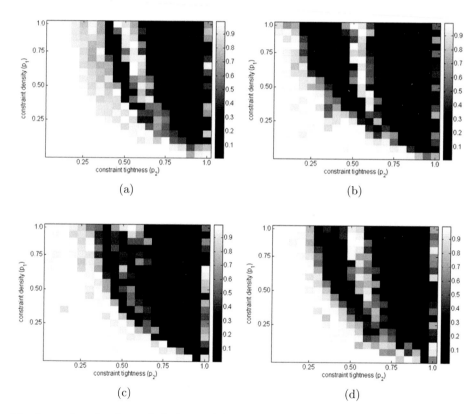

Fig. 1. p-values for the statistical test of the two branching schemes under four different heuristics: (a) MRV, (b) DEG, (c) K and (d) MXC (cells with values below 0.05 indicate significant statistical difference between both approaches with 5% of significance)

The statistical tests were performed to identify the regions where one branching scheme was statistically better than the other; not to identify which one was the best option. For this reason, once we identified the regions where there was statistical evidence that the means of the two branching schemes were different, we conducted new tests to identify which was the better branching scheme for each region. We found that, in all the cases where the statistical evidence suggests that the schemes are different, k-way branching obtained the smallest average cost. We can conclude that, for the cases where statistical evidence was found that one approach is better than the other, k-way branching is always better than binary branching on this first experiment.

Trying to understand the differences in the performance of both branching schemes, we can observe that in general (and regardless of the heuristic used), the region where unsatisfiable instances take place, k-way branching is a better option than binary branching for any of the four variable ordering heuristics. Even though there is no statistical evidence that supports that binary branching is better than k-way branching, we observed that there are regions where binary branching obtains a better mean performance than k-way branching. These regions are located just before the well known transition phase [9]; the region where the instances abruptly change from being satisfiable to being unsatisfiable. Nevertheless, no statistical evidence was found that these differences are significant. Also, it is important to mention that, in the region where loose constraints take place (low values of p_1 and p_2), both branching schemes obtain the same cost for most of the instances.

3.2   The Effect on Larger Instances

We have studied the effect of binary and k-way branching when combined with four variable ordering heuristics in the space $p_1 \times p_2$. In this experiment we study the performance of these branching schemes on a different set of instances taken from the literature. The set contains 10 random instances produced with model RB [40] (which is a revision of model B). The set includes 10 satisfiable instances, each one with 30 variables and 15 values in each domain. This instances can be obtained from `http://www.cril.univ-artois.fr/~lecoutre/research/benchmarks/frb30-15.tgz`. The idea behind this new experiment is to focus on a smaller set of larger instances with different features than the ones used before. Also, it is interesting to observe that in the previous experiment, k-way branching proved to be a better option for unsatisfiable instances and, in this case, all the instances have at least one solution.

For these instances, the interactions between variable ordering heuristics and branching schemes produce interesting behaviours. Only DEG seems to be statistically sensitive to the choice of the branching scheme. For the remaining variable ordering heuristics, the differences in the performance do not represent statistical differences with 5% of significance (this time we used a Welch's test, which is a hypothesis test based on t-distribution for small samples). These results are presented by using the box plot in Fig. 2.

The analysis of the medians confirms our conclusions obtained from the analysis of the means. For MRV, K and MXC, the statistical evidence suggests that the medians of binary and k-way branching are equal, with 95% of confidence. This can be concluded because the notches in the box plot overlap. The statistical evidence for DEG indicates that the medians of the two branching schemes may not me equal.

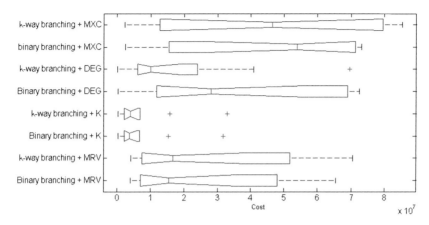

Fig. 2. Boxplot of the results obtained by each branching scheme on the set of large satisfiable instances produced with model RB when combined with each of the four variable ordering heuristics

Once again, as in the previous experiment, we observed differences in the performance of the distinct methods which are not statistical significant, but this does not mean that they may not be significant in practice. For this reason, and using only the values from the samples, we can conclude that on this set of instances:

- Binary branching is always better than k-way branching when using MRV.
- When K is used as variable ordering heuristic, binary branching always obtains the lowest costs.
- Binary branching is never worse than k-way branching when using DEG.
- When MXC is used as variable ordering heuristic, four of the 10 instances are best solved by using binary branching and the remaining ones are solved more efficiently by using k-way branching.
- We found no case where the both branching schemes produce the same cost.

Thus, we have identified that there are cases where, in combination with the variable ordering heuristic, one branching scheme should be preferred above the other. Now, the question is what occurs when we try to apply these schemes at different stages of the search. This is the purpose of the following experiment.

3.3 A Naive Random Hyper-heuristic: How Difficult Is It to Improve the Branching Schemes?

We have observed that the performance of the branching schemes is affected by the selection of the variable ordering heuristic. Then, we cannot decide when one branching scheme is better than the other without considering the variable ordering heuristic used in the search. In this experiment, we try a very simple way to analyse the probabilities of success of a naive hyper-heuristic that expands the search tree by using binary branching at some stages, and k-way branching at others. For this analysis we propose a naive random hyper-heuristic, which is defined as follows. There is a probability α that binary branching is applied (and a probability $1 - \alpha$ that k-way branching is used instead). At each stage of the search (every time a variable is to be selected) a random decision is made based on a uniform random distribution and the current value of α. For small values of α, it is very likely that k-way branching is used. Thus, when $\alpha = 0$ the search imitates k-way branching and, when $\alpha = 1$, it matches the tree produced by binary branching. This simple approach is used only to measure the probability of success of random combinations of binary and k-way branching. This is not intended to be a robust method for producing a real hybrid application; its purpose is only to show what can occur when we combine these branching approaches as the search progresses.

For each heuristic, nine values of α were tested: 0.1 to 0.9, by increments of 0.1 (values of 0 and 1.0 were taken from the results obtained in Sec. 3). For each value of α, ten runs were conducted and the average cost of these ten runs on the set of instances taken from http://www.cril.univ-artois.fr/~lecoutre/research/benchmarks/frb30-15.tgz is reported. We considered that ten runs was enough to give us an idea of the behaviour of a naive random hyper-heuristic on this set of instances. This process is performed for each of the four variable ordering heuristics. Figure 3 presents the results of this experiment.

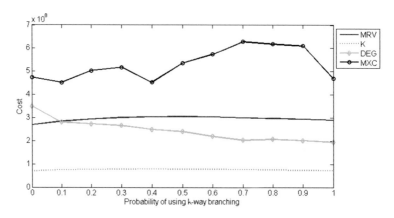

Fig. 3. Average costs of different values of α for the naive random hyper-heuristic

Before conducting this experiment, we had the hypothesis that there should be a value of α for which the average cost of solving the set of instances will be below the cost of both binary and k-way branching. Considering the results on Fig. 3 we have not found evidence to support our hypothesis. It seems that using the two branching schemes at different stages of the search by using a naive random hyper-heuristic that does not consider the current state under exploration does not help to reduce the cost. On the contrary, the results suggest that this only increases the cost of the search. We think that the selection of the branching scheme at different stages of the search requires a more refined mechanism that uses the information from the current state of the space under exploration to decide the most suitable branching scheme. The evidence suggests that relying the selection of the branching scheme to be used only on naive random decisions is not a good idea. What we can conclude from this experiment is that, if we randomly try to select between the two branching schemes as the search progresses, we are likely to increase the search cost rather than reducing it (see for example the performance of MXC on Fig. 3, where the cost of the search for $\alpha = 0.7$ is larger than the cost of both binary and k-way branching).

The development of a hyper-heuristic that exploits information from the instances to decide which branching scheme to use at each stage of the search will be part of a future investigation. At the moment, we have provided evidence that the use of different branching schemes during the search affects the search cost. The goal for future research is now to propose a mechanism that produces a mapping from problem states to branching schemes and variable ordering heuristics, similar to the work done on hyper-heuristics for variable ordering [11, 29].

4   Conclusion and Future Work

In this investigation we have analysed binary and k-way branching for CSPs together with some important and widely used variable ordering heuristics. Our analysis considers the effect of variable ordering during the search on the performance of binary and k-way branching. Based on our observations, there is no way to state that one of the branching scheme is to be preferred above the other in all cases. Not only there is much to learn from the interactions between branching schemes and variable ordering heuristics, but about the ways to exploit when these interactions produce reductions in the search cost.

One important contribution of this research is the fact that, contrary to what happens in value ordering (where binary branching is never worse than k-way branching [34, 36]), variable ordering heuristics require a deeper analysis to understand their influence on the branching schemes. At the moment, we can conclude that binary and k-way branching are variable ordering dependant and, none of them can be stated to be superior to the other for every instance. In other words, we can conclude that we cannot analyse the performance of the branching schemes without considering the variable ordering heuristic to be used with them.

We have identified regions where some heuristics are more suitable to work with one of these branching approaches. It seems that in general, it looks like for unsatisfiable instances, it is better to use k-way branching. But, more investigation is still needed to confirm whether these results apply to other classes of instances. With regard to the classes of instances used, as future work we are interested in extending the analysis of the implications in the performance of the branching schemes by considering real instances to observe how these branching schemes behave when tested on structured instances.

This investigation will be used as the basis for developing a more challenging idea, which is the generation of hyper-heuristics that control not only the heuristics used, but the branching schemes at different stages of the search. In order to produce a hyper-heuristic for this problem, we needed to justify that there are differences in the performance of the methods to be combined and that one branching scheme does not dominate the other, which is the case for the branching schemes studied when they interact with the variable ordering heuristics described in this investigation.

Acknowledgments. This research was supported in part by ITESM Strategic Project PRY075 "Intelligent Learning for Pattern Recognition and its Application in Medicine and Logistics" and the CONACyT Project under grant 99695.

References

[1] Balafoutis, T., Stergiou, K.: Adaptive branching for constraint satisfaction problems. In: Proceedings of the 2010 Conference on ECAI 2010, pp. 855–860. IOS Press, Amsterdam (2010)

[2] Balafoutis, T., Paparrizou, A., Stergiou, K.: Experimental evaluation of branching schemes for the csp. CoRR abs/1009.0407 (2010)

[3] Berlier, J., McCollum, J.: A constraint satisfaction algorithm for microcontroller selection and pin assignment. In: Proceedings of the IEEE SoutheastCon 2010 (SoutheastCon), pp. 348–351 (2010)

[4] Bessière, C., Régin, J.C.: Mac and combined heuristics: Two reasons to forsake FC (and CBJ) on hard problems. In: Freuder, E.C. (ed.) CP 1996. LNCS, vol. 1118, Springer, Heidelberg (1996)

[5] Bitner, J.R., Reingold, E.M.: Backtrack programming techniques. Communications of the ACM 18(11), 651–656 (1975)

[6] Burke, E., Hart, E., Kendall, G., Newall, J., Ross, P., Shulenburg, S.: Hyperheuristics: an emerging direction in modern research technology. In: Handbook of Metaheuristics, pp. 457–474. Kluwer Academic Publishers (2003)

[7] Burke, E.K., Hyde, M.R., Kendall, G., Ochoa, G., Ozcan, E., Woodward, J.R.: Exploring hyper-heuristic methodologies with genetic programming. In: Mumford, C.L., Jain, L.C. (eds.) Computational Intelligence. ISRL, vol. 1, pp. 177–201. Springer, Heidelberg (2009)

[8] Burke, E.K., Hyde, M., Kendall, G., Ochoa, G., Ozcan, E., Woodward, J.R.: A classification of hyper-heuristic approaches. In: Gendreau, M., Potvin, J.Y. (eds.) Handbook of Metaheuristics. International Series in Operations Research & Management Science, vol. 146, pp. 449–468. Springer, US (2010)

[9] Cheeseman, P., Kanefsky, B., Taylor, W.M.: Where the really hard problems are. In: Proceedings of International Joint Conferences on Artificial Intelligence (IJCAI 1991), pp. 331–337 (1991)

[10] Cowling, P.I., Kendall, G., Soubeiga, E.: A hyperheuristic approach to scheduling a sales summit. In: Burke, E., Erben, W. (eds.) PATAT 2000. LNCS, vol. 2079, pp. 176–190. Springer, Heidelberg (2001)

[11] Crawford, B., Soto, R., Castro, C., Monfroy, E.: A hyperheuristic approach for dynamic enumeration strategy selection in constraint satisfaction. In: Ferrández, J.M., Álvarez Sánchez, J.R., de la Paz, F., Toledo, F.J. (eds.) IWINAC 2011, Part II. LNCS, vol. 6687, pp. 295–304. Springer, Heidelberg (2011)

[12] Davis, M., Logemann, G., Loveland, D.: A machine program for theorem-proving. Communications of ACM 5(7), 394–397 (1962)
[13] Dunkin, N., Allen, S.: Frequency assignment problems: Representations and solutions. Tech. Rep. CSD-TR-97-14, University of London (1997)
[14] Epstein, S.L., Freuder, E.C., Wallace, R.J.: Learning to support constraint programmers. Computational Intelligence 21(4), 336–371 (2005)
[15] Fisher, H., Thompson, G.L.: Probabilistic learning combinations of local job-shop scheduling rules. In: Factory Scheduling Conference, Carnegie Institute of Technology (1961)
[16] Freuder, E.C.: A sufficient condition for backtrack-free search. Journal of the ACM 29(1), 24–32 (1982)
[17] Freuder, E.C., Mackworth, A.K.: Constraint-Based Reasoning. MIT/Elsevier (1994)
[18] Garey, M.R., Johnson, D.S.: Computers and Intractability; A Guide to the Theory of NP-Completeness. W.H. Freeman (1979)
[19] Geelen, P.A.: Dual viewpoint heuristics for binary constraint satisfaction problems. In: Proceedings of the 10th European Conference on Artificial Intelligence (ECAI 1992), pp. 31–35. John Wiley & Sons (1992)
[20] Gent, I., MacIntyre, E., Prosser, P., Smith, B., Walsh, T.: An empirical study of dynamic variable ordering heuristics for the constraint satisfaction problem. In: Freuder, E.C. (ed.) CP 1996. LNCS, vol. 1118, pp. 179–193. Springer, Heidelberg (1996)
[21] Haralick, R.M., Elliott, G.L.: Increasing tree search efficiency for constraint satisfaction problems. Artificial Intelligence 14, 263–313 (1980)
[22] Hwang, J., Mitchell, D.G.: 2 -way vs. d -way branching for CSP. In: van Beek, P. (ed.) CP 2005. LNCS, vol. 3709, pp. 343–357. Springer, Heidelberg (2005)
[23] Kumar, V.: Algorithms for constraint satisfaction: a survey. AI Magazine 13(1), 32–44 (1992)
[24] Lagoudakis, M.G., Littman, M.L.: Learning to select branching rules in the dpll procedure for satisfiability. Electronic Notes in Discrete Mathematics 9, 344–359 (2001)
[25] Mackworth, A.K.: Consistency in networks of relations. Artificial Intelligence 8(1), 99–118 (1977)
[26] Minton, S., Johnston, M.D., Phillips, A., Laird, P.: Minimizing conflicts: A heuristic repair method for CSP and scheduling problems. Artificial Intelligence 58, 161–205 (1992)
[27] Montanari, U.: Networks of constraints: fundamentals properties and applications to picture processing. Information Sciences 7, 95–132 (1974)
[28] O'Mahony, E., Hebrard, E., Holland, A., Nugent, C., O'Sullivan, B.: Using case-based reasoning in an algorithm portfolio for constraint solving. In: Proceedings of the 19th Irish Conference on Artificial Intelligence and Cognitive Science (2008)
[29] Ortiz-Bayliss, J.C., Terashima-Marín, H., Conant-Pablos, S.E.: Learning vector quantization for variable ordering in constraint satisfaction problems. Pattern Recogn. Lett. 34(4), 423–432 (2013)
[30] Park, V.: An empirical study of different branching strategies for constraint satisfaction problems. PhD thesis, University of Waterloo (2004)
[31] Petrovic, S., Qu, R.: Case-based reasoning as a heuristic selector in a hyper-heuristic for course timetabling problems. In: Proceedings of the 6th International Conference on Knowledge-Based Intelligent Information Engineering Systems and Applied Technologies (KES 2002), vol. 82, pp. 336–340 (2002)

[32] Purdom, P.W.: Search rearrangement backtracking and polynomial average time. Artificial Intelligence 21, 117–133 (1983)

[33] Smith, B.M.: Locating the phase transition in binary constraint satisfaction problems. Artificial Intelligence 81, 155–181 (1996)

[34] Smith, B.M.: Value ordering for finding all solutions. In: International Joint Conference on Artificial Intelligence (IJCAI 2005), pp. 311–316 (2005)

[35] Smith, B.M., Grant, S.A.: Sparse constraint graphs and exceptionally hard problems. In: Proceedings of the International Joint Conferences on Artificial Intelligence (IJCAI 1995), pp. 646–651 (1995)

[36] Smith, B.M., Sturdy, P.: An empirical investigation of value ordering for finding all solutions. In: Workshop on Modelling and Solving Problems with Constraints (2004)

[37] Soto, R., Crawford, B., Monfroy, E., Bustos, V.: Using autonomous search for generating good enumeration strategy blends in constraint programming. In: Murgante, B., Gervasi, O., Misra, S., Nedjah, N., Rocha, A.M.A.C., Taniar, D., Apduhan, B.O. (eds.) ICCSA 2012, Part III. LNCS, vol. 7335, pp. 607–617. Springer, Heidelberg (2012)

[38] Wallace, R.J.: Analysis of heuristic synergies. In: Hnich, B., Carlsson, M., Fages, F., Rossi, F. (eds.) CSCLP 2005. LNCS (LNAI), vol. 3978, pp. 73–87. Springer, Heidelberg (2006)

[39] Williams, C.P., Hogg, T.: Using deep structure to locate hard problems. In: Proceedings of AAAI 1992, pp. 472–477 (1992)

[40] Xu, K., Boussemart, F., Hemery, F., Lecoutre, C.: Random constraint satisfaction: Easy generation of hard (satisfiable) instances. Artificial Intelligence 171(8-9), 514–534 (2007)

Nash Equilibria Detection for Discrete-Time Generalized Cournot Dynamic Oligopolies

Mihai Suciu, Noémi Gaskó, Rodica Ioana Lung, and D. Dumitrescu

Babes-Bolyai University, Cluj-Napoca, Romania
mihai.suciu@ubbcluj.ro, {gaskonomi,ddumitr}@cs.ubbcluj.ro,
rodica.lung@econ.ubbcluj.ro

Abstract. The problem of equilibria detection of a discrete-time Generalized Cournot Dynamic Oligopoly is approached by using a Differential Evolution and a Particle Swarm Optimization algorithm adapted to compute and track the set of generalized Nash equilibria in a dynamic setting. Both challenges of this problem, i.e. to correctly compute the entire set of generalized Nash equilibria of the constrained (generalized) game, and also to cope with the dynamic character of the landscape, are dealt with by using a simple adaptive mechanism. Numerical experiments for settings up to 60 players are performed to illustrate the efficiency of the approach.

1 Introduction

Game theory models strategic interactions among players with conflicting interests in which the payoff of each player depends on the choices of all other partners. A constrained, generalized game, is a general model of a decision situation where the strategies of the players are not independent, i.e. they not only affect each others payoffs but also their choices can be restricted by each other.

A formalization of this idea is called Generalized Nash Equilibrium Problem (GNEP) introduced in the 50'ies in [1], [2]. A GNEP is a generalization of the classical equilibrium problem, in which players' strategies depend on the other players' strategies. From a computational point of view, one of the main challenges in solving a GNEP is to detect the entire set of generalized equilibria in a single run.

An even more realistic model is obtained by considering a dynamic environment for GNEPs, i.e. combining the GNEP with a Dynamic Game (DG). A DG is a mathematical model of the interactions between decision makers (players, agents) who are controlling a dynamical system [3] and fulfills the following characteristics [4]:

- for each period/epoch players receive a payoff;
- each player has an overall payoff - which is the sum of its payoffs obtained in each period/epoch;
- the state of the system determines the payoff of each player in a certain epoch;
- the state of the system changes in time, changes can be determined also by the actions of the players;
- a difference or a differential equation can describe the rate of the change;

G. Terrazas et al. (eds.), *Nature Inspired Cooperative Strategies for Optimization* (NICSO 2013), Studies in Computational Intelligence 512,
DOI: 10.1007/978-3-319-01692-4_26, © Springer International Publishing Switzerland 2014

In our approach a GNEP is considered in a dynamic environment, meaning that the strategies of the game are dependent within the game and that this dependence changes in time. This new class of game is called Discrete-Time Dynamic Generalized Game. In this case, the challenge is not only to detect the set of generalized equilibria but also to track the new position of this entire set within a dynamic environment.

Two heuristics for tracking the solution of a discrete-time dynamic generalized game are studied: a Differential Evolution (DE) and a Particle Swarm Optimization (PSO) algorithm that have been adapted to compute the generalized Nash equilibria and also to deal with the dynamic environment. By means of numerical experiments it is shown that a simple adaptation mechanism based on mutation is effective in tracking the set of generalized Nash equilibria for the generalized Cournot setting considered.

The paper is organized as follows: section 2 describes the Generalized Nash Equilibrium Problem. Section 3 presents the dynamic constrained version of the Cournot oligopoly. In section 4 the two proposed algorithms are described. Section 5 presents the numerical experiments and Section 6 concludes the paper.

2 Generalized Nash Equilibrium Problem

Generalized Nash equilibrium problem (GNEP) [1], [2] is a generalization of the classical Nash equilibrium problem [6], in which players' strategies depend on the other players' strategies.

GNEP can arise in some real-world situations, for example oligopoly models with shared resources, energy markets [7]. In [8] a GNEP is constructed from a spatial oligopolistic electricity model. Breton et al. [9] construct a GNEP from a game theoretic interpretation of joint implementation of environmental projects. Another application can be found in solving electrical market games formulated as a GNEP [10].

The commonly accepted solution of a GNEP is the generalized Nash equilibrium [11].

A GNEP can be described as a system $G_{GNEP} = ((N, K, u_i), i = 1, ..., n)$, where N represents a set of players, and n is the number of players.

For all players $i \in N$ the **common strategy set** is formalized as follows:

Let $s = (s_1, s_2, ..., s_n)$ a vector formed by all decision variables of the game (strategy profile); let us denote by s_{-i} the vector formed by each players strategy except of the ith player.

To accentuate that player's i strategy variables in s it can be written $(s_i, \mathbf{s}_{-i})$ instead of s.

Let $S_i \in \mathbb{R}^n$ be the strategy set of player i, $S = \prod_{j \in N} S_j$, $S_{-i} = \prod_{j \in N, j \neq i} S_j$, which means S_{-i} represents the full S set, except the ith player's set.

Let $K_i : S_{-i} \to S_i$ be a point-to-set mapping which means, that all players j can affect the feasible strategy of player i. Then $K_i(s^*) \subseteq S_i, \forall s^* \in S_{-i}$ Let $K = \prod_{i \in N} K_i(s_i)$ the mapping formed from the K_i.

For each player $i \in N$, $u_i : grK_i \to R$ represents the payoff function of i, where grK_i represents the graph of the mapping K_i.

Formally, a generalized Nash equilibrium (GNE) is a strategy profile $s^* \in S$ such that the inequality holds:

$$u_i(s_i, s_{-i}^*) \leq u_i(s^*), \forall i = 1, .., n, \forall s_i \in K_i(s_{-i}).$$

The definition of the GNE differs from the normal Nash equilibrium only in the feasible strategy of each player.

Let s and s^* be two strategy profiles; $k(s^*, s)$ denotes the number of players which benefit by deviating from s^* towards s [12]:

$$k(s^*, s) = card\{i \in N, u_i(s_i, s_{-i}^*) > u_i(s^*), s_i \neq s_i^*\}.$$

Let $s^*, s \in S$. We say that strategy s^* is better than strategy s with respect to Nash equilibrium (Nash ascends it), and we write $s^* \prec_N s$, if the following inequality holds:

$$k(s^*, s) < k(s, s^*).$$

$k(s^*, s)$ is a relative quality measure of s and s^* - with respect to the Nash equilibrium [13]. The relation $\prec_N$ can be considered as the *generative relation of Nash equilibrium*, i.e. that the set of non-dominated strategies with respect to $\prec_N$ induces the *Nash equilibrium*. Based on this property several nature inspired search methods have been adapted to compute Nash equilibria [14] by including the k operator in the fitness assignment process. However, this is not a partial order relation, as it is not transitive [15].

This relation can also be used for the GNEP taking into account the constraints imposed by the common strategy set S.

3   Constraint Dynamic Cournot Oligopoly

Let us consider a simple Cournot oligopoly model [16], where n firms produce a quantity of s_i products, $i = 1, ..., n$.

Let $P(Q)$ be the market clearing price, where $Q = \sum_{i=1}^{n} s_i$.

$$P(Q) = \begin{cases} a - Q, & \text{if } Q \leq a; \\ 0, & \text{if } Q > a. \end{cases}$$

Each firm has the common cost function $C(s_i)$. Let us assume that the total cost of company i for producing quantity s_i is $C_i(s_i) = cs_i$.

The payoff for the company i is its profit, that can be described as follows:

$$u_i(s) = s_i P(Q) - C_i(s_i),$$

where $s = (s_1, s_2, ..., s_n)$. Hence we may write

$$u_i(s) = \begin{cases} s_i(a - \sum_{j=1}^{n} s_j - c), & \text{if } \sum_{i=1}^{n} s_i \leq a; \\ -cs_i, & \text{if } \sum_{i=1}^{n} s_i > a. \end{cases}$$

The Nash equilibrium of this game is $\frac{a-c}{n+1}$ for all firms.

A generalized version of the game is obtained if we consider the following constraints on the quantity of goods produced:

$$\sum_{i=1}^{n} \alpha_i \cdot s_i \in [l, u], l < u, \alpha_i \in [0, 1], i = 1, ..., n. \tag{1}$$

All strategies that satisfy $\sum_{i=1}^{n} \alpha_i \cdot s_i = u$ are GNE if the NE of the unconstrained version of the game does not belong to the set of common strategies S (in which case the GNE is identical with the NE).

Dynamic Feature. By varying coefficients $\alpha_i, i = 1, ..., n$ in time we deal with a generalized dynamic oligopoly.

Example 1. Considering the duopoly (two players) of the Cournot game, the payoff of the two firms can be described with the following payoff function:

$$u_i(s) = s_i \cdot (a - c - (s_1 + s_2)), i = 1, 2, s_i \in [0, a - c].$$

Let us assume that $a = 24$, $c = 9$. The Nash equilibrium of the game is the strategy pair $s = (s_1, s_2) = (5, 5)$ with the payoff $(25, 25)$.

Additionally we know that the two firms can produce maxim a total quantity of 3: $s_1 + s_2 \leq 3$. This condition makes from the normal Cournot duopoly a GNEP problem.

Considering the constraint $s_1 + s_2 \leq 3$ it is easy to verify that $(5, 5)$ can't be a Nash solution of the game. The GNEP has an infinity number of GNE with every strategy pair $s_1 + s_2 = 3$, $s_1, s_2 \in [0, 15]$ being a GNE of the game.

4   Dynamic Generalized-Equilibrium Tracking Algorithms

In order to compute the GNEs in a constrained dynamic environment an Evolutionary Algorithm for Equilibria Detection (EAED) [17] can be endowed with a mechanism for tracking and adapting to environment changes.

An EAED uses a selection for survival operator that guides the search using the generative relation described in Section 2 (an offspring replaces its parent only if it is better than it in *Nash* sense).

Two EAEDs are adapted to detect and track the GNEs: the Dynamic Generalized Equilibrium Tracking Differential Evolution (*DGET-DE*) based on Differential Evolution [18], and the Dynamic Generalized Equilibrium Tracking Particle Swarm Optimization (*DGET-PSO*) based on SMPSO [19]. Both methods use the same mechanism to cope with the the dynamic features of the environment, presented in the following.

We address the equilibrium detection in a GNEP using DE and PSO based algorithms because they are computationally effective. These simple and efficient algorithms require less payoff function evaluations than a standard EA would. Using the generative relation to guide the search towards the Nash equilibrium of a game is a computationally expensive step, due to the selection for recombination operators a standard Ea would require a significantly larger number of payoff functions evaluations.

Detection of an Environment Change. A sentinel is used to detect a change in the game. The sentinel is randomly generated at the beginning of the search and it does not change during the evolutionary process. In each generation its fitness and constraint are evaluated, if these values differ from the last generation a change it is inferred that in the environment occurred. Based on the new and old values the algorithm estimated the amplitude of the change and adapts the search accordingly.

Coping with Changes. The following steps are taking for dealing and adapting to changes in the dynamic game:

1. (*DGET-DE* and *DGET-PSO*) When a change in the game is detected the magnitude of change m is estimated:

$$m = \frac{|constraint_{Old} - constraint_{New}|}{\max(constraint_{Old}, constraint_{New})} \quad (2)$$

   where $constraint_{Old}$ and $Constraint_{New}$ represent the old and new constraint violation values for the sentinel. If the sentinel is a feasible solution, the value of the *constraint* is taken the middle of the interval $[l, u]$.

2. (*DGET-DE* and *DGET-PSO*) In order to increase population diversity uniform mutation, with $p_m \in [p_{min}, p_{max}]$, where p_{min} and p_{max} are parameters of the algorithm, is applied to each individual. The probability of mutation is computed using the following relation:

$$p_m = p_{min} + m \cdot (p_{max} - p_{min})$$

   which ensures that p_m is directly proportional to m. Thus a large change in the environment will produce a higher mutation rate inducing population diversity, while a smaller change will keep a low mutation rate maintaining the search in the equilibrium region.

3. (*DGET-DE* and *DGET-PSO*) Apply uniform mutation with probability p_m and mutation step $N(0, \sigma)$ where σ is linearly correlated to the estimated magnitude of the change m:

$$\sigma_m = \sigma_{min} + m \cdot (\sigma_{max} - \sigma_{min}).$$

   Thus an adaptive mutation step is used to further cope with the change in the environment.

4. (*DGET-DE*) The value of of the scaling factor F is increased (in order to better explore the search space), and linearly decreases at each generation until it reaches a certain threshold. The value of C_r (crossover probability) is decreased (in order to promote exploration) and linearly increases at each generation until it reaches a predefined threshold.

In the generation in which a change is detected in the game we introduce diversity in the population by mutating each individual in the population. A mutation factor scaled to the amplitude of the detected change is used. This step is needed because the population converges to the Nash equilibrium (which is a point/front depending on the game) and all individuals have the Nash strategy. When the game changes because there is no diversity, the population does not converge to the new Nash equilibrium. This step is used in order to promote diversity and give individuals in the population a chance to find the new equilibrium.

Outline of the Methods. The general outlines of the *DGET-DE* and *DGET-PSO* algorithms are presented in Algorithm 1 and Algorithm 2 respectively by emphasizing only the changes to their standard versions. Both algorithms are compared with a simple DE version, called Adaptive Equilibrium Differential Evolution Algorithm (*AEDE*), that only uses step 4 of the adaptation mechanism presented here.

We compare our algorithms with a simple adaptive version of Differential Evolution (AE-DE) (modified for detecting the Nash equilibrium of a game). We use this algorithms for comparison because there are no other evolutionary approaches for detecting game equilibrium, and especially Nash equilibrium in a constrained game.

Algorithm 1. Dynamic Generalized Equilibrium Tracking Differential Evolution Algorithm (*DGET-DE*)

Randomly generate initial population of game situations;
repeat
    Create offspring by mutation and recombination (DE/rand/1/bin);
    Evaluate offspring (compute payoff functions for all players);
    if the offspring is better (in Nash sense) than parent **then**
        Replace the parent by offspring in the next generation;
    end if
    if change detected **then**
        Apply uniform mutation with the adapted value of p_m and step $N(0, \sigma)$
        according to Steps 1-3;
    end if
    Apply adapted Differential evolution (using Step 4. for varying F);
until termination condition is met;

Constraints Handling. An individual is deemed feasible if it satisfies the constraints defined by (1). When comparing two individuals, selection for survival is based on the following rules [20]:

1. If both individuals are infeasible choose the one that violates restriction less;
2. If one individual is feasible and the other is not then choose the feasible one;
3. When both individuals are feasible choose the one that Nash ascends the other. If these vectors are indifferent (no individual is dominated) then keep the parent.

Algorithm 2. Dynamic Generalized Equilibrium Tracking Particle Swarm Optimization Algorithm (*DGET-PSO*)

Randomly generate initial population of game situations;
repeat
    Perform a SMPSO iteration*;
    if change detected **then**
        Apply uniform mutation with the adapted value of p_m and step $N(0, \sigma)$,
        according to Steps 1-3;
    end if
until termination condition is met;
* In the archiving step of SMPSO the Pareto domination relation is replaced with the Nash ascendancy relation presented in Section 2.

5   Numerical Experiments

Numerical experiments are conducted for a constrained dynamic Cournot duopoly. The dynamic version of the game (described in Section 3) is obtained by randomly changing α_i from relation (1) at the beginning of each epoch. We use a real value representation, each individual in the population consists of a vector of n real numbers where the genes of our genome represent the strategy for each player: $g_i = s_i, i = 1, ..., n$.

The performance of each algorithm is evaluated by computing the Nash Generational Distance ($Nash - GD$) indicator, which is similar to the multiobjective GD [21], between obtained strategies and true Nash front composed of the theoretical strategies computed analytically for each change of the fitness landscape.

We consider 20 independent runs for each algorithm with each run consisting of 50 epochs. For all simulations we use a population of $N = 150$ individuals.

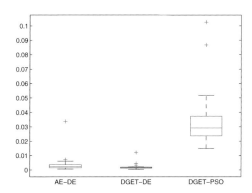

Fig. 1. Box-plots for the average *Nash-GD*. *DGET-DE* has the lowest mean value and standard deviation for the *Nash-GD* indicator.

For *DE* based algorithms initial parameter values are: $C_r = 0.8$ and $F = 0.2$. For *PSO* variant, based on [22], [23],[24], parameter values are: $c_1 = 1.4962, c_2 = 1.4962, w_{initi} = 0.7968, w_{final} = 0.7968$. Game parameters are: $a = 24$, $c = 9$, $s_i \in \{0, 15\}, l = 2, u = 4$ and the ones used for the adaptation mechanism are $p_{min} = 0.02, p_{max} = 0.1, \sigma_{min} = 0.1$, and $\sigma_{max} = 1$. After 200 generations (1 epoch) the constraints of the game change by generating new α_i values following a uniform distribution.

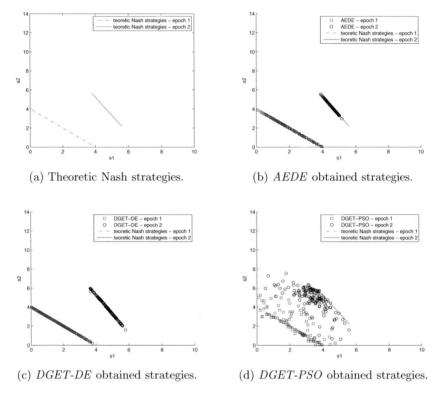

(a) Theoretic Nash strategies.   (b) *AEDE* obtained strategies.

(c) *DGET-DE* obtained strategies.   (d) *DGET-PSO* obtained strategies.

Fig. 2. Theoretical Nash strategies for 2 epochs (epoch 0 : $\alpha_1 = 1, \alpha_2 = 1$ and epoch 1 : $\alpha_1 = 0.6, \alpha_2 = 0.3$) and obtained Nash strategies for *AEDE, DGET-DE,* and *DGET-PSO. DGET-DE* algorithm seems to better approximate the true Nash strategy front.

When a change is detected, in order to promote diversity, F is set to 0.4 and decreased down to 0.2 with a step of 0.02 per generation for *AEDE* and *DGET-DE* algorithms.

Figure 2 illustrates a theoretical Nash front of strategies for 2 epochs (epoch 0 : $\alpha_1 = 1, \alpha_2 = 1$ and epoch 1 : $\alpha_1 = 0.6, \alpha_2 = 0.3$). It can be observed that *DGET-DE* is able to cope well with constraints changes of the Cournot dynamic game, Figure 2(c). The simpler *AEDE* version is able to find strategies that lie on the true Nash front but is unable to ensure a good diversity when a change is detected, figure 2(b) and the *PSO* variant is not able to find a good approximation of the true Nash strategy front when the game changes, Figure 2(d).

In order to illustrate the effect of step 3 of the adaptation mechanism on the results (variation of the mutation step σ) a set of experiments that take into account fixed values for σ were performed. Boxplots represented in Figure 3 indicate that the variation of σ yields better results in terms of standard deviation and the Wilcoxon sum-rank test indicate that results obtained by using the adaptive version are significantly better than those obtained for a fixed value for this parameter.

Figure 1 presents box-plots of the average *Nash-GD* indicator computed over all 20 independent runs for all generations. It can be observed that *AEDE* and *DGET-DE* outperform *DGET-PSO* algorithm. A Wilcoxon sum ranked test was performed in order to asses the differences in between the results of the two

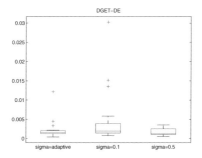

(a) DGET-DE *Nash-GD* boxplots. Comparison of results obtained for different values of mutation step σ.

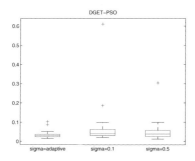

(b) DGET-PSO *Nash-GD* boxplots. Comparison of results obtained for different values of mutation step σ.

Fig. 3. The effect of the mutation step σ size

algorithms and indicated that there is no statistical difference between $AEDE$ and $DGET\text{-}DE$ methods for the mean values of the $Nash\text{-}GD$ indicator.

On a simple duopoly, according to the $Nash\text{-}GD$ indicator, $DGET\text{-}DE$ outperforms the other tested approaches. Next we wanted to see if this observation is also true for a many player version of the constrained game. For this we run $DGET\text{-}DE$ and $DGET\text{-}PSO$ algorithms on a constrained Cournot game with $n = \{3, 5, 10, 20, 40, 60\}$ players. Figure 4 presents box-plots of the average $Nash\text{-}GD$ indicator over 20 independent runs. It can be observed that also for $n > 2$ players algorithm $DGET\text{-}DE$ outperforms $DGET\text{-}PSO$.

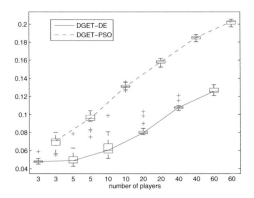

Fig. 4. Box-plots of the average $Nash\text{-}GD$ for a $n = \{3, 5, 10, 20, 40, 60\}$ player constrained Cournot. $DGET\text{-}DE$ has the lowest mean value and standard deviation for the $Nash\text{-}GD$ indicator.

6 Conclusions

Two methods for computing the Generalized Nash equilibrium of a discrete-time dynamic game are presented: one based on Differential Evolution - Dynamic Generalized Equilibrium Tracking Differential Evolution ($DGET\text{-}DE$), and one based on particle swarm optimization ($DGET\text{-}PSO$). Using a generative relation the proposed methods are able to find the GNE of the game in a dynamic environment.

Numerical experiments are conducted for the 2, 3, 5, 10, 20, 40 and 60 players version of the constrained dynamic Cournot game. We evaluate the performance of each algorithm by computing the Nash Generational Distance indicator between obtained strategies and the analytically computed true Nash front over 20 independent runs for each algorithm. Considering that a 60 palyers game can be compared with a 60 objectives multicriteria optimization problem, the results presented indicate the competitivness of this simple approach.

The proposed $DGET\text{-}DE$ is able to cope well with constraints changes finding strategies that lie on the true Nash front and ensures good solution diversity.

The simpler *AEDE* version is able to find strategies that lie on the true Nash front but is unable to assure good diversity when a change is detected. *DGET-PSO* is not able to find a good approximation of the true Nash strategy front when the game changes.

Acknowledgements. The first would like to thank for the financial support provided from program co-financed by the Sectoral Operational Programme Human Resources Development, Contract POSDRU/107/1.5/S/76841 with the title Modern Doctoral Studies: Internationalization and Interdisciplinarity. This project was supported by the national project code TE 252 financed by the Romanian Ministry of Education and Research CNCSIS-UEFISCSU.

References

[1] Debreu, G.: A social equilibrium existence theorem. Proc. Nat. Acad. Sci. U.S.A. (1952)
[2] Arrow, K.J., Debreu, G.: Existence of an equilibrium for a competitive economy. Econometrica 22, 265–290 (1954)
[3] Haurie, A., Krawczyk, J.: An introduction to dynamic games (2001)
[4] Ngo, V.L.: A Survey of Dynamic Games in Economics. World Scientific Books, vol. 1. World Scientific Publishing Co. Pte. Ltd. (2010)
[5] Nguyen, T.T., Yang, S., Branke, J.: Evolutionary dynamic optimization: A survey of the state of the art. Swarm and Evolutionary Computation 6, 1–24 (2012)
[6] Nash, J.: Non-Cooperative Games. The Annals of Mathematics 54(2), 286–295 (1951)
[7] Cardell, J.B., Hitt, C.C., Hogan, W.W.: Market power and strategic interaction in electricity networks. Resource and Energy Economics 19(1-2), 109–137 (1997)
[8] Jing-Yuan, W., Smeers, Y.: Spatial oligopolistic electricity models with cournot generators and regulated transmission prices. Oper. Res. 47(1), 102–112 (1999)
[9] Breton, M., Zaccour, G., Zahaf, M.: A game-theoretic formulation of joint implementation of environmental projects. European Journal of Operational Research 168(1), 221–239 (2006)
[10] Contreras, J., Klusch, M., Krawczyk, J.: Numerical solutions to nash-cournot equilibria in coupled constraint electricity markets. IEEE Transactions on Power Systems 19(1), 195–206 (2004)
[11] Han, D., Zhang, H., Qian, G., Xu, L.: An improved two-step method for solving generalized nash equilibrium problems. European Journal of Operational Research 216(3), 613–623 (2012)
[12] Dumitrescu, D., Lung, R.I., Mihoc, T.D.: Generative relations for evolutionary equilibria detection. In: Proceedings of the 11th Annual Conference on Genetic and Evolutionary Computation (2009)
[13] Lung, R.I., Dumitrescu, D.: Computing Nash equilibria by means of evolutionary computation. Int. J. of Computers, Communications & Control 3, 364–368 (2008)
[14] Lung, R.I., Mihoc, T.D., Dumitrescu, D.: Nash equilibria detection for multi-player games. In: IEEE Congress on Evolutionary Computation, pp. 1–5 (2010)
[15] Mihoc, T., Lung, R., Dumitrescu, D.: Notes on a fitness solution for nash equilibria in large games. In: 2010 11th International Symposium on Computational Intelligence and Informatics (CINTI), pp. 53–56 (2010)

[16] Cournot, A.: Recherches sur les Principes Mathématique de la Théorie des Richesses. Hachette, Paris (1838)

[17] Suciu, M., Lung, R.I., Gaskó, N., Dumitrescu, D.: Differential evolution for discrete- time large dynamic games. In: CEC 2013 (accepted, 2013)

[18] Storn, R., Price, K.: Differential evolution - a simple and efficient adaptive scheme for global optimization over continuous spaces. Journal of Global Optimization 11, 341–359 (1997)

[19] Nebro, A.J., Durillo, J.J., García-Nieto, J., Coello Coello, C.A., Luna, F., Alba, E.: SMPSO: A New PSO-based Metaheuristic for Multi-objective Optimization. In: 2009 IEEE Symposium on Computational Intelligence in Multicriteria Decision-Making, MCDM 2009, pp. 66–73. IEEE Press (2009)

[20] Deb, K.: An efficient constraint handling method for genetic algorithms. Computer Methods in Applied Mechanics and Engineering 186(2-4), 311–338 (2000)

[21] Van Veldhuizen, D., Lamont, G.: On measuring multiobjective evolutionary algorithm performance. In: Proceedings of the 2000 Congress on Evolutionary Computation, vol. 1, pp. 204–211 (2000)

[22] Parsopoulos, K.E., Vrahatis, M.N.: Recent approaches to global optimization problems through particle swarm optimization 1(2-3), 235–306 (June 2002)

[23] Trelea, I.C.: The particle swarm optimization algorithm: convergence analysis and parameter selection. Inf. Process. Lett. 85(6), 317–325 (2003)

[24] Shi, Y., Eberhart, R.: Empirical study of particle swarm optimization. In: Proceedings of the 1999 Congress on Evolutionary Computation, CEC 1999, vol. 3. p. 1950 (1999)

Author Index